Lake Sediments and
Environmental History

Winifred Tutin has made many notable contributions to palaeolimnology, palaeoecology, and vegetational history, particularly in the English Lake District and north-west Scotland. The papers presented to her in this volume by her colleagues are a reflection of her own wide-ranging scientific interests and research activities. (Photograph: Trevor Furnass.)

Lake Sediments and Environmental History

Studies in palaeolimnology
and palaeoecology
in honour of Winifred Tutin

*edited by Elizabeth Y. Haworth and
John W. G. Lund*

Leicester University Press   1984

First published in 1984 by Leicester University Press.
Published in North America by University of Minnesota Press

Designed by Douglas Martin
Phototypeset in Linotron 202 Molior
Printed in Great Britain at The Pitman Press, Bath

British Library Cataloguing in Publication Data
Lake sediments and environmental history.
1. Palaeolimnology
I. Haworth, Elizabeth Y. II. Lund, John W. G.
III. Tutin, Winifred
551.48'2   QE39.5.P3

ISBN 0–7185–1220–0

Contents

Illustrations

Tables

Note   Places of publication are given only for works published outside the United Kingdom. Journal abbreviations follow the style of the *World List of Scientific Periodicals* (4th edition).

Editors' note

THIS BOOK was conceived at the time of the retirement of Professor Winifred Tutin (née Pennington) from the Freshwater Biological Association, to honour her achievements in the field of palaeolimnology and environmental history. The response of the contributors has allowed the editors to achieve their own ambition to produce a fitting gift that contains much that is new science or which provides a new perspective of some particular topic. These contributions reflect the wide scope of Professor Tutin's own research interests, as is evident from the introductory chapter. They also provide a very wide-ranging view of the topics and the literature that is very relevant to the present concept of environmental history.

The editors would like to extend their thanks to all those who responded to the original request and have contributed to this volume as authors. The same thanks go to Dr John Birks, who organized the volume and did all the preliminary work, and also to the reviewers. We are grateful to the Leicester University Press for their enthusiastic extension from their more usual range of publications in order to confer this distinction on a member of their University.

The editors would also like to offer their own appreciation of the scientific contribution made by Professor Winifred Tutin and wish her a long, successful continuation of her work.

Elizabeth Y. Haworth
John W. G. Lund

Contributors

Allen, Peter V.

Freshwater Biological Association, The Ferry House, Ambleside, Cumbria LA22 0LP, England

Andersen, Svend T.

Geobotanical Department, Geological Survey of Denmark, Thoravej 31, DK-2400, Copenhagen NV, Denmark

Appleby, Peter G.

Department of Applied Mathematics and Theoretical Physics, University of Liverpool, P.O. Box 147, Liverpool L69 3BX, England

Birks, Hilary H.

22 Rutherford Road, Cambridge CB2 2HH, England

Bonny, Anne P.

8 Firfield Road, Addlestone, Weybridge, Surrey KT15 1QU, England

Cambray, Roger S.

Environmental and Medical Sciences Division, A.E.R.E. Harwell, Oxfordshire OX11 0RA, England

Chambers, Kevan C.

NPK Landscape Architects, 47 Stephens Road, Withington, Manchester M20 9XB, England

Cranwell, Peter A.

Freshwater Biological Association, The Ferry House, Ambleside, Cumbria LA22 0LP, England

Davis, Margaret B.

Department of Ecology and Behavioral Biology, University of Minnesota, Church Street, Minneapolis, Minnesota 55455, U.S.A.

Deevey, Edward S., Jr

Department of Natural Sciences, Florida State Museum, Museum Road, Gainesville, Florida 32611, U.S.A.

Eakins, John D.

Environmental and Medical Sciences Division, A.E.R.E. Harwell, Oxfordshire OX11 0RA, England

Engstrom, Daniel R.   Limnological Research Center, University of Minnesota, Pillsbury Hall, Minneapolis, Minnesota 55455, U.S.A.

Ford, Jesse   Department of Ecology and Behavioral Biology, University of Minnesota, Church Street, Minneapolis, Minnesota 55455, U.S.A.

Haworth, Elizabeth Y.   Freshwater Biological Association, The Ferry House, Ambleside, Cumbria LA22 0LP, England

Lally, Arthur E.   Environmental and Medical Sciences Division, A.E.R.E. Harwell, Oxfordshire OX11 0RA, England

Livingstone, David   9 Springhill, Barby Road, Rugby, England

Lund, John W. G.   Freshwater Biological Association, The Ferry House, Ambleside, Cumbria LA22 0LP, England

Moeller, Robert E.   Kellogg Biological Station, Michigan State University, Hickory Corners, Michigan 49060, U.S.A.

Oldfield, Frank   Department of Geography, University of Liverpool, P.O. Box 147, Liverpool L69 3BX, England

Thompson, Roy   Department of Geophysics, James Clerk Maxwell Building, University of Edinburgh, Mayfield Road, Edinburgh, EH9 3JZ, Scotland

Turner, Judith   Department of Botany, University of Durham, South Road, Durham DH1 3LE, England

Watts, William A.   School of Botany, Trinity College, University of Dublin, Dublin 2, Ireland

Wright, Herbert E.   Limnological Research Center, Pillsbury Hall, University of Minnesota, Minneapolis, Minnesota 55455, U.S.A.

Winifred Tutin – a personal note

John W. G. Lund

IT NEEDS someone experienced in the fields concerned to write a full and critical appreciation of Professor Winifred Tutin's great contributions to the study of post-glacial vegetation history and lake development. Particularly in writing something about her early career and about the events which and people who influenced it, I am at a disadvantage in that I knew nothing about them before I joined the staff of the Freshwater Biological Association (hereafter FBA), to the renown and growth of which she has made such a large contribution.

In 1936, possibly while I was attending the annual Easter course, there arrived at Wray Castle a very remarkable student called Winifred Pennington. Probably on this first occasion she travelled, as she did later, by train from Barrow-in-Furness to Coniston and then by bicycle to the laboratory. Then at Reading University, she had spoken to Professor T.M. Harris of her wish to spend part of her summer vacation at the FBA's laboratory which was in Wray Castle, a Victorian monstrosity ill-designed for living or, it seemed, working in. However, it and its boathouse were conveniently sited on the western shore and near the head of Windermere and there were keen young biologists on its small staff. The FBA was then only seven years old and its laboratory less than five years old, but it had the active support and the encouragement of several outstanding scientists. Professor Harris supported her idea but told her that she herself should go and ask if she could work in the laboratory. She has told me that it was with great trepidation that she went to see the Naturalist-in-Charge, Mr R.S.A. Beauchamp. She was received in a friendly manner and became a regular visitor during vacations, and so came into contact with two people who were to play a large part in determining future events. They were Dr (now Professor) C.H. Mortimer, a member of the staff, and the late Professor W.H. Pearsall, then a lecturer at Leeds University. Waiting, as it were, in the wings was Dr (now Professor Sir Harry) H. Godwin at Cambridge, and at Reading there was Professor Harris.

I have said that she was a remarkable student. Apart from hearsay, it would be remarkable, even in these days of massive publication, for an undergraduate to carry out research worthy of publication over such a wide field as the anatomy of *Fucus* (Pennington 1937), the air-pores of *Preissia* (Walker and Pennington 1939) and the ecology of mosses in and around a Lakeland stream (Pennington 1949). Moreover, at Wray Castle, she was also studying the freshwater red alga *Hildenbrandia*, entering into lake studies and, for all I know, into other investigations as well.

It is hardly surprising that, having obtained her B.Sc. in 1938, she was awarded a postgraduate studentship to work at Reading under the direction of Professor Harris. Harris had a keen interest in freshwater algae, as his notebooks in the FBA's Windermere library attest. Moreover, he was a palaeobotanist and had been at Cambridge where Dr Godwin was carrying out pioneer studies on the history of our vegetation. Later, Godwin was to initiate Winifred Pennington into the use of pollen analysis in elucidating such matters and so, in turn, for studying lake development. Part of her research was concerned with the ecology of present-day algae. This investigation merits mention, even if it is not strictly germane to this book. Part of the work was an experimental study of the interaction between phyto- and zooplankton (Pennington 1941a, 1941b), a subject which is now being vigorously pursued and is fundamental to an understanding of the seasonal and spatial changes in the composition of the phytoplankton, just as, for example, supplies of nitrogen and phosphorus are to its quantity; and, of course, as she showed in her study, grazing also affects total algal abundance. In the observations and the results of the experiments in 100l tubs can be seen a microcosm of virtually all that is generally accepted or is argued about over 40 years later. A small offshoot of another part of her thesis, concerned with the supply of nitrogen to algae, its utilization and possible sources of loss during their later decomposition (Pennington 1942), was an early example of her common sense, in this case in relation to one of the various wartime ideas for the use of algae (Pennington 1941c). She also discovered in her tubs the aptly named new coccoid green algal genus *Diogenes* (Pennington 1941d). The other part of the thesis concerned lake sediments and the sedimentation of seston, based on work done or samples taken in the Lake District. Although the title of her thesis (Pennington 1941a) suggests a unified plan of research, I have the feeling that the developments at Wray Castle, which had started before she graduated, were already attracting her away from algae, apart, for a time, from diatoms. In her final year as an undergraduate she attended one of the FBA's annual Easter Courses, which were the seedbed for the flowering of British limnology.

Mortimer was a zoologist, but a condition of his appointment to the FBA's staff was that he carried out routine chemical analyses of lake water. 'This unsought task and the inspiration of Professor Pearsall laid the foundations of a life's study of the aquatic environment' (C.H. Mortimer, personal communication). He and Pearsall were, in their turn, to have a profound influence on Winifred Pennington's career. In 1937, the Admiralty made a bathymetric survey of Windermere by echo-sounding in which Mortimer participated. Afterwards he continued the work, in the course of which he was intrigued by the layering of the traces obtained, then a novel observation (Mortimer 1940, Mortimer and Worthington 1940). He thought that these layers might represent interfaces between different kinds of post-glacial deposits. He had also made contact with the late Mr B.M. Jenkin whose daughter, Dr Penelope Jenkin, was the FBA's first research worker. Mr Jenkin was an ingenious engineer with wide interests. In April 1936 he was holidaying in the nearby hamlet of Wray and 'doing preliminary tests of his sampler' (P. Jenkin, personal communication). This first sampler was simply a steel tube which, when driven into shallow water deposits near the Wray Castle boathouse, brought up a three-layered core, with clay at the bottom and top and brown mud in the middle. Both clays were laminated and the mud contained traces of plants. Mr Jenkin then devised a corer (Jenkin and Mortimer 1938) which overcame the compression and so distortion of a core obtained by ramming a tube into the deposits. In 1940, after a delay caused by the outbreak of war, his improved version which collected a 6m long core came into use (Jenkin, Mortimer and Pennington 1941).

The original, relatively crude three-layered core mentioned above, in which Winifred Pennington found parts of birch catkins, had already 'suggested comparison with the threefold division of the late-glacial period originally recognized at Allerød in Denmark by Hartz and Milthers (1901) and subsequently recognized at many sites in North West Europe and Ireland (Godwin 1947)' (Pennington 1948). In the FBA Annual Report for 1937–8 (FBA 1938) it is recorded that Dr H. Godwin had made a preliminary examination of pollen and at Reading Mr L.R. Wager's examination of the clay particles supported the view that they were of glacial origin. Also at Reading, Professor Harris and, at the Natural History Museum, Mr R. Ross were examining the diatoms present. Dr G.H. Mitchell of the Geological Survey was studying the relationships with other glacial deposits. In this flurry of activity, as yet there is no mention of Miss Pennington. However, she was soon to appear in her rightful place, for next year, 1938–9 (FBA 1939), we read that 'the study of plant remains . . . has been undertaken by Miss Winifred Pennington' and, later in the same report, that she had made 'considerable progress' in analysing the diatom flora, the full results of which were to become part of her Ph.D. thesis. She had also found roots, moss leaves (see also Pennington 1962), catkin scales, and birch and pine pollen.

If, as it would seem, she had been temporarily ignored in the 1938 report, the progress made a year later suggests that the impression she made on one observer was not far from the mark. This observer noted 'what a determined character W. Pennington is and how well she knows what she wants and goes on firmly till she gets it'. Those who have been privileged to know her and her work for a long time will be aware that there have been occasions and periods when she has had much need of these alleged characteristics. However, the determination which is so necessary for any worthwhile research work is not usually so cool and calculating as that anonymous remark may suggest. It is enthusiasm, often delight, and wonderment which, despite laborious, even tedious work and setbacks, drives her or him to elucidate the, in this case, literally muddy record of the past.

Exciting though these early discoveries were, the Development Commission, responsible for informing and advising the government about the infant Freshwater Biological Association, felt that these post-glacial investigations were not within its remit. Possibly the Commission thought that this work was nearer geology than fisheries. However, Winifred Pennington's work was able to continue, thanks to grants from the Royal Society. From 1941, when she obtained her Ph.D., she worked partly at Wray Castle, where she held the position of research student, and partly with Dr Godwin at Cambridge in order to receive training in pollen analysis. The results of her investigations were two now classic papers, one on the diatoms and the other on the pollen (Pennington 1943, 1947) in the late- and post-glacial deposits of Windermere.

In the Annual Report for 1942–3 (FBA 1943) it is noted that a Mr T.G. Tutin 'was working from time to time' at Wray Castle. He was to produce some valuable data and considerations concerning the aquatic vegetation of the Lake District as part of his paper on the concept of climax in plant ecology (Tutin T.G. 1941) and it was in this year that they were married. Talented though they clearly were, it may be doubted if anyone then foretold the election of both into the Royal Society.

The birth of her first child and the claims on her of a growing family restricted the time Winifred Tutin could give to research for many years. However, University College, Leicester (after 1957 the University of Leicester), where Tom Tutin was a lecturer and later professor, made her a temporary lecturer in 1947, part-time lecturer in 1948, and special lecturer in 1961. By about 1955, she was once more making regular visits to the FBA's laboratory which was now at The Ferry House, also on the western shore of Windermere and at the junction between the two biologically and bathymetrically distinct lake basins. She was now to regenerate the FBA's post-glacial studies and, indeed, inspire a major development in the history of the Association. From 1956 to 1967, when she joined the staff though still based at Leicester, she was continuously on the FBA's Council and apart from the statutory two-year period between each three-year period of office, for most of the time on its Scientific Advisory Committee.

It must have been some time before she was first on the Council that she aroused the interest of the FBA's chemist, the late Mr F.J.H. Mackereth, in her field – a field, moreover, which clearly needed more detailed chemical studies than those which she had carried out almost from the earliest days. Mackereth was a man of outstanding ability and ingenuity over a remarkable range of professional and personal interests, as can be seen in part from obituary notices (FBA 1972, Macan 1973); it was a great tragedy that he died at the early age of 51. At this point I would also like to mention Mr G.J. Thompson, the laboratory steward and longest-serving member of the staff of the FBA, who died at almost the same time, for I feel sure that Winifred Tutin would wish, even in this partial record of her career, tribute be paid to him, since from the early days of the Jenkin corer he gave her great assistance and support.

The corers then available either needed, like the Jenkin corer, a platform such as a pontoon or ship or, if easily portable, could only be used in relatively shallow water. In 1957, Mackereth (1958) devised and himself largely made a pneumatic corer which, needing neither weights nor lifting gear, was portable and could be operated by two people in a rowing boat. Its robustness is attested by the fact that the first one made was used by Winifred Tutin for about 20 years. Now she could extend the area of her coring with relative ease, though not with complete ease since she often worked in difficult terrain. As might be expected, she took full advantage of this technical advance. With the aid of the Mackereth corer she was to make detailed and comparative studies all over the Lake District (e.g. Pennington 1964, 1965) and further afield (e.g. north-west Scotland, Pennington 1975a, 1975b, 1977a, 1977b, Pennington et al. 1972). Mackereth (1965, 1966) collaborated in the Lake District investigations, making a detailed study of the inorganic chemistry of cores from lakes of diverse trophic status.

The important work already done by her enabled the FBA, after the passing of the Science and Technology Act and the setting up of the Natural Environment Research Council in 1965, to offer her a post on the staff and to set up the Quaternary Research Unit under her direction. It included assistants working at Leicester or Ferry House on pollen and chemical analysis. A few months later a postgraduate student from the University of Wales, Miss (now Dr) Elizabeth Y. Haworth, came to study fossil diatoms in a Lakeland tarn and, after a year in the U.S.A., was to join the group as diatomist. In 1969, Dr P.A. Cranwell joined the group to study the origin and fate of the organic matter in deposits (e.g. Cranwell 1974, 1976, Pennington et al. 1977).

Mackereth's ingenuity once again helped when he made a magnetometer (Mackereth 1971) and showed that the oscillations in the horizontal direction of the remanent magnetism in cores could be used for dating purposes. In addition he devised a short form of pneumatic corer

(Mackereth 1969) in which the disturbance of the mud–water interface in cores collected by the long corer was minimized. Winifred Tutin used this corer to establish dates for recent (0–400 years) sediments in collaboration with workers at the Atomic Energy Research Establishment at Harwell (Pennington *et al.* 1973, Pennington *et al.* 1976) estimating ^{137}Cs and [210]Pb. Dr Haworth's studies made possible a comparison between weekly records of diatom abundance in the plankton since 1945 with the frequency of their occurrence in ^{137}Cs-dated deposits in one of the lakes (Haworth 1980).

As early as 1940 (Pennington 1941b, 1955), Professor Tutin was making direct measurements of sedimentation rates using traps, the design of which she showed affected the nature of the results obtained. Later, using estimates of sedimentation from trap catches and dating of deposits by measurements of remanent magnetism, ^{137}Cs, and ^{210}Pb, she was able to compare and contrast the rates of accretion of sediment in oligotrophic and eutrophic lakes and the degree to which resuspension of sedimentary material took place. In relation to later studies on eutrophication (i.e. man-made or cultural eutrophication), about which so much has been written in recent years, it is interesting to note that, more than 40 years ago, Pennington (1941a, 1943) showed that there was a great increase in the abundance of the diatom *Asterionella* in Windermere 'from 20 centimetres below the mud surface'. She suggested that this change in abundance might be 'correlated with the opening up of the Lake District at the beginning of the nineteenth century'. Though there may be 'some changes in our views ... the general thesis proposed by Pennington remains unchallenged' (J.W.G. Lund in Pearsall and Pennington 1973).

Professor Tutin has expressed her debt to Sir Harry Godwin in the preface to her very readable and highly regarded book *The History of British Vegetation* (Pennington 1969). More than 25 years ago she and Professor W.H. Pearsall published a paper on the ecological history of the Lake District (Pearsall and Pennington 1947). Perhaps one of the highest tributes I can pay to her is to say that she herself has shown truly Pearsallian width of vision, combined with the meticulous attention to detail so necessary in the elucidation of the past. She too, in fact, pays tribute to Pearsall in *The Lake District* (Pearsall and Pennington 1973). Though the facts are clear from the preface, it may not be generally known or remembered that she herself wrote the book, basing it on the plan and notes left by Pearsall and with some help from colleagues. She also revised Pearsall's *Mountains and Moorlands* (1950, revised 1968).

This account, because of my lack of competence in this field, gives little or no idea of one of the major aspects of her work, namely investigations on late-glacial or early post-glacial time (e.g. Pennington 1975a, 1975b, 1975c, 1977a, 1977b). As an observer, working at the most on some of the fringes of her subject, I have been impressed by her ability 'to see the wood despite the trees' and, I would add, the trees despite the wood. Professor Tutin has an extremely keen instinct for the possible or probable relevance to her research of a wide variety of subjects and investigations. She has, alone or in collaboration, entered such diverse fields as, for example, diatom analysis; pollen analysis, its origin, transport and fate; prehistory; recorded history; chronological markers; regional and local vegetation; climate and soil history (e.g. Pennington and Lishman 1971), soil development and destruction; accumulation rates of sediment by direct and indirect measurement, and various aspects of limnology.

Of course, it may be said by those who are workers in palaeoecology and palaeolimnology that there is nothing novel in all this; it represents the essential approach to the elucidation of the past. If so, I can only say that, in my view, she has and continues to be, outstanding in vision, controlled by check and counter-check from every available angle. The German chemist Kekulé, elucidator of the arrangement of the atoms in the benzene ring, told how the idea came to him while he was dozing before the fire and, as it were, saw the atoms gambolling before his eyes. The following remark which he made about his dream may, I feel, be applied to Professor Tutin's career.

'Let us learn to dream, then perhaps we shall find the truth, but let us beware of publishing our dreams before they have been put to the proof by our waking understanding' (Japp 1897).

References

PAPERS BY PROFESSOR TUTIN

[Pennington W. (Mrs T.G. Tutin) or Tutin W. – the last asterisked]
Note: This is not a complete list of her publications.

1937.　The secondary thickening of *Fucus*. *New Phytol. 36*: 267–79.
1941a.　An investigation of some problems of the ecology of freshwater algae, with special reference to the process of sedimentation (Ph.D. thesis, pp.1–191, University of Reading).
1941b.　The control of the numbers of freshwater phytoplankton by small invertebrate animals. *J. Ecol. 29*: 204–11.
1941c.　Plankton as a source of food. *Nature 148*: 314.
1941d.　*Diogenes rotundus* gen. et sp. nov. – a new alga. *J. Bot. Lond. 79*: 83–5.
1942.　Experiments on the utilisation of nitrogen in fresh waters. *J. Ecol. 30*: 326–40.

1943.   Lake sediments: the bottom deposits of the North Basin of Windermere, with special reference to the diatom succession. *New Phytol.*
42: 1–27.

1947.   Studies on the post-glacial history of British Vegetation VIII. Lake
sediments: pollen diagrams from the bottom deposits of the North
Basin of Windermere. *Phil. Trans. R. Soc. B,* 233: 137–75.

1948.   The lake sediments of Windermere. *Br. Sci. News* 1: 17–21.

1949.   The moss ecology of a Lakeland stream. *Trans. Br. Bryol. Soc.* 1:
166–71.

*1955.   Preliminary observations on a year's cycle of sedimentation in
Windermere, England. *Memorie Ist. Ital. Idrobiol. Suppl.* 8:
447–84.

1962.   Late-glacial moss records from the English Lake District. Data for the
study of post-glacial history. *New Phytol.* 61: 28–31.

1964.   Pollen analyses from the deposits of six upland tarns in the Lake
District. *Phil. Trans. R. Soc. B,* 248: 205–44.

1965.   The interpretation of some post-glacial vegetation diversities at
different Lake District sites. *Proc. R. Soc. B,* 161: 310–23.

1969.   *The History of British Vegetation.*

1969.   The usefulness of pollen analysis in interpretation of stratigraphic
horizons, both late-glacial and post-glacial. *Mitt int. Verein. theor.
angew. Limnol.,* 17: 154–64.

1970.   Vegetation history in the north-west of England: a regional synthesis.
In *Studies in the Vegetational History of the British Isles.* ed. D.
Walker and R.G. West, 41–79.

1973.   The recent sediments of Windermere. *Freshwater Biol.* 3: 363–82.

1973.   Absolute pollen frequencies in the sediments of lakes of different
morphometry. In *Quaternary Plant Ecology,* ed. H.J.B. Birks and R.G.
West, 79–104.

1974.   Seston and sediment formation in five Lake District lakes. *J. Ecol.* 62:
215–51.

1975a.   A chronostratigraphic comparison of Late-Weichselian and Late-
Devensian subdivisions, illustrated by two radiocarbon-dated
profiles from western Britain. *Boreas* 4: 157–71.

1975b.   Climatic changes in Britain as interpreted from lake sediments,
between 15,000 & 10,000 years ago. In *Palaeolimnology of Lake Biwa
and the Japanese Pleistocene,* ed. S. Horie, 3: 536–69.

1975c.   The effect of Neolithic man on the environment in north-west
England: the use of absolute pollen diagrams. In *The Effect of Man on
the Landscape: the Highland Zone* (CBA Res. Rep.), ed. J.G. Evans,
S. Limbrey, H. Cleere, 11: 74–86.

1977a.   The Late Devensian flora and vegetation of Britain. *Phil. Trans. R.
Soc. B,* 280: 247–71.

1977b.   Lake sediments and the late-glacial environment in northern Scotland. In *Studies in the Scottish Late-glacial Environment*, ed. J.M. Gray and J.J. Lowe, 119–41.

1978.   The impact of man on some English lakes: Rates of change. *Polskie Archwm Hydrobiol. 25:* 429–37.

1978.   Responses of some British lakes to past changes in land use on their catchments. *Verh. int. Verein. theor. angew. Limnol. 20:* 636–41.

1979.   The origin of pollen in lake sediments: an enclosed lake compared with one receiving inflow streams. *New Phytol. 83:* 189–213.

1980.   Modern pollen samples from west Greenland and the interpretation of pollen data from the British late-glacial (Late Devensian). *New Phytol. 84:* 171–201.

1981.   Sediment composition in relation to the interpretation of pollen data. IV Int. Palynol. Conf., Lucknow (1976–7) 3: 188–213.

1981.   Records of a lake's life in time: the sediments. *Hydrobiologia 79:* 197–219.

1981.   The representation of *Betula* in the Late Devensian deposits of Windermere, England. *Striae 14:* 83–7.

PAPERS, ETC. WRITTEN WITH OTHER AUTHORS

Pearsall, W.H. and Pennington, W., 1947. The ecological history of the English Lake District. *J. Ecol. 34:* 137–48.

Pearsall, W.H. and Pennington, W., 1973. *The Lake District.*

Pennington, W. and Bonny, A.P., 1970. Absolute pollen diagram from the British late-glacial. *Nature 226:* 871–3.

Pennington, W., Cambray, R.S. and Fisher, E.M., 1973. Observations on lake sediments using fallout ^{137}Cs as a tracer. *Nature 242:* 324–6.

Pennington, W., Cambray, R.S., Eakins, J.D. and Harkness, D.D., 1976. Radionuclide dating of the recent sediments of Blelham Tarn. *Freshwater Biol. 6:* 317–31.

Pennington, W., Cranwell, P.A., Haworth, E.Y., Bonny, A.P. and Lishman, J.P., 1977. Interpreting the environmental record in the sediments of Blelham Tarn. *Rep. Freshwat. Biol. Assoc. 45:* 37–47.

Pennington, W. and Frost, We., 1961. Fish vertebrae and scales in a sediment core from Esthwaite Water (English Lake District). *Hydrobiologia 17:* 183–90.

Pennington, W., Haworth, E.Y., Bonny, A.P. and Lishman, J.P., 1972. Lake sediments in northern Scotland. *Phil. Trans R. Soc. B, 264:* 191–294.

Pennington, W. and Lishman, J.P., 1971. Iodine in lake sediments in northern England and Scotland. *Biol. Rev. 46:* 279–313.

Pennington, W. and Sackin, M.J., 1975. Numerical methods in Quaternary palaeoecology. An application of principal components analysis and the zonation of two Late-Devensian profiles. *New Phytol. 75:* 419–53.

Walker, R. and Pennington, W., 1939. The movements of the air pores of *Preissia quadrata* (Scof.). *New Phytol. 38:* 62–8.

OTHER REFERENCES

Cranwell, P.A., 1974. Monocarboxylic acids in lake sediments: indicators derived from terrestrial and aquatic biota, of palaeoenvironmental trophic levels. *Chem. Geol. 14:* 1–14.

Cranwell, P.A., 1976. Organic geochemistry of lake sediments. In *Environmental Biogeochemistry*, ed. J.O. Nriagu, 75–88 (Michigan).

Godwin, H., 1947. The late-glacial period. *Sci. Prog. 138:* 185–92.

FBA (1938), (1939), (1943), (1972). *Rep. Freshwat. Biol. Assoc.* 1937–8, 1938–9, 1942–3 and 1972–3 respectively.

Hartz, N. and Milthers, V., 1901. Det senglaciale ler i Allerød teglvacrksgrav. *Meddr dansk geol. Foren. 8:* 7–12.

Haworth, E.Y., 1980. Comparison of continuous phytoplankton records with the diatom stratigraphy in the recent sediments of Blelham Tarn. *Limnol. Oceanogr. 25:* 1093–1103.

Japp, F.R., 1897. The Kekulé Memorial Lecture. *J. Chem. Soc. 73:* 97–138.

Jenkin, B.M., Mortimer, C.H. and Pennington, W., 1941. The study of lake deposits. *Nature 147:* 496–500.

Jenkin, B.M. and Mortimer, C.H., 1938. Sampling lake deposits. *Nature, Lond. 142:* 834–5.

Macan, T.T., 1973. Frederic John Haines Mackereth. *Arch. Hydrobiol. 72:* 270–2.

Mackereth, F.J.H., 1958. A portable corer for lake deposits. *Limnol. Oceanogr. 3:* 181–91.

Mackereth, F.J.H., 1965. Chemical investigations of lake sediments and their interpretation. *Proc. R. Soc. B. 161:* 295–309.

Mackereth, F.J.H., 1966. Some chemical observations on post-glacial sediments. *Phil. Trans. R. Soc. B, 250:* 165–213.

Mackereth, F.J.H., 1969. A short core sampler for subaqueous deposits. *Limnol. Oceanogr. 14:* 145–51.

Mackereth, F.J.H., 1971. On the variation in direction of the horizontal component of remanent magnetisation in lake sediments. *Earth Plant. Sci. Lett. 12:* 332–8.

Mortimer, C.H., 1940. Echo-lotungen von Seeablagerungen: eine neue Methode in der regionalen Limnologie. *Verh. int. Verein. theor. angew. Limnol. 9:* 334–40.

Mortimer, C.H. and Worthington, E.B., 1940. A new application of echo-sounding. *Nature, Lond. 145:* 212.

Pearsall, W.H., 1950. *Mountains and Moorlands.*

Tutin, T.G., 1941. The hydrosere and current concepts of the climax. *J. Ecol. 29:* 268–79.

1 Chemical stratigraphy of lake sediments as a record of environmental change

D. R. Engstrom and H. E. Wright, Jr

Introduction

STRATIGRAPHIC ANALYSIS of pollen and plant macrofossils in lake sediments commonly provides a wealth of information about regional and local vegetational history and a certain amount about lake history, and from the interpretations it is possible in many cases to reconstruct the past climate and even the palaeohydrology of the area. Developmental changes in lakes themselves are further elucidated by analysis of diatoms and cladocerans, although inadequate ecological and taxonomic knowledge of these groups may hamper the development of detailed interpretations. Stratigraphic studies of other lake-sediment fossils (e.g. molluscs, ostracods, rhizopods, fish scales, algae other than diatoms) have not been extensively pursued, primarily because of their uncommon occurrence, poor preservation, or undiagnostic ecological distribution.

The matrix for all these fossil components, however, usually constitutes the bulk of lake sediment, and its analysis may have the potential for revealing further details about the lake development or drainage-basin history. The sediment matrix includes organic materials that cannot be recognized as specific fossils, along with a fine-grained inorganic component that likewise cannot easily be distinguished as to mineral species. Chemical analysis provides a possible method for further environmental reconstruction.

A chemical approach involves several difficulties, however, and geochemical stratigraphy has not yet reached the point where routine analyses can lead to generalities about lake or catchment history. These difficulties derive from the varied and not easily identified sources for the particulate material that constitutes most of the matrix, as well as from diagenetic chemical changes that take place in the sediment after deposition. Sources of the organic component include both the autochthonous algal production of the lake water and the allochthonous detritus from terrestrial or littoral vegetation. The inorganic component includes mineral particles washed into the lake from the drainage basin by streams or hill-wash, or blown into it from beyond as dust, as well as inorganic ions dissolved in the inflowing water or the regional precipitation. During the depositional process more interactions between the lake water and the organic and inorganic components may take place, commonly related to the changing physical and chemical conditions that accompany the annual cycle of stratification of the water column in dimictic lakes.

Initial deposition of either the organic or inorganic particles on the lake bottom does not end the interactions. In shallow water the particles are commonly re-suspended and carried to deep water, where water currents may not further disturb them. There they are subject to chemical action at the contact with overlying water, where chemical gradients may be very steep and where bacterial action is lively. Burrowing organisms may mix the bottom sediment to a depth of a decimetre or more. Interstitial water reacts with particulate matter, and ions may diffuse upward or downward In the sediment column. Water with its contained ions moves upward (or sidewards) as the sediment compacts, providing further opportunity for chemical interactions.

In view of all these potential chemical changes in the lake water, and in the particulate matter during and after initial deposition, chemical profiles of lake sediments are difficult to interpret. Yet there remains the challenge to develop further understanding of the stratigraphic variations in chemical components for environmental reconstructions and attempts continue to be made by various investigators, among whom Professor Winifred Tutin, to whom this review is dedicated, has been prominent. The present synthesis is an effort to draw some generalities from these attempts and to point out some of the more promising lines of research.

Any discussion of stratigraphic studies of lake sediment chemistry must start with Mackereth's (1966) monograph on the English Lake District. His basic thesis was that the inorganic component of lake sediments, as calculated from loss-on-ignition and by the proportions of Na and K, reflects erosion intensity in the catchment. Thus if the landscape is stabilized by vegetation cover the rocks will weather deeply, and the runoff waters will carry the dissolved ions into the lake. Those that are chemically conservative pass through the system and may be removed at the lake outlet. But nutrient ions like phosphorus enhance algal production in the lake, and the algal detritus accumulates as organic matter in the sediment. On the other hand, if the landscape is not fully stabilized, erosion may cut through the mantle of soil and weathered rock and may bring to the lake unweathered minerals, which are recorded in the lake sediments by a higher content of Na and K and a higher proportion of mineral v. organic matter.

Mackereth's work in the Lake District involved relatively oligotrophic lakes, in which the content of mineral matter in post-glacial sediments averages about 70%, as measured by loss-on-ignition. He interprets the variable thickness of lake sediments in the area as reflecting primarily the degree of catchment erosion and thus the topography of the area. Because of low nutrient supply the organic productivity in these lakes is low, and most of what is produced is oxidized in the water before burial in the sediments. Consequently most of the organic component that the sediments do contain is derived from the catchment, where oxidation also destroys a large part of the organic matter produced, resulting in the delivery of only the most refractory organic materials to the lake. He concludes that the stratigraphy of lake sediments in the Lake District represents a 'series of samples of soils eroded from the drainage basin and deposited chronologically in the lake bed.'

The interpretations of Mackereth in the English Lake District have been extended by Tutin and her colleagues not only in this area but to northern Scotland as well, with the elaboration of many stratigraphic studies other than sediment chemistry: pollen analysis to decipher the vegetational history of the catchments before and since the inception of agricultural disturbance, diatom analysis to reveal the trends in alkalinity, nutrient supply, and related chemical conditions in the lakes, and isotope dating to refine the chronology of the stratigraphic changes and to permit the computation of annual influx of various components in the sediment. An integrated summary of this research (Pennington 1981a) shows the synergistic values of interdisciplinary approaches to palaeolimnology.

In these British studies the contributions of sediment chemistry to the integrated picture are substantial, for they yield an estimation of the history of landscape stability as it responds to climatic change, soil development, and human disturbance. Under conditions of cool but fluctuating climate of late-glacial time 15,000–10,000 years ago, mineral sediments dominated, although during the main late-glacial interstadial interval the content of organic matter temporarily increased, implying humus accumulation in the catchment in response to revegetation. Ca, Mg, Na, and K all have high values in late-glacial sediments compared to those in post-glacial sediments, but the slightly lower values of Mg and K in the interstadial zone of the profiles are attributed to the loss of these elements in the mineral fraction during interstadial weathering of feldspars to clay minerals in the catchment (Pennington et al. 1972).

The post-glacial history involved rapid stabilization of the landscape, with deep leaching of the soils, limited erosion in the catchment, and maximum organic production within the lake. There followed in the second half of post-glacial time a gradual increase in the mineral matter and in the content of Na and K, as a manifestation of increased erosion resulting from deforestation, attributed in turn to climatic deterioration or to agricultural land clearance.

Of particular help in relating sediment chemistry to lake history and the history of land use are short-core studies, in which the stratigraphic record can be correlated with well-known historical events. The time scale in these cases can be determined by identifying the increase of pollen from crop plants or agricultural weeds, by the increase in [137]Cs (a product of atmospheric bomb tests starting in 1954), or by ^{210}Pb dating (applicable for the last 100–150 years). The short-core method is particularly successful in the eastern half of the United States, where agricultural land clearance and other catchment disturbances started abruptly, leaving sharp stratigraphic changes in the microfossil assemblages, sediment lithology, and sediment chemistry. In some cases land clearance was so extensive that erosion resulted in an increase in mineral sediment, detectable not only through a shift in the loss-on-ignition curve but also by the increase in Na, K, and Mg (Likens and Davis M. 1975, Davis R. and Norton 1978). In other cases the major effect of catchment disturbance was lake pollution, caused by input of sewage waters or agricultural wastes, and resulting in increased algal productivity and thus an increase rather than a decrease in the organic component of the lake sediment.

In western Europe short-core stratigraphy is not so easy to interpret: cultural disturbance started thousands of years ago, and its chronology as documented by archaeological or historical records is less clear. On the other hand, the multitude of stratigraphic changes provides a greater challenge for palaeolimnological reconstructions. Of particular interest are the stratigraphic profiles for Blelham Tarn in the English Lake District, where land disturbance dating from Viking times resulted in distinctive changes in the sediment chemistry, and where even those since the Neolithic 5000 years ago are detectable in the record (Pennington 1981a). Investigations of a similar nature have been undertaken in Finland (e.g. Huttunen et al. 1978, Huttunen and Tolonen 1977) and Sweden (e.g. Digerfeldt 1972, 1975, Renberg 1976).

The discussion of sediment chemistry that follows centres on the interpretation of elemental stratigraphy in late-glacial and Holocene (post-glacial) lake deposits. Because of limitations of space, consideration is restricted largely to inorganic elements, and among these little mention is made of calcium and magnesium carbonates, which have been well treated in recent reviews by Dean (1981) and Kelts and Hsü (1978). Nor is there any discussion of trace metals or contaminants from atmospheric deposition. Organic materials are considered to a certain degree, however, because inorganic ions are commonly complexed with organic compounds in lacustrine sediments. Thus we have largely confined our review to those areas of geochemical research that Mackereth

(1966) addressed in his pioneering work on the English Lake District. Like Mackereth, we view elemental stratigraphy as a potential record of environmental history, which may be read by applying an understanding of modern geochemistry and limnology to the past. We have tried in particular to integrate recent experimental studies of sediment-water interactions with palaeolimnological investigations of chemical stratigraphy from both European and North American sites.

Sediment composition and fractionation methods

EXISTING PROCEDURES

Lacustrine sediments consist of a heterogeneous mixture of materials that can be roughly classified according to their origin. In this review, those components derived from outside the lake proper are termed *allogenic*, while those deposited directly from aquatic solution through biological uptake or chemical sorption and precipitation are designated *authigenic*. This latter fraction also includes materials formed *in situ* within the sediments. Authigenic components include biochemically precipitated carbonate minerals, amorphous and cryptocrystalline Fe and Mn oxyhydroxides, sulphides, and phosphates, biogenic silica (opal), and sorbed or coprecipitated elements. Organic matter, which may be either allochthonous or autochthonous, possesses only a small structural complement of inorganic cations but may complex or sorb large amounts of metallic ions from solution within the lake. Allochthonous organic matter also chelates cations from soil solutions and may enter the lake in both dissolved and particulate form. Strictly speaking, metals bound to the particulates are allogenic; however, in practice the distinction between the different organic components is difficult to make. Thus in this review organically bound metals are considered authigenic, regardless of the provenance of their organic ligands. Elements in the authigenic fraction are ultimately derived from the decomposition of crystalline rock through terrestrial weathering within the catchment except in the case of anthropogenic effluents. Rain and snow may also be an important source for certain elements in marine aerosols. In contrast, the allogenic fraction consists entirely of clastic mineral particles resulting from erosion of the catchment soils or from dust.

Other classifications have been proposed that differentiate between authigenic components of chemical origin (hydrogenous) and biological origin (biogenous) (Goldberg 1954), or between *endogenic* materials strictly derived from the water column and *authigenic* components formed through diagenetic changes within the sediments (Jones and Bowser 1978). However, these distinctions are often difficult to make in practice, because many sedimentary components are the net product of a combination of authigenic processes; biogenic silica formed from diatom frustules is an important exception. The terms *minerogenic* or *mineral matter* are also frequently encountered in the palaeoecological literature; they generally are used to designate the ash content of sediments from loss-on-ignition determination. Mineral matter is thus a composite of inorganic materials of mixed origin, including both amorphous and crystalline compounds, and the term is conceptually less useful than the authigenic-allogenic classification. Fractionation procedures for sedimentary phosphorus frequently distinguish between organic and inorganic phases of authigenic P (respectively nonapatite inorganic phosphorus and organic phosphorus; Williams *et al.* 1976a, 1976b), while another common measure in geochemical studies is that of *extractable* cations. The extractable fraction is a measure of the ions held by clays and organic colloids and is defined by the chemical extractant utilized.

Because different environmental information is contained within each sedimentary component, analytical separation of different fractions can help elucidate geochemical history from heterogeneous lake muds. This detailed information is obscured in bulk analysis, which has been the normal procedure in palaeolimnological investigations. Sediment fractionation techniques, which have been used primarily in the analysis of marine deposits and terrestrial soils, separate different materials by selective chemical attack on the total sediment. Because of the chemical similarity of various components, a precise division of conceptually distinct fractions is difficult to attain; analytical separations are only estimates of actual composition. Some of the more commonly used methods for selective fractionation are outlined below, along with techniques for the bulk digestion of sediments.

1. *Bulk (total) sediment digestion.* A simple analysis of total sediment composition is most frequently employed in palaeolimnological studies because of the time-consuming aspect of fractionation procedures and because a reliable protocol for sediment partitioning has not been clearly designed for limnetic deposits. In bulk analysis, sediment samples are fused with various fluxes or digested with strong acids, and the resulting materials are analysed by atomic absorption, emission spectroscopy, colorimetric measurements, or other techniques. Results from these methods are most easily interpreted for elements that are readily ascribed to a single sedimentary component (e.g. Na to allogenic minerals) but are less reliable for elements such as Si or Fe, which may be derived from several different components. For fusion techniques, anhydrous sodium carbonate (Jackson 1958) or lithium metaborate (Suhr and Ingamells 1966) are the most widely used fluxes. $LiBO_2$ may be preferred over Na_2CO_2 flux if Na determination is desired; also, inexpensive graphite

crucibles may be used rather than platinum. A mixture of hydrofluoric and perchloric acids is employed in acid digestion procedures (e.g. Allen *et al.* 1974). Fusion is generally more rapid than acid digestion and does not employ potentially dangerous reagents such as HF and $HClO_4$. In some geochemical studies a strong-acid leaching with various mixtures of $HClO_4$, H_2SO_4, and HNO_3 is preferred over total digestion of the sediments (e.g. Brugam 1978, Bengtsson 1979, Elner and Happey-Wood 1980). Although these reagents effectively destroy organic matter they cannot break down certain minerals, particularly silicates, so that total concentrations of elements such as K, Al, and Si cannot be readily determined by such methods. Other techniques that do not rely on solutions such as X-ray fluorescence or electron-probe analysis are also used.

   2. *Authigenic materials.* Various chemical separation techniques have been used to differentiate among elements incorporated into the sediments from solution and those occurring in pre-existing mineral lattices. These methods range from mild exchange reactions to rather vigorous leaching procedures that also attack silicate mineral lattices. Those techniques that best seem to approximate an ideal separation include (in approximate order of increasing extraction efficiency): citrate-dithionite-bicarbonate or CDB (Mehra and Jackson 1960), 0.2M acid ammonium oxylate (Saunders 1965), 1.0M hydroxylamine hydrochloride in 25% acetic acid (Chester and Hughes 1967), and 0.3N HCl (Malo 1977). Generally, these procedures extract absorbed metals, most carbonates (CDB less so than other reagents), Fe and Mn oxyhydroxides and a variable portion of structural cations from clay minerals. Detailed testing of these methods (except ammonium oxylate) by Malo (1977) showed that the 0.3N HCl extraction had greater operational efficiency and recovered more trace metals and oxide coatings from a wide variety of sediments. Lahann (1976) evaluated the effect of these procedures on the X-ray diffraction patterns of clay minerals and found slightly greater structural degradation with the 0.3N HCl treatment than with citrate-dithionite or hydroxylamine-hydrochloride. Clay destruction by 0.3N HCl was minimal, however, as shown by the small percentage of structural Al measured in the extractant. Similar studies by Dudas and Harward (1971) showed no structural alteration of clays by ammonium oxalate.

   Lake sediments are usually treated to remove organic matter in chemical fractionation procedures so that organically bound metals may be analysed. The most commonly used reagents for the destruction of organic matter include 30% hydrogen peroxide, 5% sodium hypochlorite (bleach), and 0.1M sodium pyrophosphate. Peroxide is clearly the most vigorous of these methods and, according to Douglas and Fiessinger (1971), can lead to some clay degradation if samples are not suitably buffered to prevent acid conditions from the oxidation of organic matter.

However, tests on a variety of river and lake sediments by Malo (1977) indicate that only a small proportion of total silicate Al is dissolved by peroxide pre-treatment.

3. *Biogenic silica.* Amorphous silica, composed primarily of diatom frustules, often forms a substantial fraction of the sediment found in small temperate lakes. Because silica is not dissolved by extractants such as dilute acids or citrate-dithionite, it is possible to measure this important biogenic component independently of other authigenic materials. Alkaline dissolution techniques are most commonly used to selectively remove amorphous SiO_2 from allogenic minerals in analyses of both sediments and water samples. In a laboratory comparison of different extractants, Krause *et al.* (1983) conclude that hydrolysis with 0.2N sodium hydroxide (Werner 1966) or 0.5% sodium carbonate (Tessenow 1966) quantitatively digests diatom silica and causes only minor degradation of clays and other mineral silicates. In these experiments less mineral silica was solubilized from clay and silt samples by NaOH than by Na_2CO_3 because of the longer extraction time needed for the latter (10mins v. 2hrs), although for both extractants this amounted to less than 1.5% of the total silica in the mineral samples.

Biogenic silica may also be measured without chemical extraction by quantitative X-ray diffraction analysis of cristobalite formed by heating amorphous silica (Goldberg 1958), with infrared spectroscopy (Chester and Elderfield 1968), or through a normative calculation technique by which total silica is corrected for mineral silica on the basis of the Al and Mg content of the sediments (Leinen 1977).

4. *Interstitial and extractable ions.* The ion content of sediment pore-water and the complement of metals absorbed to sediment particles are often analysed as distinct authigenic components in geochemical studies involving sediment-water interactions. Interstitial fluids are typically extracted from sediments by squeeze-filtration (e.g. Manheim 1966, Sasseville *et al.* 1974), but such methods require the exclusion of oxygen, lest precipitation on exposure to air lowers the concentration of redox-sensitive elements in the extracted water (Bray *et al.* 1973). Temperature change in cores following collection is another important variable in pore-water analysis as is the length of time between collection and sampling (Jones and Bowser 1978 and references therein). One approach that may eliminate the problems associated with sediment handling utilizes a compartmentalized probe that operates by equilibration of pore-waters across a dialysis membrane (Hesslein 1976).

Adsorbed ions are displaced from the sediment matrix with mild chemical extractants such as 1.0N ammonium acetate (Jackson 1958) or 0.5N acetic acid (Allen *et al.* 1974). Pore fluids and some ions solubilized from labile particulates are also extracted by these reagents. Ammonium acetate at pH 7 is the most widely used extractant for the major cations, whereas acetic acid is employed when phosphorus determinations are also desired. These techniques are largely borrowed from the soils literature, where they are used to assess the level of 'bio-available' or 'exchangeable' nutrients; because concepts of nutrient availability are not defined for lake sediments, the term 'extractable' is more appropriate for palaeolimnological studies.

PROPOSED FRACTIONATION PROCEDURE FOR LACUSTRINE SEDIMENTS

A step-wise protocol for selective chemical dissolution of sediments was used by Engstrom (1983) to analyse cores from two sites in Labrador, Canada. This fractionation procedure, assembled from a selection of the aforementioned extraction methods, was chosen according to three important criteria: efficiency of operation, consistency for sediments of varying composition, and accuracy of the separation between components. According to this scheme, which is outlined below, sediment samples are first treated with 30% hydrogen peroxide to destroy organic matter and release organically sorbed cations, followed by 0.3N HCl to dissolve Fe and Mn oxyhydroxides and associated metals. Biogenic silica is then selectively dissolved with 0.2N NaOH, and the remaining mineral residue is fused with lithium borate to complete the digestion. The basic division in this procedure is between authigenic and allogenic fractions, with an additional separation of biogenic silica. Although this is a very simple model of sediment composition, only approximated by the actual extractions, it represents a substantial improvement over conventional bulk analysis and so affords a greater resolution of chemical stratigraphy and environmental change.

1. React 1.0g wet sediment with 25ml 30% H_2O_2 at 90–95°C to destroy organic matter (some refractory organics may not be totally oxidized).
2. When oxidation reaction ceases (usually less than 1hr) add 5ml 3.0N HCl and bring volume to 50ml with pure H_2O; react at 90–95°C for 30 minutes to extract carbonates and amorphous oxides.
3. Filter sample through 0.45μm HA-Millipore filter to separate extract from residue; rinse and bring filtrate to volume (1000ml) with pure H_2O.
4. Wash residue from filter into 50ml 0.2 NaOH and react at 90–95°C for 15 minutes to dissolve biogenic Si; filter sample and dilute as in (3) above.
5. Fold filter and residue and place in graphite crucible, add 0.5g $LiBO_2$ and fuse at 950°C in muffle furnace for 15 minutes.

6. Dissolve molten bead in 100ml 0.5 N HCl on ultrasonic mixer, and dilute 1 : 10 for analysis.

This fractionation procedure requires few manipulations, yields easily analysed solutions, and is operationally efficient, although it is clearly more time-consuming than one-step bulk digestion. The reagents used in this scheme are also among the more vigorous extractants that we consider suitable for selective dissolution of fine-grained sediments, so that inadvertent degradation of clastic components is a potential source of error.

The accuracy and consistency of the authigenic extraction procedures may be determined from the fractionation results for conservative elements like potassium, which occur predominantly in the lattices of silicate minerals. If silicate degradation is minimal, then the concentration of K measured in the allogenic fraction after extraction should be nearly equal to the K content of the total sediment. In the Labrador study mentioned, this relation is shown by the absence of any significant correlation between the potassium ratio (allogenic K : total K) and those variables representing sediment composition, such as organic content, total K concentration, sample dry weight, or age (depth) of the sample (Engstrom 1983).

By similar reasoning, in the extraction of biogenic silica the degradation of allogenic minerals was also minimal, whereas visual scans of the mineral residue following hydroxide extraction showed virtually total dissolution of diatom frustules, confirming the efficacy of this method.

Despite some inherent inaccuracy in the separations and the additional time required for these procedures, the value of separate analyses for individual fractions is clearly indicated in the Labrador results (figs. 1.1 and 1.2). Major changes in overall sediment composition are evident in both profiles; the lower sediments, derived from a tundra landscape, are predominately minerogenic, whereas the upper sediments, beginning with the invasion of conifer forests about 7000 BP, are more organic. Across this transition the partitioning of several elements of mixed origin, notably Fe, Mn, Al, and Si, shifts markedly, such that authigenic (or biogenic) components largely replace their allogenic counterparts. A subsequent reversal of this trend is evident in the Moraine Lake diagram. If these sediments had not been fractionated a very different picture of sediment composition would have emerged from chemical analysis, the authigenic and allogenic signals would have been confounded, and serious errors in interpretation could result.

Data representation

The results from geochemical analyses may be numerically represented either in units of concentration or as rates of accumulation. Concentration calculations are the conventional means of representing stratigraphic data, and in geochemical studies they are expressed as relative measures of sediment composition (e.g. mg/g dry sediment, weight-percent of mineral matter). In contrast, accumulation rates are a direct measure of the annual net deposition of sedimentary materials per unit area at the coring site. The interpretations derived from these two methods are fundamentally different, as are the assumptions upon which they are based, so that some workers (e.g. Likens and Davis M. 1975, Digerfeldt 1972) have argued in favour of accumulation calculations, while others (e.g. Pennington 1981b) suggest that concentration determinations are preferred.

The primary difficulty in interpreting concentration units arises from the interdependence of chemical constituents of the sediments, whereby variations in the level of each component influences the concentration profiles of all other components. If elemental concentrations are expressed per gram dry sediment, for example, increases in the sedimentation of organic matter or biogenic silica will dilute the sedimentary concentration of elements associated with clastic minerals.

However, even relative changes in elemental concentrations may provide important palaeoecological information regarding the composition of the source material for the sediments. In oligotrophic lakes both organic and mineral matter are predominantly allochthonous, so that variations in total sediment composition represent changes in the structure and composition of catchment soils. For other lakes a constant deposition rate can be assumed for certain sedimentary components against which relative variations in the accumulation of other materials may be measured. Moreover, the elemental composition of a particular sedimentary fraction may be calculated separately to correct for variations in the input of unrelated materials. Thus Mackereth (1966), Koljonen and Carlson (1975), Davis R. and Norton (1978), and others have determined elemental composition per gram mineral matter (ash weight) to remove the influence of variable organic content. Similarly, Engstrom (1983) calculated the weight-percent composition of the allogenic fraction alone to eliminate the influence of authigenic and biogenic components.

Accumulation rates, on the other hand, are normalized to time, and this procedure eliminates the problem of covariation among different sedimentary components. Furthermore, the rate of deposition of various elements may represent important palaeoenvironmental signals that cannot be deciphered from concentration calculations alone. For

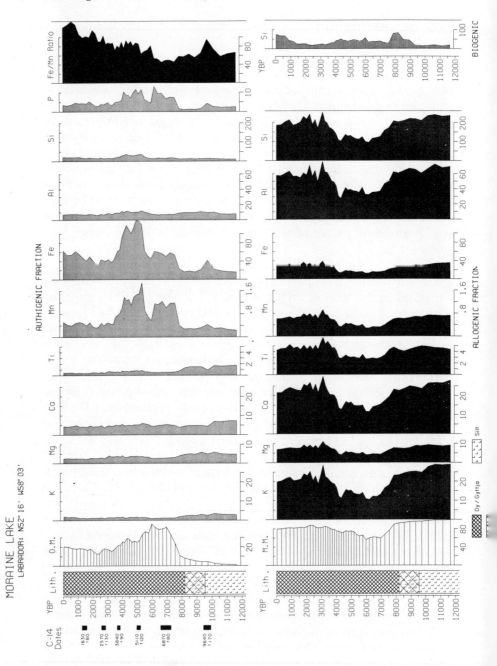

Fig. 1.1   Moraine Lake, Labrador, sediment chemistry diagram. Elemental concentrations in mg/g dry sediment. O.M. = % organic matter and M.M. = % mineral matter (ash) in dry sediment based on loss-on-ignition. From Engstrom 1983.

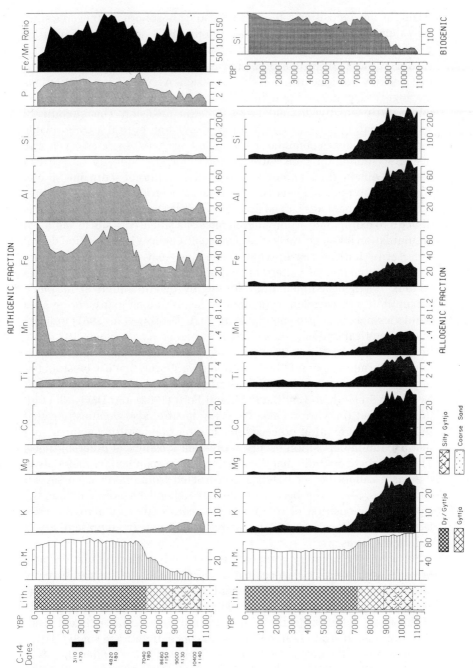

Fig 1.2 Lake Hope Simpson, Labrador, sediment chemistry diagram. Elemental concentrations in mg/g dry sediment. O.M. = % organic matter and M.M. = % mineral matter (ash) in dry sediment based on loss-on-ignition. From Engstrom 1983.

example, net deposition rates for authigenic phosphorus or biogenic silica may represent historic levels of lake productivity, and the accumulation rate for allogenic components may indicate erosional intensity in the past. However, there are several problems associated with accumulation rates that severely limit the utility of this approach. Accumulation calculations are highly dependent upon reliable and close-interval dating and are numerically sensitive to small errors in dating precision. The reliability of sediment dating is usually difficult to assess, and even if a conformable sequence of dates is obtained the sediment age between dated levels and hence accumulation rates must be interpolated. Typically only a few dates are obtained for an entire post-glacial section of sediment, so that accumulation rates are necessarily averaged over long intervals of time.

Even if sediment age can be precisely defined throughout a profile, it is difficult to extrapolate accumulation rates at a single core site to sediment accumulation for an entire lake. Site-specific changes in deposition do not necessarily indicate basin-wide events but may result instead from changes in the spatial pattern of sediment accumulation across the lake bottom. Detailed sedimentation studies by determination of magnetic susceptibility to correlate synchronous levels among many coring sites within a single lake (Bloemendal et al. 1979, Dearing et al. 1981) indicate that theoretical models of sediment focusing (cf. Lehman 1975) are insufficient to describe actual patterns of sediment deposition, even in relatively simple basins. Other multiple-core studies that use biostratigraphy to correlate among sites such as those by Battarbee (1978a) on Lough Neagh, N. Ireland, and by Davis M. and Ford (1982) and Davis M. et al. (this volume) on Mirror Lake, New Hampshire, also demonstrate that sediment accumulation proceeds in a complex and shifting manner, and that maximum deposition occurs at different locations at different times. These authors conclude that the average flux of sedimentary materials to the basin cannot be accurately reconstructed from one or even several coring sites. Moreover, because of density-dependent sedimentation, the chemical composition of sediments also varies spatially across a basin (e.g. Sly 1976, Håkanson 1977), which further limits the representivity of individual cores.

For some sedimentary components, even reliable accumulation rates provide little useful palaeoecological information and instead actually obscure important changes in sediment composition. This occurs because accumulation rates for individual components are the numerical product of their concentration in the sediments and the overall sediment-accumulation rate. If stratigraphic changes in concentration are small relative to variations in the rate of total sediment deposition (as is often the case), the compositional changes will be difficult to discern from accumulation diagrams. For example, the elemental composition of the allogenic fraction may provide information on the degree of mineral leaching in catchment soils (cf. Mackereth 1966), but because these changes are very subtle they are easily confounded in accumulation profiles by attendant changes in sedimentation rates. Furthermore, all elements of the allogenic fraction are delivered to the sediments bound together as detrital mineral particles, so that individual element accumulation rates for this fraction are all highly correlated and provide little independent information about the rate of clastic-mineral sedimentation that is not already available from the accumulation rate for the allogenic fraction as a whole.

In conclusion, because of the different interpretational problems associated with each method, both concentration and accumulation calculations can provide complementary information and should be used together whenever good time control is available. This is particularly true for short-core studies where ^{210}Pb and ^{137}Cs techniques are used for fine-scale dating or in the case of laminated sediments where accurate varve chronology is obtainable.

Sodium, potassium, magnesium, and calcium

The alkali and alkaline-earth elements are major constituents of common silicate minerals and occur in most lake sediments primarily in allogenic clastics eroded from catchment soils and rocks. Dissolution of silicate minerals in fresh waters is extremely slow, and authigenic silicates are rarely formed in non-arid climates. Although some investigators have argued for the authigenesis of clay minerals in lake sediments (e.g. Carmouze *et al.* 1976, Johnson and Eisenreich 1979), Jones and Bowser (1978) conclude that there is no definitive evidence (with the exception of nontronite, an iron smectite). Thus because detrital minerals are little altered in the lake environment, their distribution in sediments is particularly useful in assessing weathering, soil development, and erosion in the catchment.

As dissolved ions in surface waters, Na, K, and Mg are relatively conservative and generally are not appreciably sedimented through biological uptake or chemical precipitation. However, sorption onto suspended particulates and exchange across the sediment-water interface may incorporate small amounts of these elements into the sediment. Calcium has a strong affinity for organic ligands such as humic and fulvic acids, and organic sediments may thus contain substantial amounts of calcium not associated with allogenic minerals. A general correspondence between calcium and organic content has been noted in sediment stratigraphy from a wide variety of lakes (e.g. Mackereth 1966, Tolonen 1972, Brugam 1978). A portion of the sorbed complement of sediment particles may also originate from exchange processes within the soil profile prior to erosion and transport.

In alkaline lakes, Ca and occasionally Mg and Fe may precipitate as authigenic carbonates and, in more saline environments, as sulphates. While carbonates in particular constitute an important fraction of many lake sediments, they are not considered further in this review.

PALAEOSALINITY

Investigators in a number of studies have attempted to reconstruct lake salinity in the past by measuring levels of adsorbed ions within the sediments or their concentrations in interstitial waters, under the assumption that these phases represent exchange equilibria with the bottom waters at the time of sediment deposition. This research includes questions about salinity changes associated with isolation of coastal lakes from marine influence (Tolonen 1972, Ericsson 1073, Kjensmo 1968), climatic and hydrological effects on palaeosalinities in tropical and arid regions (Hecky 1976, Lewis and Weibezahn 1981), and ionic composition of dilute temperate lakes in relation to trophic development (Digerfeldt 1972, 1975, Sasseville and Norton 1975). Results from these studies show that geochemical evidence for palaeosalinity is generally difficult to interpret, because many variables in addition to lake-water composition may influence the retention of ionic species within the sediments, e.g. high rates of ionic diffusion.

Profiles for cations in interstitial water commonly show patterns that result from diagenesis of solid phases within the sediment column. In cores from several Maine lakes (Sasseville and Norton 1975), interstitial levels of K, Na, Ca, and Mg show *increasing* concentrations with depth, which the authors attribute to mineral dissolution following progressively deep burial. A 'reactive' zone enriched in interstitial cations that results from a more rapid diagenesis of labile materials also occurs near the surface of these profiles. Similarly, post-depositional solution of carbonates may be responsible for the high levels of Ca and Mg measured in sediment pore-waters from Lake Valencia, Venezuela (Lewis and Weibezahn 1981).

Diffusion of solutes between zones of different ion concentrations is sufficiently rapid in lake sediments to eliminate stratigraphic variation in pore-water chemistry unless solid phases are participating in the reaction. Hecky (1976) found in Lake Momela, Tanzania, that the stratigraphy of interstitial Na was totally unrelated to salinity changes as determined from diatom analysis. Apparently diffusion between sediment layers laid down during different salinity phases of the lake resulted in a nearly featureless Na profile. Smoothing of irregularities in pore-water concentrations for K and Mg was also recognized by Sasseville *et al.* (1975). The stratigraphy of extractable metals is less subject to alteration by diffusion than are profiles of metals in interstitial waters, because sorbed metal phases are less mobile. However, different results for extractable ions may be obtained with various extraction techniques, depending on sediment structure and composition (cf. Digerfeldt 1972).

The exchange capacity of the sediments has an important effect on the composition of interstitial waters and on the amount of ions absorbed to sediment particles. This capacity depends on the proportions of organic matter, clays, and amorphous hydroxides, on grain size, and on the pH and redox conditions within the mud. Changes in sediment composition frequently accompany major shifts in lake salinity during deposition, so that it is usually difficult to separate in the sedimentary record the influence of exchange capacity from that of water chemistry. In a relatively homogeneous sediment sequence deposited during the Ancylus and Littorina phases of the Baltic, Ericsson (1973) found good agreement between palaeosalinity interpretations based on diatom analyses and concentration profiles of extractable Ca and Mg. My contrast, at a second site, Rässan Mire, where marine clays underlie organic sediment of freshwater origin, concentrations of extractable ions decrease upward and are better correlated with physical characteristics of the sediments such as dry matter and carbon content. In lower sections of the profile, where sediment composition is more constant, Ca and Mg concentrations have a relationship to salinity as interpreted from diatoms. Extractable K and Na at both sites show no relationship to former salinity conditions.

SOIL HISTORY

The interpretation of soil development from sediment geochemistry was first clearly formulated by Mackereth (1966) in his classic investigation of chemical stratigraphy of the English Lakes. Mackereth envisaged that both the processes of leaching of mineral matter *in situ* and the erosive transport of soil particles could be reconstructed from sedimentary profiles of the alkali and alkaline earth elements. He reasoned that these two processes should produce contrasting patterns in the sediments. During periods of active erosion, the mass transport of raw unleached soils should increase the level of Na, K, and Mg in both the mineral fraction and in the sediments as a whole, whereas during episodes of relatively stable soils deep weathering of mature soil profiles should diminish the base content of mineral material prior to its erosive removal and sedimentation. Mackereth concluded that sedimentary Na, K, and Mg, being primarily associated with detrital minerals, directly reflect the intensity of weathering and erosion within the catchment.

Since Mackereth's initial observations, many studies have demonstrated the utility of these elements as indices for soil erosion. In late-glacial sediments from highland lakes in the Lake District and in Scotland, Pennington (1981b) attributes decreased levels of Na and K in interstadial deposits (13,000–11,000 BP) to more-favourable climatic conditions leading to vegetated and stabilized soils. A return to glacial conditions and intense solifluction is recorded in subsequent deposits of the Younger Dryas stadial (11,000–10,000 BP), in which Na and K again increase and organic content declines. In early post-glacial sediments the proportions of sodium and potassium are substantially lower throughout than in the underlying late-glacial deposits, whereas periods of accelerated erosion corresponding to climatic and anthropogenic events commencing around 5000 BP are inferred from the increased sedimentary content of Na and K. Palynological investigations of these same sites indicate a close correspondence between vegetational history and the presumed sequence of soil changes based on the alkali metals.

Likewise Huttunen et al. (1978) recognize intense erosion during the early post-glacial history of the lake Hakojärvi in southern Finland from high sedimentary concentrations of Na, K, and Mg. A subsequent decline in these elements at the close of the Preboreal represents the gradual revegetation of barren deglaciated soils and the closing of the pine–birch forest. Accelerated erosion resulting from agricultural activities is indicated in the chemical profiles from the lake Prästsjön near the Swedish coast of the Gulf of Bothnia (Renberg 1976) and from Lovojärvi in southern Finland (Huttunen and Tolonen 1977), corresponding to the appearance of cultural pollen in the sediments. Sasseville and Norton (1975) report geochemical evidence for increased erosion associated with the arrival of European settlers in northern New England. In two of the four Maine lakes they studied, elevated levels of K and Mg accompany this increase in disturbance pollen types in the sediments, but at the other two sites these elements do not increase upward across the disturbance horizon. This individualized response in the sediments may result from differences in erosive transport or deposition among the four basins.

Engstrom's (1983) chemical and pollen profiles from Lake Hope Simpson and Moraine Lake in Labrador also demonstrate the simultaneous impact of environmental change on soil and vegetational development (figs. 1.1 and 1.2). During the first 2000 years of post-glacial history, maximum levels of all allogenic components, including K and Mg, characterize sediments derived from inorganic tundra soils. A rapid invasion of spruce-fir forests between 7500 and 7000 BP dramatically increased the organic content of catchment soils and decreased mineral erosion and the allogenic component of the sediments. Chemical stratigraphy at Hope Simpson remains relatively constant after about 6000 BP, but a return to minerogenic sediments with high K and Mg occurs at Moraine Lake after 4000 BP, possibly because of increased erosion under a more open lichen woodland. While this last development at Moraine Lake is difficult to detect in the pollen spectra of the depauperate vegetation, similar patterns in sediment composition from nearby sites suggest a regional climatic cause.

In the discussion thus far it has been assumed that variations in the intensity of erosion are directly related to changes in soil composition as reflected in the mineral content of the sediments. Increases in sedimentary Na, K, and Mg, for example, represent changes in soil composition that one might expect to accompany accelerated erosion of catchment soils. The dissection of soil horizons by surface flow and increased channel scour associated with intense erosion should transport to the sediments greater quantities of allochthonous materials, which are also more minerogenic. However, as Mackereth clearly recognized, variations in the rate of deposition of other sedimentary components, particularly organic matter, might alter sediment composition independently of the rate of mineral erosion. Although it is intuitively logical that densely vegetated organic soils should be more resistant to erosive forces than newly deglaciated barren landscapes, this hypothesis cannot be tested on the basis of total sediment composition alone.

Mackereth (1966) surmised that the relative intensity of soil erosion might be determined from sediment composition by measuring the extent of chemical weathering of the allogenic minerals. Weathering intensity, he felt, should depend upon the length of time that the mineral matter was exposed to leaching before its removal to the sediments, so that detrital minerals derived from stable soils should be more thoroughly leached of alkali and alkaline-earth elements as compared to rapidly eroded material. Thus if total ash weight of the sediments was directly related to erosional intensity, as hypothesized, then it should also be inversely proportional to the degree of chemical leaching of the mineral matter. This is precisely what Mackereth found when he graphed the sodium and potassium content of the mineral matter against the percent mineral matter in the sediments from Esthwaite, Ennerdale, and Buttermere. These results, reproduced in fig. 1.3, show progressively higher concentrations of Na and K in the mineral matter (less leaching) with increasing total mineral content of the sediments (greater erosion). He concluded that major changes in chemical weathering had accompanied variations in soil stability throughout the post-glacial history of the English Lake District and that these changes had strongly influenced trophic development in the lakes themselves.

However, Mackereth's analysis of soil leaching may not apply to all sediments because his determination of 'mineral matter' is based on

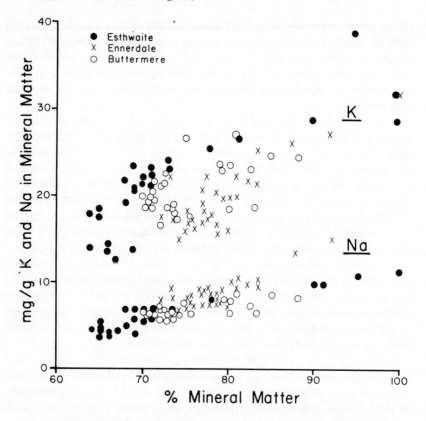

Fig. 1.3   Relationship between the sodium and potassium content of the mineral matter and the total mineral content of the sediments of Esthwaite, Ennerdale, and Buttermere. From Mackereth 1966.

ash weight (loss-on-ignition) and thus includes variable quantities of authigenic oxides and diatom silica. These components can constitute a large fraction of the toal dry weight of freshwater sediments, as seen, for example, in Engstrom's (1983) profiles from Lake Hope Simpson (fig. 1.2), in which authigenic Fe or biogenic Si alone exceed the combined allogenic complement of the dry sediments in the upper half of the core. While Mackereth's analyses do not include Si, his measurements of iron (primarily authigenic) exceed 10% of the total dry sediment (including organics) in sections of the Esthwaite and Windermere cores.

It is also apparent from the Labrador diagrams that biogenic Si and amorphous Fe oxides are proportionally more abundant in organic sediments, while muds highest in inorganic matter include the least amounts of these materials. Therefore the dilution of alkali and alkaline earth elements by non-mineral inorganics is greatest in highly organic deposits and least in highly minerogenic sediments. This effect can be clearly seen in figs. 1.4 and 1.5, where the Labrador data are graphed in the same manner as Mackereth's English results. Fig. 1.4 shows the relationship between the K and Mg per gram inorganic matter (i.e. mineral matter according to Mackereth) and the total inorganic content of the sediments from Moraine and Hope Simpson. Calculated in this way, K and Mg exhibit the same trend that Mackereth reported from the English lakes (cf. fig. 1.3), i.e. a diminishing base-metal content of the inorganic matter with progressively more-organic sediments. However, if the K and Mg contents are calculated as a weight percent of the allogenic fraction alone (all elements represented as their common oxides), these data are totally uncorrelated with the inorganic content of the sediments (fig. 1.5). Thus when the influence of non-clastic inorganics is removed the predictive relationship shown in fig. 1.4 is eliminated. While Mackereth's results cannot be tested in this same manner, the Labrador data suggest that dilution of allogenic minerals by authigenic and biogenic materials also may be responsible for the trend in the Na and K content of the 'mineral matter' from the English lakes.

Alternatively, it is possible that the composition of the allogenic minerals in the Labrador sediments is an artifact of the fractionation procedures, and that variations in the elemental content of eroded clastics were eliminated during the laboratory extraction of authigenic and biogenic components. This explanation seems unlikely in view of the evidence presented in a previous section, which demonstrates that extraction efficiency was relatively constant throughout all levels in both sediment profiles. Nevertheless, wet-chemical separations only approximate the true fractional composition of heterogeneous sediments, so that the validation of interpretations will depend upon similar results from other sediments, with refinements in the techniques outlined here. Unfortunately, of those palaeoecological investigations that have presented geochemical evidence for soil leaching (e.g. Pennington *et al.* 1972, Guppy and Happey-Wood 1978) most are based on bulk chemical analysis, and only a few have expressed their results per gram inorganic matter to eliminate the influence of organic matter, as did Mackereth (e.g. Davis R. and Norton 1978, Rippey *et al.* 1982, Koljonen and Carlson 1975).

The progressive diminution of nutrients in soil by chemical weathering is frequently cited in palaeological studies as a potential driving force for limnological changes in lake productivity and alkalinity as well as autogenic succession in catchment vegetation. Declining lake productivity (meiotrophication, cf. Quennerstedt 1955) and a gradual replacement of alkaline-water diatoms by those tolerant of acid conditions are characteristic of many north-temperate lakes and are commonly attributed to the

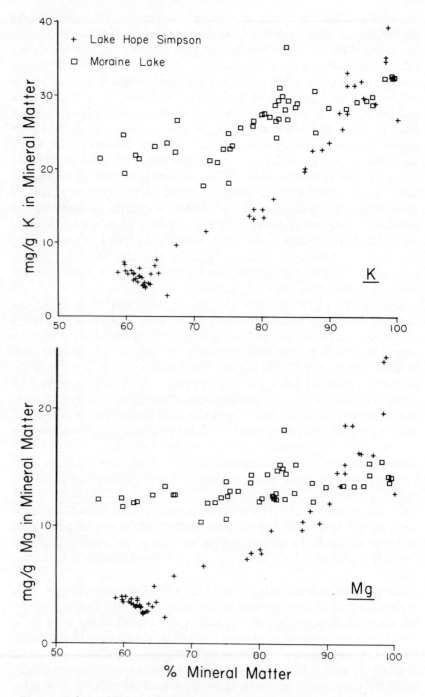

Fig. 1.4   Relationship between potassium and magnesium content of the mineral matter and the total mineral content of the sediments of Lake Hope Simpson and Moraine Lake.

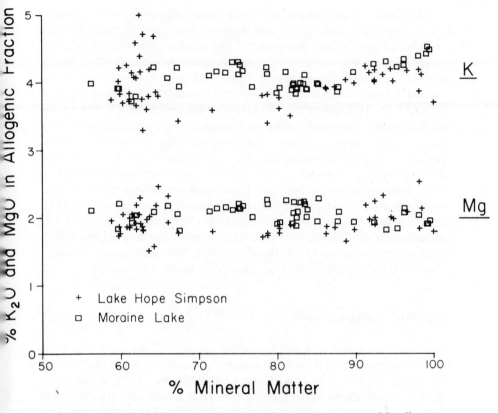

Fig. 1.5 Relationship between potassium and magnesium content of the allogenic fraction (calculated as common oxides) and the total mineral content of the sediments of Lake Hope Simpson and Moraine Lake.

progressive leaching of catchment soils (e.g. Pennington *et al.* 1972, Pennington 1981a, Digerfeldt 1972, 1977, Renberg 1976, Crabtree 1969, Walker 1978). Likewise, Iversen (1958, 1964) and Andersen (1964, 1966) emphasized the gradual development of nutrient-poor acid soils as an important factor in the interglacial vegetation cycles of north-west Europe.

These patterns in lake ontogeny and vegetation development offer indirect evidence for progressive leaching of soil minerals, but they may also result from other changes in the physical and chemical properties of catchment soils, such as organic content, cation exchange capacity, and water retention. Some investigators have suggested, for example, that the limnological transition from alkaline to acidic conditions resulted from the invasion of conifer forests (Huttunen *et al.* 1978, Tolonen 1972) or the development of blanket peat in the watershed (Round 1956, 1961).

Nevertheless, the progressive chemical degradation of primary minerals is a well-known process in soil genesis, and investigations of soil chronosequences have demonstrated the gradual leaching of mineral constituents such as carbonates from the solum (e.g. Crocker and Major 1955, Ugolini 1966, Jacobson and Birks 1980).

In the Labrador profiles (Engstrom 1983), mineral weathering may not be detectable because of the slow rate of decomposition of the native granites under the cool subarctic climate of the region. Sediments representing warmer climates, more soluble minerals, or longer time-spans may record progressive leaching of catchment soils. However, it is also possible that gradual changes in soil mineralogy are too subtle to detect with the crude fractionation techniques available to palaeoecologists. If the bulk of detrital clastics are unaltered primary minerals, as in the Labrador sediments, small variations in the proportion of weathered minerals may be difficult to detect, particularly because leaching occurs primarily at the surface of soil particles, while the entire particle volume, which is largely unweathered, is digested in sediment analysis.

Iron and manganese

These two components of lake sediments are perhaps the most important for palaeolimnology, for their abundance in the sediments is determined by conditions both in the catchment and in the lake. They are found in lake sediments bound in the mineral lattices of allogenic clastics and also as components of authigenic oxides, sulphides (Fe), carbonates, and organic complexes derived from the weathering of catchment soils. Because a number of independent environmental factors control the delivery and sedimentation of these various materials, Fe and Mn profiles are difficult to interpret.

Authigenic forms of Fe and Mn in lake sediments, unlike the highly stable clastic components, are potentially labile, and factors controlling preservation can exert considerable influence on the final quantity of these materials ultimately incorporated into the bottom muds. Thus any explanation of iron and manganese stratigraphy must take into account historical changes in both rate of supply to the lake and in those processes that control the pattern and degree of preservation. While the supply of these elements is largely controlled by environmental processes occurring in the catchment, preservation in the sediments is a function of limnological conditions within the lake.

SUPPLY

Fe and Mn may reach the lake as constituents of mineral particles, and thus they may represent the severity of geologic erosion in the same way as do the alkaline and alkaline-earth elements. On the other hand, they may be weathered from the rocks in the catchment and retained as hydrated oxides within the mineral soils (primarily the B horizon) or in organic soils (complexed or adsorbed on organic matter). Both elements exhibit very low solubility under oxidizing conditions, but under reducing conditions that come from waterlogging or from the build-up of raw humus on the soil surface they may be released from the soil – Mn more readily than Fe because of its greater solubility – and travel to the lake in solution as organic complexes or as colloidal particulates.

Mackereth (1966) clearly recognized that the supply of these elements to the lake sediments could be controlled by changes in soil composition, and he hypothesized that the differential mobility of these two elements could be used to reconstruct palaeo-redox conditions in catchment soils. On the basis of his analyses of cores from the English lakes he argued that the low Fe:Mn ratio in the post-glacial sediments of Windermere resulted from reducing conditions in Windermere soils, and that the higher Fe:Mn ratio in the sediments of Ennerdale Water reflected the more barren mineral soils of the Ennerdale watershed. Presumably the mildly reducing conditions in the Windermere catchment permitted Mn to be selectively transported relative to Fe, thereby lowering the Fe:Mn ratio below the lithospheric average.

Pennington *et al.* (1972) supported this hypothesis with their work on Loch Sionascaig and other sites in northern Scotland. Here pollen profiles and peat sections indicate that blanket peat and water-logged organic soils became widespread during the mid-Flandrian (c. 5000 BP). At the same time Fe and Mn concentrations rose dramatically in lake sediments, indicating increased solutional transport of these elements as reducing conditions developed within catchment soils.

While these studies emphasize the role of redox conditions in the mobilization of iron and manganese from catchment soils, it seems likely that soil humic materials may play an even more important role in the transport of these elements. Humic and fulvic acids produced by the microbial decay of terrestrial plant materials form strong soluble complexes with multivalent cations such as Fe and Mn (Gjessing 1976). These organometallic complexes are carried by surface water or groundwater to the lake, where they are slowly polymerized and precipitated as particles of dy. Enriched iron levels are often found in lakes that have high water colour caused by dissolved organics, and several studies have demonstrated a close correlation between iron content and dissolved humics in surface waters (e.g. Shapiro 1966, Clair and Engstrom, in press). In addition correlations between sedimentary organics and Fe and Mn, which have been noted in cores from such widely contrasting environments as tropical Lake Valencia in Venezuela (Lewis and Weibezahn 1981) and temperate lakes in southern Finland (Alhonen 1971, Koljonen and Carlson 1975, Kukkonen 1973), affirm that soil organics play an important role in the transport of dissolved metals to the sediments.

Soils that form beneath coniferous vegetation are particularly prone to the build-up of thick organic horizons, because of the slow rate of microbial decay in conifer litter. These organic soils are rich in acid humic materials, so their development should increase the mobility of Fe and Mn. Sedimentary profiles from the two sites in south-eastern Labrador studied by Engstrom (1983) provide striking examples of the close relationship between vegetational change and Fe and Mn stratigraphy (figs. 1.1, 1.2). Concentrations of authigenic Fe and Mn, together with allochthonous sedimentary organics, increase sharply at both sites at the same time as the vegetation shifts from shrub tundra to conifer forest at 7500–7000 BP. Such changes are taken to represent increase mobilization of organometallic complexes from waterlogged soils produced through humus accumulation under coniferous vegetation.

Because of the low solubility of oxidized forms of Fe and Mn, groundwater may be important for the transport of these elements to lakes. A progressive decrease in the flux of subsurface water has been hypothesized to explain the decline in Fe and Mn in sediment profiles from several Swedish and Norwegian lakes. In southern Sweden at Lake Trummen (Digerfeldt 1972), Ranviken Bay of Lake Immeln (Digerfeldt 1975), and Lake Flarken (Digerfeldt 1977), the Fe content of early post-glacial sediments is very high (as is Mn in Lake Flarken) but decreases dramatically at the close of the Boreal period. Digerfeldt attributes the high content of Fe (up to 15–25% of dry matter at Ranviken Bay) to early leaching from glacial soils in the catchment and transport by groundwater through the lake bottom. As impermeable organic sediments accumulated on top of the permeable glacial sands and gravels, the hypolimnetic environment became progressively sealed from the influence of upwelling groundwater. This decreased the supply of Fe and other dissolved ions to the lake water and ultimately the sedimentary record. A similar account has been offered by Kjensmo (1978) for Vilbergtjern Lake in south-eastern Norway.

The role of accumulating fine-grained organic sediment in 'sealing' the bottom from inflowing subsurface water, although intuitively logical, is speculative. Flow of shallow groundwater tends to emerge at a lake shore regardless of the permeability of the bottom sediments, simply because of the geometry of the topography and the equipotential flow lines (Winter

1978). With the development of a 'seal' the inflow of groundwater will merely be more concentrated near the sandy lake shore rather than reduced in amount. Moreover, lakes from other regions of similar topography and lithology exhibit Fe and Mn maxima during periods other than the early post-glacial, contrary to the predictions of the basin-sealing hypothesis. Hope Simpson and Moraine Lake in Labrador exhibit low authigenic Fe and Mn during the first 3000 years of their post-glacial history (Engstrom 1983). This is similarly true in southern Finland for Lovojärvi (Huttunen and Tolonen 1977), Hakojärvi (Huttunen et al. 1978), and several of the lakes studied by Alhonen (1971).

In Vilbergtjern the early post-glacial peaks in Fe and Mn occur in sediments that are highly minerogenic, in which maxima in Na and K also occur (Kjensmo 1978). Kjensmo's analyses are based on bulk digestion of the sediments, and thus the Fe and Mn maxima probably come in part from allogenic clastics derived from erosion rather than only solutional transport by groundwater. Digerfeldt (1972) suggests that a decrease in minerogenic Fe may have contributed to the decline of Fe deposition during the early Boreal period of Lake Trummen. It is also possible that the early Fe peaks in these Swedish sediments are associated with catchment vegetation. Notable in this regard is the shift from pine–birch forest to mixed deciduous forest during the early Atlantic period, corresponding to the decline in Fe and Mn at Lakes Trummen and Flarken (Digerfeldt 1972, 1977). This may be purely coincidental, but alternatively it may represent a decrease in Fe and Mn mobilization as soil humus changed when the vegetation shifted.

SEDIMENTATION AND PRESERVATION

In aerated waters ionic species of Fe are normally in very low concentrations, and most iron occurs either as colloidal suspensions of ferric hydroxide or bound to organic ligands. Mn is more soluble than Fe and is also more complexed with sulphates, carbonates, and organic acids. The precipitation of these various forms of Fe and Mn is governed by the ionic composition of the water, redox conditions, pH, and (in the case of humic complexes) temperature, light penetration, and microbial activity. It is possible that long-term changes in these limnological conditions could alter the proportions of Fe and Mn reaching the sediments and the amount being flushed from the lake. However, aside from redox, which is discussed below, little serious consideration has been given in palaeolimnological investigations to other factors controlling Fe and Mn solubility.

The Fe and Mn that precipitate from the water column and accumulate in the sediments may be redissolved. Diagenetic changes and migration of ions within the sediment column can also alter the patter of Fe and Mn distribution, as material that was once present on the surface passes downward through a steep gradient in redox potential and pore-water concentration.

In oxygenated bottom waters Fe and Mn are deposited largely as hydrated oxides and coagulates of humic organic complexes. Mn may be sedimented by sorption onto less soluble Fe oxides. In surface sediments further microbial decay releases organically bound cations, and the Fe and Mn are quickly immobilized as hydrous oxides. Apparently most iron exists in oxygenated surface muds as amorphous ferric hydroxide (Coey et al. 1974). However, Nriagu and Coker (1980) determined that large concentrations ($1000-2000\mu g/g$) of iron occur tightly bound to humic acids in Lake Ontario sediments. Delfino and Lee (1971) suggest that Mn is present in sediments primarily as amorphous hydrous Mn oxides. In addition to these fine-grained sediments, enriched deposits of Fe and Mn occur as ferromanganese nodules and crusts (Jones and Bowser 1978).

Because of the fine texture and low density of Fe and Mn precipitates, wave and current actions tend to winnow these particles from coarse-grained materials and selectively transport them to deeper, more protected regions of a lake bottom. Thus Fe and Mn concentrations tend to increase with increasing depth of the overlying water column (Syers et al. 1973 and references therein). In large water bodies a further separation between Mn and Fe can occur spatially across a lake bottom. In Green Bay of Lake Michigan the Fe : Mn ratio decreases with distance from major river sources (Jones and Bowser 1978). This occurs because Mn tends to remain in solution long after iron has precipitated. Thus changes in lake hydrodynamics over time could be expected to alter the pattern of Fe and Mn deposition over a lake bottom. In single-core studies, such variations are assumed to be minor relative to changes in overall inputs and preservation. However, this assumption has yet to be tested by multiple core studies, in which spatial variation in sedimentary Fe and Mn profiles can be evaluated.

In most lakes only a thin layer of the uppermost sediment is oxidized, typically a few centimetres or less in thickness. Below this layer oxygen consumption by microbial activity and the diagnosis of inorganic chemical species lead to reducing conditions. As sediments are moved downward across this boundary by progressive burial, oxidized forms of Fe and Mn undergo reduction, for they are no longer stable at the lower redox potential. While the exact chemical configuration of Fe and Mn below the oxidized microzone is uncertain, Williams et al. (1971a) suggest that most Fe exists as a hydrated form of amorphous ferrous hydroxide. Presumably manganese also exists as Mn (II) sorbed into ferrous complexes or as hydrous oxides (Delfino and Lee 1971). In their reduced states both Fe and Mn are considerably more mobile and enter into solution when sediments are buried below the redox-cline. Interstitial fluids below the oxidizing microzone are enriched in these elements relative to the surface muds, and upward diffusion may occur along this concentration gradient generated by precipitation in the oxidized surface sediments. This post-depositional migration can result in Fe and Mn peaks at the top of a sediment profile that have little to do with historical changes in sediment deposition. This process is clearly illustrated for iron in fig. 1.6, taken

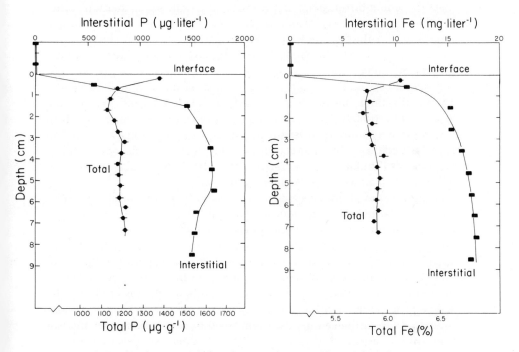

Fig. 1.6   Total and interstitial phosphorus and iron profiles in lake sediments five weeks after experimental homogenization. From Carignan and Flett 1981.

from Carignan and Flett's (1981) experimental studies of Lake Memphremagog sediments. The strong downward increase of interstitial Fe and the surface peak in total Fe were developed within five weeks of incubation of homogenized sediments under oxygenated water.

Surface maxima in both Fe and Mn have been noted in undisturbed sediment cores from a wide variety of lakes. Examples include those found by Sasseville and Norton (1975) in a number of Maine lakes, by Kemp *et al.* (1976) in several cores from Lake Erie, and by Engstrom (1983) in Lake Hope Simpson, Labrador (fig. 1.2). Surface enrichment in Mn has long been known from deep-sea sediment cores, and Lynn and Bonatti (1965) have related the thickness of this zone to rates of accumulation and organic content of the sediments. Unlike Mn, Fe does not exhibit an upward flux in marine sediments, presumably because sulphide is sufficiently abundant that Fe forms stable sulphides insoluble under reducing conditions.

Because of the potential for post-depositional migration of Fe and Mn, trends in these elements near the top of cores must be interpreted with caution. This is particularly true for short-core studies that attempt to assess recent anthropogenic changes in Fe and Mn sedimentation. Older profiles, however, may also preserve enriched surface zones through rapid burial following formation, as shown in cores from Hovvatn and Blåvatn, Norway (Norton and Hess 1980), and Moraine Lake, Labrador (Engstrom 1983).

If the surface muds remain oxidized, most of the sedimentary Fe and Mn remains immobile, even in reduced forms below the redox-cline. The fact that interpretable patterns of manganese and iron do occur at all in lake sediments supports this conclusion. Stability of reduced Fe and Mn may be enhanced by interaction of Fe(II) with organic and sulphur compounds (Shukla *et al.* 1971, Syers *et al.* 1973) or through equilibrium of sorbed Fe and Mn with enriched interstitial waters.

As long as the sediment surface remains oxidized, iron and manganese that enter the sediments remain permanently entombed within the mud. If hypolimnetic waters become deoxygenated through temporary or permanent stratification, however, the oxidized microzone quickly disappears, and Mn followed by Fe and other chemical species diffuse from the sediments into the bottom waters. Fe and Mn are highly enriched in interstitial fluids relative to the lake water and migrate upward along the sediment-water gradient when the oxidized barrier is removed. Continued loss from the sediments is sustained by desorption of Fe and Mn from solid phases and dissolution of their hydrated oxides. Mayer *et al.* (1982) suggest that in some eutrophic lakes the release of Fe and Mn may be mediated by the initial formation of carbonate phases (siderite, rhodochrosite) from the high-valence oxyhydroxides in the sediments.

Mortimer (1941–2) demonstrated this process in a set of experiments, in which he incubated Windermere sediments under both oxic and anoxic conditions. He noted a flush of Mn and Fe into the waters above his deoxygenated sediments when redox conditions dropped to the point where the 0.20 isovolt (Eh) emerged above the sediment interface. The release of Mn preceded that of Fe.

Mackereth (1966) recognized that long-term changes in hypolimnetic oxygen conditions could affect the distribution of redox-sensitive elements within sediment cores, and he demonstrated that Mn and Fe profiles could be used to reconstruct the history of such changes. He reasoned that Mn, by virtue of its greater mobility, could be selectively depleted in sediments under moderately anaerobic hypolimnia. Presumably some of the dissolved Mn would be mixed into the water column during overturn and flushed from the lake or resedimented in shallower water above the thermocline. If seasonal anoxia became more severe or prolonged, Fe as well as Mn should be lost from the sediments. On the basis of these arguments he determined that among the English lakes Esthwaite Water has experienced seasonal anoxia over most of its post-glacial history, while Windermere and Ennerdale Water retained oxic hypolimnia. More recently Davison (1981) has shown that the separation of manganese from iron may also result from dissolution of Mn precipitates as they descend into anoxic waters.

Mackereth also realized that changes in supply of Fe and Mn from the watershed could produce sedimentary patterns that might be difficult to distinguish from those produced by changes in redox of the bottom waters. He reasoned that, if peaks in the Fe : Mn ratio were correlated with minima in iron concentration, then changes in hypolimnetic redox were responsible, because Fe should be lowest during strongly reducing conditions at the sediment surface and Mn should be even lower relative to iron. On the other hand, if iron and the Fe : Mn ratio were positively correlated then changes in supply (soil redox) were controlling, because high Fe concentrations and Fe : Mn ratios both indicate low soil redox and hence higher rate of supply.

However, as previously mentioned, Fe and Mn may occur in enriched crusts and buried layers that result from post-depositional migration below the redox-cline. Such patterns readily develop without bottom-water anoxia and may alter the initial Fe : Mn ratio because of the different mobility of these two elements. This process may result in a poor correspondence between the Fe : Mn ratio and other evidence for hypolimnetic oxygen conditions.

In the hypolimnion of very productive lakes or in the monomolimnion of meromictic lakes redox potential may be lowered below 100mv, and H_2S may be formed through sulphate reduction and the breakdown of organic matter. Because ferrous iron is released at higher potential (Eh = 200 − 300mv), Fe appears in the water before H_2S does but is quickly re-sedimented as insoluble iron sulphides if reducing conditions become severe enough for H_2S to form (Mortimer 1941–2, Einsele 1937). Manganous sulphide, on the other hand, is much more soluble, so that H_2S does not inhibit Mn losses from the sediments. Thus the final stages of increasing anoxia may be marked in the sediments by an enrichment of iron and an increase in the Fe : Mn ratio (Hutchinson 1957).

This is illustrated in Kjensmo's (1968) study of post-glacial sediments from meromictic Lake Svinsjøen in south-eastern Norway. The development of permanent stratification is indicated by a sharp increase in Fe and a concomitant drop in Mn beginning about 2000 BP. Digerfeldt (1975) suggests that iron sulphide deposition during the Preboreal and Boreal periods of Ranviken Bay of Lake Immeln may have occurred under prolonged or permanent stratification caused by meromixis. Unusually high iron levels and only modest Mn concentration are present in these sediments. A similar case is also made for the early post-glacial of Lake Flarken (Digerfeldt 1977).

In most palaeolimnological studies the identification of iron sulphides is either inferred from the black colour of the sediments or is based on stratigraphic association of high iron and sulphur concentrations in the core. Because iron sulphide occurs in various mineralogic forms $(FeS_{0.9} - FeS_2)$ in lake sediments and because other sulphur-bearing compounds may be present, sulphur analysis provides only a rough index of sedimentary FeS (Jones and Bowser 1968). Nevertheless, major increases in total sedimentary sulphur are often related to iron-sulphide deposition, as Nriagu (1968) showed for Lake Mendota sediments. In Lovojärvi in southern Finland a sharp rise in Fe and S marks the onset of meromixis, which in turn is correlated with the beginning of slash-and-burn agriculture in the watershed c. 1700 BP (Huttunen and Tolonen 1977). Because only a slight increase in Mn occurs at this time, a selective precipitation of iron sulphide in the permanently anoxic monomolimnion is indicated.

Iron sulphides are also commonly found in sediments deposited during the early stages of the isolation of Swedish lakes from both freshwater and marine phases of the Baltic. According to Ingmar (1973) conditions favourable for FeS deposition resulted from decreased water circulation and hypolimnetic anoxia associated with the formation of a closed lake from an open bay. Alternatively, these FeS deposits may have formed under oxygenated waters through iron and sulphate reduction below the redox-cline, as commonly occurs in marine sediments (Hallberg 1972) and occasionally in lakes (e.g. Kemp et al. 1976, Vallentyne 1963). Sediments deposited at the time of isolation are typically low in organic matter, particularly humic materials that tend to stabilize reduced phases of iron oxides so that muds rich in authigenic FeOOH or $Fe(OH)_3$ may have undergone diagenesis during burial to form FeS.

In certain meromictic lakes that are unproductive or low in sulphate, H_2S production may be insufficient to precipitate all of the Fe generated from the bottom, so that sediments may become depleted in Fe rather than enriched. For example, the recent meromictic sediments of Brownie Lake in Minneapolis, Minnesota, are low in Fe relative to older holomictic muds (E. Swain, unpublished data). In turn the monomolimnion contains very high levels of dissolved Fe, which probably contributes to the stability of meromixis (Kjensmo 1968).

Other factors in addition to redox can affect the flux of Mn and Fe across the sediment-water interface. Delfino and Lee (1971) determined that the release of Mn from Lake Mendota sediments was pH-dependent. In laboratory experiments where sediment suspensions were incubated under anoxic conditions, Mn was released as pH was lowered, and was taken up as pH was increased. Bioturbation of surface sediments by benthic organisms should also influence the preservation of Mn and Fe. Burrowing and respiratory activities of oligochaetes and midge larvae greatly increase the flux of dissolved solutes into and out of the sediments. Oxygen penetration is also enhanced, thus affecting both microbial activity and redox conditions within the mud (Petr 1976).

Because numerous factors control the final stratigraphy of Fe and Mn, these elements should be used with caution to reconstruct palaeo-redox conditions in lakes. Independent supporting evidence should be sought whenever possible. In this regard Eh measurements of the sediments might be useful. While in practice, redox potential may be measured by inserting a platinum electrode into the sediments (taking care not to introduce oxygen), interpretation of the data is considerably more difficult (cf. Mortimer 1971). Because of the electrochemical irreversibility of many of the redox reactants in lake sediments and because of continual diagenesis during burial, Eh measurements are different from the redox potential that existed at the time the sediments were laid down. However, major shifts in sediment Eh may still provide a useful proxy for the direction of palaeo-redox changes. For example, Huttunen and Tolonen (1977) use sediment Eh to confirm the historical development of meromixis in Lovojärvi. They found that Eh drops precipitously to negative values at the same level where meromixis is indicated by the iron and sulphur profiles. Likewise, Vuorinen (1978) employed Eh measurements to indicate the onset of meromictic conditions in Hännisenlampi, Finland.

Fossil evidence from benthic midge larvae have also been used to reconstruct past hypolimnetic oxygen conditions. Brugam (1978) determined that head-capsules from the subfamily Chironominae dominated throughout his core from Linsley Pond, Connecticut. Because this group of midges is characteristic of anaerobic environments, Brugam concluded that the hypolimnion had been seasonally anoxic throughout the lake's history, which was also indicated by the Fe:Mn ratio. Yet in Esthwaite Water, Goulden (1964) found an oxygen-demanding midge fauna throughout all but the last 1000 years of the lake's sediments, whereas Mackereth (1966), on the basis of Fe and Mn profiles, had argued for hypolimnetic anoxia during this same period of time. This may indicate, as Pennington (1981a) suggests, that chemical and biological responses to oxygen regimes do not always coincide. But it may also mean that one or the other of these two interpretations is actually incorrect.

Phosphorus

One of the primary tasks of palaeolimnology has been to reconstruct the trophic status of lakes in the past. Because of close relationship between phosphorus concentration and primary productivity (e.g. Vollenweider 1968, Smith 1979) and because sediments are an effective sink for P, numerous attempts have been made to use the phosphorus content of sediments as a proxy for palaeo-productivity. Some of these studies have found a strong correlation between sedimentary P and other historical evidence of trophic conditions (e.g. Shapiro *et al.* 1971, Bradbury 1978), whereas others have found quite the opposite (e.g. Brugam 1978, Brugam and Speziale 1983). The inconsistency of these results arises from historical variation in other limnological factors that control the extent to which P in the water is incorporated within the sediments. Like Fe and Mn, the sedimentary signal from phosphorus is commonly mixed and difficult to interpret.

Lake sediments may be greatly enriched in phosphorus relative to catchment soils. While concentrations of sedimentary P are normally less than 0.25% (by dry weight as phosphate), in organic muds this level can be 4–5 times that found in unweathered bedrock (Koljonen and Carlson 1975). Phosphorus occurs in lake sediments in both authigenic and allogenic materials, although in non-calcareous deposits P bound in detrital minerals is a small portion of the total. The sedimentation of authigenic P occurs through the biological uptake of dissolved inorganic P and its subsequent deposition as particulate organic P. Sorption by humic complexes and iron oxides, precipitation as iron phosphates, and possibly coprecipitation with carbonates (Otsuki and Wetzel 1972) may also transport dissolved P to the bottom sediments. Stripping of dissolved P from the water column is relatively rapid, so that surface waters are usually depleted and the sediments enriched in this limiting nutrient. Some of the particulate organic P that rains to the bottom is incorporated directly into the sediments, whereas more labile materials are degraded either in the hypolimnion or in the surface sediments through microbial activity that liberates orthophosphate and soluble organic compounds. These in turn may become fixed in the sediments primarily by sorption to hydrated ferric oxides or by complexation with refractory organic materials such as humic and fulvic acids.

FORMS OF PHOSPHORUS

Phosphorus within the sediments may be classified into three functional categories, according to Williams *et al.* (1976a): (1) constituents of discrete mineral phases such as apatite or vivianite, (2) sorbed or coprecipitated components of amorphous inorganic phases, and (3) P bound to organic compounds. Detrital apatite may comprise an appreciable portion of the total P in calcareous muds (e.g. Williams *et al.* 1971b, 1976b), but in oligotrophic lakes of low alkalinity apatite P is only a very minor component of the total sedimentary P (Syers *et al.* 1973). Among authigenic phosphate minerals, vivianite $(Fe_3(PO_4)_2 \cdot 8H_2O)$ has been occasionally reported from reduced sediments rich in silt or clay; the diagenetic formation of other discrete-phase phosphates in lake sediments has not been demonstrated. Although several authors have postulated the formation of hydroxyapatite on the basis of thermodynamic principles, evidence for its formation in lake sediments is decidedly lacking (Syers *et al.* 1973 and references therein). No mineralogical evidence exists for the occurrence of other discrete-phase ferric or aluminum phosphates such as strengite $(FePO_4 \cdot 2H_2O)$ or variscite $(AlPO_4 \cdot H_2O)$ in lake sediments (Jones and Bowser 1978), although Nriagu and Dell (1974) suggest that at least cryptocrystalline Ca, Al, and Fe hydroxyphosphates might form at pH and Eh conditions encountered at the sediment-water interface.

Inorganic phosphorus in most sediments occurs primarily as sorbed components of amorphous iron oxides. This iron complex is largely responsible for the exchange of dissolved P in interstitial waters and ultimately for the levels of P that accumulate in the sediments. While P is often more closely correlated with manganese than with iron in lake sediments, the low Mn:P ratios in most cases argue against Mn controlling the retention of inorganic P. Electron-probe analyses of ferromanganese nodules typically show high levels of P in Fe-rich zones but not in Mn-rich zones (Jones and Bowser 1978). Thus the relationship between Mn and P occurs indirectly through the independent association of both elements with Fe.

Shukla *et al.* (1971) and Williams *et al.* (1971a) postulate that a gel complex of the type envisaged by Mattson *et al.* (1950), largely consisting of hydrated Fe oxide along with smaller amounts of organic matter, P, and Al and associated SiO_2, is the major contributor to the sorption of inorganic P by both calcareous and non-calcareous sediments. Shukla *et al.* (1971) showed that Fe extraction by acid ammonium oxalate or citrate-dithionate-bicarbonate (CDB) eliminated or greatly reduced the ability of sediments to sorb added P. Williams *et al.* (1971a) determined that most of the variability in total P found in surface sediments from 14 Wisconsin lakes resulted from variation in extractable inorganic P (as oxalate or CDB). In turn inorganic P and total P were highly correlated with extractable Fe. Both studies showed P sorption to be inversely related to the $CaCO_3$ content of the sediments, indicating that carbonates are much less important than Fe in controlling phosphorus retention.

Adsorption process rather than a precipitation reaction is indicated by the ability of sediments to retain added inorganic P and by the rapidity of the exchange between solid and solution phases (Syers *et al.* 1973, Theis and McCabe 1978).

Organic phosphorus includes all sedimentary P associated with carbon atoms through C-O-P or C-P bonds in either particulate or dissolved form. The relationship with carbon was verified by Sommers *et al.* (1972) and Williams *et al.* (1971a), who showed a strong correlation between organic C and organic P in both calcareous and non-calcareous sediments. Organic P is largely associated with high-molecular-weight humic-fulvic complexes and may constitute from 10 to 70% of the total P in lake sediments (Sommers *et al.* 1972).

The fractionation of sedimentary P into different functional categories has clearly allowed a better understanding of the various processes that control phosphorus retention in modern sediments. Although palaeolimnological studies may also benefit from phosphorus fractionation, in these investigations the separation of allogenic mineral P from authigenic forms, both inorganic and organic, is most critical. Because authigenic P passes to the sediments primarily through biotic cycling, it is a potential proxy for trophic conditions in the past, unlike detrital phosphate minerals, which do not contribute to primary productivity. The importance of this distinction for sediments rich in detrital apatite is illustrated by the study of Williams *et al.* (1976b) on the history of phosphorus deposition in Lake Erie (fig. 1.7). Six cores from widely spaced locations in this lake show constant levels of apatite P throughout. In pre-cultural sediments this amounts to 60–80% of the total phosphorus, so that the magnitude of recent increases in P of anthropogenic origin would be underestimated (and conversely the relative productivity in the past would be overestimated) if only total P had been analysed. In other situations the changes in the flux of detrital P could totally obscure any pattern in authigenic forms. For sediments with low detrital apatite, total P or acid-leached P are probably suitable measures of authigenic phosphorus.

PHOSPHORUS RETENTION

As long as a constant proportion of dissolved P in the water is permanently deposited at the bottom, sedimentary phosphorus profiles should reflect levels of past lake productivity. However, because phosphorus retention in the sediments is strongly controlled by sorption onto iron oxides, variations in both Fe content and redox conditions might change phosphorus accumulation independently of its concentration in the water. This is clearly shown in P profiles from Moraine Lake, Labrador

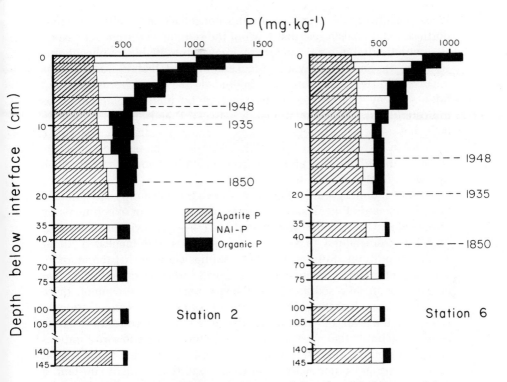

Fig. 1.7   Profiles of apatite P, non-apatite inorganic P, and organic P at two coring stations in Lake Erie. From Williams *et al.* 1976b.

(fig. 1.1), in which authigenic Fe (and Mn) concentration closely parallels non-detrital phosphorus levels throughout; correlation coefficients of P with Fe and Mn are 0.82 and 0.93, respectively (Engstrom 1983). The highest P levels occur between 4200 and 7500 BP, during which independent evidence from sedimentary pigments indicate a progressive decline in lake productivity. Rather than representing past phosphorus levels in the lake, the P curve rises and falls in apparent response to variations in iron concentrations.

Mackereth (1966) found a close correspondence between P and Fe or Mn profiles from Windermere and Esthwaite Water and attributed the P stratigraphy to variation in Fe and Mn deposition. He assumed that most of the P in these sediments was deposited by coprecipitation with Fe and Mn and that this abiotic mechanism precluded any interpretation of past trophic conditions from sedimentary phosphorus. Bortleson and Lee (1974) obtained similar results from cores of six Wisconsin lakes and noted that the transition from pre- to post-cultural sediments did not affect the relationship between P and Fe or Mn. Only in oligotrophic

Weber Lake were the P and Fe profiles not related, and in this instance Bortleson and Lee suggest that most of the sedimentary iron was associated with detrital minerals and therefore was unavailable for phosphorus sorption. In Ennerdale Water, Mackereth found that P was likewise uncorrelated with Fe, and he concluded that phosphorus was not influenced by the iron cycle but instead by biotic precipitation, as indicated by the strong relationship between P and organic carbon. In both these studies only total iron was analysed (HF digestion), so that a correlation between P and authigenic Fe could not be tested without the confounding influence of allogenic iron, which is abundant in Ennerdale as well as in Weber Lake sediments.

Several laboratory studies have also shown that the ability of sediments to sorb inorganic P is closely related to iron content. In experiments by Shukla *et al.* (1971) and Bortleson and Lee (1974), known concentrations of PO_4-P were added to sediment suspensions and incubated under aerobic conditions. Results from both studies showed that the more Fe was present in the sediments, the more P was sorbed. Because the content of native P in lake sediments is likewise related to Fe content, these studies indicate that the ability of sediments to sorb inorganic P strongly influences the levels of P that accumulate under natural conditions. These results also imply that most sediments have the capacity to sorb additional amounts of phosphorus in response to an increased flux to the lake. If iron content and other sedimentological conditions remain constant, a record of changing P levels in the water may be preserved in the sediments.

The phosphorus content of sediments taken from a selection of different lakes is invariably uncorrelated with measures of lake productivity. This is true for both organic P (Sommers *et al.* 1972) and inorganic P (Williams *et al.* 1971a, Bortleson and Lee 1974), and it is not surprising given the great variation among sediments in their capacity to sorb and retain phosphorus. Spatially within a lake both the amount and form of phosphorus in the sediments is also highly variable, because of density-dependent settling of the fine particulates with which P is associated. Phosphorus deposition, like Fe and Mn, is relatively greater with depth (e.g. Delfino *et al.* 1969, Frink 1967). Nevertheless, Birch *et al.* (1980) report that contemporary rates of P accumulation in four Washington State lakes are correlated with primary production. However, these results are based on single cores from each lake and may be fortuitous rather than broadly applicable.

If hypolimnetic waters become anoxic under prolonged stratification, phosphorus is released from the sediments to the overlying water in much the same way as Fe and Mn. Mortimer's (1941–2) laboratory experiments with Windermere sediments demonstrate that an oxidized surface layer is critical for the retention of P in the sediments. Under reducing conditions hydrous ferric oxides are converted to more soluble and highly dispersed ferrous forms, from which P is easily mobilized into interstitial waters, where upward diffusion occurs. Phosphorus release may continue even after concentrations of P in the overlying water exceed levels in interstitial fluids, the flux apparently being driven by a concentration gradient between the surface layers of the ferrous gel and the bottom water (Theis and McCabe 1978).

Because redox conditions strongly influence the retention of P in the sediments, the reconstruction of past trophic conditions from phosphorus profiles is usually possible only if redox conditions are constant over time. If not, stratigraphic trends in sedimentary P may be quite contrary to actual historical variations in P influx to the water. For example, if sediments become reduced so that P is released from the bottom, P flux to the surface waters will increase while the actual record preserved in the sediments may be just the opposite.

Mackereth (1966) recognized this problem in the phosphorus profile from Esthwaite Water and concluded that changes in retention of P within the sediments were largely responsible for the stratigraphy that he observed. Phosphorus levels were lowest in those sections of the core where iron and manganese profiles indicated strongly reducing conditions. Digerfeldt (1975, 1977) also found that low P levels in Lake Flarken and Ranviken Bay coincided with apparent episodes of severe anoxia in the bottom waters as indicated by the deposition of iron sulphides. Apparently both P and Fe were liberated through the reduction of ferric oxides, but Fe was reprecipitated as FeS while P remained in solution.

In Linsley Pond, Connecticut, Livingstone and Boykin (1962) discovered that sedimentary phosphorus told a strikingly different story about trophic development than that revealed by faunal remains and pigments. High P levels characterized sediments in the early history of the lake, during which time biotic evidence suggested low productivity, whereas low P and fossil evidence for high productivity were found in the upper half of the core. Mackereth (1966) pointed out that the iron profile for Linsley Pond reported by Hutchinson and Wollack (1940) follows a pattern very similar to the P profile of Livingstone and Boykin. The sharp mid-core decline in both Fe and P strongly suggests that reducing conditions eventually developed in the lake and mobilized both Fe and P from the sediments. Thus, the discrepency between biotic and chemical evidence is explained by the fact that the P curve represents palaeo-redox conditions and not palaeo-productivity.

Progressive changes in redox during sediment burial may also alter phosphorus stratigraphy in a manner similar to that described for Fe and Mn. Phosphorus, mobilized from reduced sediments at depth, may migrate upward along a concentration gradient generated by precipitation of interstitial P in the oxidized microzone. As long as this process continues, an 'artificial' maximum in P concentration is maintained at the sediment surface. This process was clearly demonstrated by both Tessenow (1972) and Carignan and Flett (1981), who found significant P enrichment in the surface sediments of homogenized lake muds following incubation under oxic conditions. The interstitial and total P profiles from Carignan and Flett, shown in fig. 1.6, developed within five weeks of complete sediment mixing, illustrating the rapid alteration of phosphorus levels that may occur following deposition. Phosphorus mobilization peaks unrelated to the known history of P loading have also been observed in undisturbed sediment cores from several lakes (e.g. Williams *et al.* 1976b, Carignan and Flett 1981). However, the phosphorus maxima are often less pronounced than those of Mn or Fe. Williams *et al.* (1976b) determined that post-depositional migration of P had altered phosphorus stratigraphy from two of their six stations in Lake Erie, but they noted that this was less evident than the upward mobilization of Mn. Actual phosphorus deposition was differentiated from vertical migration by comparison of P profiles with historical records of phosphorus loading to the lake. This type of approach is necessary whenever P stratigraphy is used in palaeolimnology to document recent trends in anthropogenic eutrophication.

Other factors in addition to sediment composition and redox are known to influence the retention and release of phosphorus in limnetic sediments. The transport of P between sediments and overlying lake water is regulated by rates of diffusion and turbulent mixing. Sediment mixing increases the rate of physical contact between mud and water and may accelerate the release of P under both anoxic and oxygenated bottom waters (Lee *et al.* 1976, Stevens and Gibson 1976, Holdren and Armstrong 1980). The churning of surface muds by wave and current actions and the activity of benthic organisms also enhances oxygen penetration under aerobic conditions and increases Eh at depth in the sediments. This may decrease phosphorus mobility by reducing interstitial P concentrations as shown in bioturbation experiments by R. Davis *et al.* (1975). However, Holdren and Armstrong (1980) found that large populations of chironomids actually increased the rate of P flux from intact sediment cores. Water temperature in the hypolimnion may also affect release rates by regulating the microbial activity, which in turn influences redox potential. Theis and McCabe (1978) noted maximum P flux from anoxic sediments during periods when hypolimnetic temperatures were highest, while in laboratory tests Holdren and Armstrong (1980) also measured greater P release at higher temperatures. Other studies indicate that pH interacts with redox to control P solubility. In anoxic sediments peak rates of phosphorus exchange occur near a pH of 7 (Kamp-Nielsen 1974), while under aerobic conditions P release from ferric oxides may result from competition by hydroxl ions at elevated pH (Lijklema 1976). While these different factors may vary throughout the developmental history of a lake and affect the final stratigraphy of sedimentary P, they are generally assumed to be minor relative to changes in P influx, redox, and sediment composition.

PHOSPHORUS SUPPLY

Despite the numerous problems associated with phosphorus sedimentation, several palaeolimnological studies have found good agreement between sedimentary P and other historical evidence for trophic development. Williams *et al.* (1976b) compared recent trends in total municipal phosphorus loading to Lake Erie with P concentration profiles taken from six coring stations. They observed that increases in non-apatite inorganic P in four of the cores closely paralleled the sharp rise in P influx to the lake from 1935 to 1971 (fig. 1.7). The remaining two cores were appreciably influenced by post-depositional migration, presumably caused by the slow rates of sediment accumulation at these sites. Likewise, Bradbury (1978) correlated sedimentary phosphorus stratigraphy from Shagawa Lake, Minnesota, with historical records of sewage discharge from the nearby town of Ely. Known cultural events such as settlement of the town (1887), the introduction of phosphate detergents (1948), and the installation of sewage treatment facilities (1954) were clearly evident in both the P profile and the diatom stratigraphy from Shagawa sediments.

In their investigation of the recent eutrophication of Lake Washington, Shapiro *et al.* (1971) compared phosphorus profiles from cores taken both before (1958) and after (1968, 1970) the diversion of sewage effluent from the lake. Although all cores showed recent enrichment of P near the surface, maximum P concentrations that were evident at the top of the 1958 profile had been displaced downward in the 1968 and 1970 cores by subsequent deposition of sediment with lower P content. These results indicate that a record of high P deposition from the period of maximum enrichment had been permanently fixed in the sediments. Ulén (1978) reported similar results from Lake Norrviken in central Sweden, where sediment cores taken in 1972 and 1976 record the progressive decline in phosphorus loading with the cessation in 1969 of sewage discharge into the lake. The P content of organic material caught in sediment traps declined over this same period of time, thus accounting for the change in phosphorus profiles.

Other studies that have attempted to reconstruct trophic history from sedimentary phosphorus have usually been less successful than these examples. Often equally plausible explanations for P profiles based on redox conditions or sediment composition cannot be differentiated from the effects of P loading, and frequently the supporting evidence from other palaeolimnological indices either contradicts the phosphorus stratigraphy or is ambiguous. This is particularly true of investigations into natural lake ontogeny during which gradual changes in phosphorus inputs are probably small and easily obscured by attendant variations in phosphorus retention. Those investigations in which P stratigraphy has been useful are characterized by dramatic changes in P influx that create decisive patterns in the sediments. Under these circumstances even decreased P retention caused by hypolimnetic anoxia may be of less importance than phosphorus loading in determining the final stratigraphy (cf. Bradbury 1978, Ulén 1978). The anthropogenic enrichment of lakes from the discharge of municipal or industrial sewage appears to be the one event that is sufficiently dramatic to leave conclusive evidence in the phosphorus record.

Silica and aluminium

Silicon and aluminium are typically the most abundant inorganic elements in lacustrine sediments, with the largest fraction of both contributed by clastic silicate minerals. In many lakes significant amounts of biogenically derived SiO_2 are also sedimented as diatom frustules, chrysophyte cysts, sponge spicules, and other silicified remains of aquatic organisms. Both SiO_2 and Al may be complexed with other sedimentary components, silica as an anionic ligand of hydrated Fe and Mn oxides and aluminium chelated with high-molecular-weight humic materials. Adsorption of dissolved silica from sediment pore waters by clay minerals may also contribute to SiO_2 retention in lake sediments (Johnson and Eisenreich 1979 and references therein). In addition, Nriagu (1978) has presented evidence for the authigenic formation of amorphous ferroaluminium silicates from the reaction of biogenic silica with ferric oxyhydroxides. The direct inorganic precipitation of silica is rare in lakes and may be restricted to the cooling of geothermal solutions and the dilution of alkaline (pH>10) brines (Jones and Bowser 1978).

Because both allogenic and biogenic forms of SiO_2 are usually abundant in lake sediments, the separation of these two fractions is essential for palaeoecological interpretations of silica stratigraphy. The chemical composition of detrital silicates is commonly used to reconstruct the history of mineral leaching in the development of catchment soils. As previously discussed, the depletion of alkali and alkaline-earth elements and the increase of 'resistate' elements such as Si in the sediment column is used to document this process. However, for sediments rich in amorphous silica, variations in biogenic deposition can confound these interpretations if only bulk chemical composition is determined.

Investigations of sedimentary biogenic silica include studies of silica cycling and the role of lake sediments in regulating dissolved SiO_2 (e.g. Nriagu 1978, Parker *et al.* 1977) as well as palaeolimnological reconstructions of diatom productivity and trophic development in the past. Biogenic silica determinations are considerably less time-consuming than diatom counts and in addition remove the problem of unidentifiable pieces of diatoms that are not included in counts yet represent some of the diatom production of a lake. However, chemical analysis of amorphous silica does not distinguish between SiO_2 from diatom production and contributions from other sources. Selective extraction techniques may remove some detrital SiO_2 from labile clay minerals or alternatively fail to dissolve all frustules completely, although in the presence of abundant diatom silica these errors should be relatively small. Opaline silica from other biologic sources such as chrysophyte cysts may occasionally be important, and inorganic amorphous silica can also be proportionally abundant in some sediments.

An additional problem with both diatom counts and silica determinations is the dissolution of diatom frustules within the sediment or water column. Most lakes are undersaturated with respect to amorphous silica, and diatom frustules are potentially soluble in such waters. Laboratory experiments by Lewin (1961) demonstrate that frustule dissolution commences following cell death, although cell walls of living diatoms are apparently stabilized by either organic coatings (Cooper 1952) or multivalent cations (Lewin 1961). Many investigations of diatom stratigraphy have documented poor preservation of frustules at selected depths or throughout a sediment profile, while studies of the silica budgets of North American Great Lakes indicate significant internal loading of dissolved SiO_2 from diatom dissolution (Parker *et al.* 1977, Nriagu 1978, Johnson and Eisenreich 1979). Even in those instances where microscopic examination reveals good preservation some silica losses may be sustained by partial dissolution of valve surfaces and labile structures (Battarbee 1978b). Biogenic silica in the sediments represents the net balance between diatom production and dissolution, so that changes in preservation must be considered in assessing palaeo-productivity from silica determinations, as well as in reconstructing water chemistry in general.

The dissolution of biogenic silica in lacustrine environments is a complex function of several variables, such that it is difficult to predict how well diatoms may be preserved in a particular lake. Although experimental studies by Jørgensen (1955) show that greater dissolution occurs with increasing pH, preservation in calcareous sediments is not always poor (Hecky and Kilham 1973), nor are diatoms in acidic lakes always well preserved (Round 1964). Lewin (1961) presents experimental data indicating that metal oxyhydroxide coatings may retard dissolution, while a number of workers suggest that silica content of the cell wall may also be an important factor in preservation (e.g. Meriläinen 1973, Parker and Edgington 1976). In deep lakes considerable dissolution may occur as diatoms descend through the water column (Parker et al. 1977, Johnson and Eisenreich 1979), and this process may be enhanced by mechanical breakage from zooplankton grazing and other processes (Cooper 1952). Within the sediments silica losses are regulated by the interstitial concentration of SiO_2 and the gradient between the sediments and overlying water (Tessenow 1966). High pH and temperatures at the sediment-water interface tend to enhance outflow of dissolved silica (Rippey 1977), as do bioturbation and turbulent mixing (Nriagu 1978). Hypolimnetic redox conditions may also regulate silica losses from the sediments of some lakes, as shown in experimental studies by Mortimer (1941–2) and Kato (1969). An increased flux of dissolved SiO_2 from the sediments occurs when bottom waters become anoxic; however, the loss is sustained by the solution of amorphous ferroaluminium silicate—a diagenetic product of biogenic silica (Nriagu 1978). The sediment-accumulation rate and hence the amount of time diatom frustules remain in the active surface sediments can also influence silica dissolution; more rapid burial should enhance preservation (Bradbury and Winter 1976, Riedel 1959).

Despite the potential for silica dissolution, diatom frustules seem to be well preserved in the sediments of most lakes. A relative assessment of preservation may be obtained for stratigraphic work with the scanning or transmission electron microscope to detect visually subtle evidence of frustule corrosion (Battarbee 1980). Moreover, those studies that have employed both 'absolute' diatom counts and biogenic silica determinations have demonstrated a good correlation between frustule abundance and SiO_2 concentrations in the sediments. Flower (1980) found that diatom counts, expressed either as bio-volume or valves per gram dry sediment, show a strong linear relationship with alkali-extractable silica in sediment traps from Lough Neagh, Northern Ireland. Renberg (1976) found a similar correspondence in stratigraphic work on the lake Prästsjön, Sweden, except that a decline in frustule abundance in the upper sediments did not occur in the profile for biogenic silica. He attributed this discrepancy to a greater quantity of uncountable fragments in the upper part of the core. Surprisingly, the number of diatom valves equivalent to one gram SiO_2 is roughly the same in both studies (c. 3×10^9), however, this value is about twice that reported by Parker and Edgington (1976) for plankton hauls in Lake Michigan. This difference may result from partial dissolution of the diatoms in the sediment studies, differences in taxonomic composition of the floras, or simply analytical error in silica determinations or diatom counts.

Palaeolimnological studies of biogenic silica stratigraphy have met with limited success in reconstructing lake productivity in the past. Digerfeldt (1972) found a sharp increase in the accumulation rate for alkali-soluble SiO_2 at the onset of recent cultural eutrophication in the sediments of Lake Trummen, Sweden. However, in the early post-glacial sediments low levels of biogenic silica do not agree with other geochemical evidence for relatively high productivity and nutrient availability. To explain this discrepancy Digerfeldt suggests that frustule dissolution was more prevalent during the early post-glacial, possibly because of higher pH of the water or mechanical breakage from reworking of minerogenic sediments. Renberg (1976) reports a similar problem in the biogenic silica profiles from Lake Prästsjön, in which SiO_2 accumulation and concentrations are unexpectedly low during the early phases of lake isolation from the Baltic (c. 3200 BP), when nutrient conditions were supposedly more favourable than during subsequent periods. Poor preservation may also be responsible for the lack of agreement between silica and other measures of lake productivity.

Likewise, the concentration profiles for biogenic silica reported by Engstrom (1983) from Labrador show little relationship to the probable history of trophic development. Data for sedimentary pigments and humic content of these cores strongly suggest higher productivity during the earlier tundra phases of landscape development, while silica content is consistently highest subsequent to this period in Lake Hope Simpson (fig. 1.2) and shows no interpretable pattern in Moraine Lake (fig. 1.1). In the case of Hope Simpson, dilution of silica inputs by allogenic clastics during the early post-glacial may be the cause of this discrepancy. Accumulation rates for biogenic silica correspond more closely with the hypothesized trophic history; SiO_2 accumulation is highest during earlier times (9000–7000 BP). However, accumulation calculations do not improve the 'fit' of the Moraine Lake silica profile, so that loss through dissolution may also be important in these sediments. Finally, it is important to remember that diatom populations account for only a portion of the primary productivity in lakes and that silica determinations may not accurately reflect trophic conditions in the past if the proportion of non-siliceous algae in the lake has changed.

Aluminium in the sediments of most lakes is primarily allogenic and is usually viewed as a resistate element in investigations of erosional intensity and soil leaching. Several palaeolimnological studies have also utilized sedimentary Al as a conservative element against which variations in the accumulation of bio-active elements and anthropogenic input are compared (e.g. Kemp et al. 1976). Titanium concentrations are employed in a similar manner under the assumption that clastic minerals are the principal source for these metals and that post-depositional diagenesis is minor (e.g. Cowgill and Hutchinson 1966). However, in humic-rich dystrophic lakes of the boreal region significant amounts of both Al and Ti may be authigenically deposited in the sediments. Koljonen and Carlson (1975) found highly enriched concentrations of titanium in the organic sediments of Järvenkylänjärvi and Telkko, Finland, and aluminium levels exceeding that which would could be present in silicate minerals according to CIPW-norms for bedrock samples. They conclude that both elements are chelated in acid soils by humic substances and are sedimented as organic colloids of dy. Alternatively, these patterns could result from changes in the sedimentation pattern for different minerals (with different densities) and the elements associated with them. However, similar results are reported by Engstrom (1983) where fractionation procedures should have eliminated the influence of changing mineral composition. For Lake Hope Simpson, Labrador (fig. 1.2), authigenic Al, and to some extent Ti, increase together with the organic content of the sediments.

Aluminium solubility in soils and lakes is strongly dependent on pH and rises markedly below a pH of 5. Elevated levels of dissolved Al have been recently noted in natural waters in large areas of eastern North America and north-western Europe (Wright et al. 1980). These increases are usually attributed to Al mobilization from poorly buffered soils as the result of industrially acidified precipitation (Cronan and Schofield 1979). Experimental studies by Schindler et al. (1980) further indicate that aluminium may be leached from sediments as the result of lowered pH of lake water. These changes are apparently not reflected in the sedimentary record, presumably because of the large reservoir of Al in the sediments. Conversely, palaeolimnological studies by Fritz and Carlson (1982) demonstrate that increased sedimentation of Al may accompany the recovery of previously acidified waters. Their stratigraphic investigation of the sediments from an acid strip-mine lake revealed a dramatic rise in sedimentary Al resulting from the elevated pH that accompanied lake restoration.

Summary

The discussions in this chapter emphasize the palaeoecological reconstruction of past environments from the inorganic chemistry of late-glacial and Holocene lake sediments. Recent experimental studies of sediment-water interactions are integrated with palaeolimnological investigations of chemical stratigraphy from both European and North American sites. A brief synopsis of the most important points follows.

1. Sedimentary components may be functionally classified according to their origin as *allogenic* if formed outside the lake proper, or *authigenic* if precipitated from aquatic solution or formed diagenetically within the sediments. Authigenic components include biochemically precipitated carbonate minerals, metal oxyhydroxides, sulphides, and phosphates, biogenic silica, and sorbed or coprecipitated elements. The allogenic fraction consists entirely of mineral particles resulting from the erosion of catchment soils.

2. Because different environment information is contained within each sedimentary component, analytical separation of different fractions can help elucidate geochemical history from heterogeneous lake muds. This detailed information is obscured through bulk analysis, which has been the normal procedure in palaeolimnological investigations. Sequential extraction procedures, adapted from investigations of soils and marine sediments, are recommended for the chemical analysis of lake sediments.

3. Geochemical data may be numerically represented in units of concentration or as rates of accumulation. Although accumulation calculations eliminate the problem of covariation among different sedimentary components that arises with concentration data, additional difficulties with sedimentation rates and core representivity severely limit the reliability of accumulation data in reconstructing basin-wide events.

4. Chemical evidence for palaeosalinity is usually difficult to interpret because many variables in addition to lake-water composition may influence the retention of ionic species within the sediments.

5. The relative intensity of soil erosion is generally reflected in the concentrations of elements primarily associated with clastic minerals (Na, K, and Mg). However, estimates of soil weathering based on the level of these elements in the mineral matter (inorganic ash) may be seriously confounded by variations in the sedimentation of biogenic silica and authigenic oxides. Definitive evidence for progressive soil leaching may be difficult to obtain from sediment chemistry because changes in soil mineralogy are too subtle to detect.

6. Humic materials produced by the microbial decay of terrestrial plant materials play an important role in the mobilization of Fe and

Mn from catchment soils. Soil redox and water retention may also be important in some cases, so that changes in catchment vegetation can strongly influence the stratigraphy of authigenic Fe and Mn in lake sediments.

7. The deposition and retention of authigenic phases of Fe and Mn in lacustrine sediments is highly dependent on the oxidizing–reducing condition of the lake system. However, the interpretation of palaeo-redox conditions from iron and manganese profiles requires careful evaluation of potential changes in catchment soils as well as post-depositional diagenesis and migration. Other factors such as pH, sediment mixing, and the spatial patterns of sediment deposition across the basin can also influence Fe and Mn profiles.

8. Phosphorus occurs in lake sediments as sorbed components of amorphous iron oxides, as constituents of discrete mineral phases, and as organically bound P. The inorganic phosphorus-iron complex is largely responsible for the exchange of dissolved P in interstitial waters and ultimately for the levels of P that accumulate in most sediments. If iron content and other sedimentological conditions remain constant, a record of changing P levels in the water may be preserved in the sediments.

9. Redox conditions strongly influence the retention of P in lake sediments, while sediment mixing, pH, and temperature may also affect phosphorus release rates. In addition, alteration of the phosphorus profile can result from post-depositional migration of interstitial P from reduced sediments at depth.

10. The reconstruction of past trophic conditions from phosphorus profiles is usually difficult because gradual changes in phosphorus inputs are usually obscured by attendant variations in phosphorus retention. Those investigations in which P stratigraphy has been useful are characterized by major changes in P influx that create decisive patterns in the sediments. The point-source discharge of industrial or municipal waste is one event that is sufficiently dramatic to leave conclusive evidence in the phosphorus record.

11. The analysis of amorphous silica may be used as a proxy for relative diatom abundance in lake sediments. However, problems of frustule dissolution and primary production by non-siliceous algae limit the direct reconstruction of trophic development from biogenic silica determinations.

12. Elevated levels of authigenic aluminium and titanium may occur in the sediments of polyhumic lakes because of chelation of these metals from humic-rich acid soils. Aluminium retention in sediments is also pH-dependent, such that increased sedimentation of Al may accompany the recovery of previously acidified lakes.

13. In oligotrophic lakes the primary environmental signal represented in the sediments is the composition of catchment soils. In more productive lakes autochthonous production of carbon, bacterial activity, and redox variations can produce internal modification of the sediment composition that may be read from chemical stratigraphy as a history of limnological development. But in most instances the chemical and biological conditions in the lake itself are difficult to discern from geochemical analyses because of changes in the composition and supply of eroding soils and because of the numerous factors that influence the retention of chemical species within the mud. Despite recent advances in our understanding of sediment-water interactions, chemical stratigraphy has not yet reached the point where it can be used independently of microfossil and other biotic evidence to reconstruct lake history.

Acknowledgments

This review is in part an outgrowth of a graduate seminar organized at the University of Minnesota for the purpose of reviewing recent literature on this subject. Regular participants included E.J. Cushing, Margaret B. Davis, Mary S. Ford, Nancy Eyster-Smith, Robert Moeller, Nancy Radle, and Edward B. Swain, all of whom contributed to the discussions and the development of the ideas summarized here. Subsequent consultations particularly with Swain and Ford led to the identification of many of the problems and to refinement of this manuscript. Final review and comment by Walter E. Dean of the U.S. Geological Survey and Stephen A. Norton of the University of Maine are also gratefully acknowledged. Radiocarbon dating for the Labrador sediments was generously provided by Weston Blake of the Geological Survey of Canada, and support for the field work in Labrador was provided by grants from the U.S. National Science Foundation (DEB–7913419 and DEB–8004286), the Geological Society of America (2445–79), the National Geographic Society, the U.S. Department of Energy (DE–AC02–79EV 10097), and the Smithsonian Institution. This paper is contribution 264 from the Limnological Research Center, University of Minnesota.

References

Alhonen, P., 1971. On the vertical distribution of iron and organic matter in some Finnish lake sediments. *Aqua Fennica* 1971: 98–104.

Allen, S.E., Grimshaw, H.M., Parkinson, J.A. and Quarmby, C., 1974. *Chemical Analysis of Ecological Materials.*

Andersen, S.T., 1966. Interglacial vegetational succession lake development in Denmark. *Palaeobotanist 15:* 117–27.

Andersen, S.T., 1964. Interglacial plant successions in the light of environment changes. *Report of VI International Congress on Quaternary, Warsaw (1961) 2:* 359–68.

Battarbee, R.W., 1978a. Biostratigraphical evidence for variations in the recent pattern of sediment accumulation in Lough Neagh, N. Ireland. *Verh. int. Verein. theor. angew. Limnol. 20:* 624–9.

Battarbee, R.W., 1978b. Observations on the recent history of Lough Neagh and its drainage basin. *Phil. Trans. R. Soc. B. 281:* 303–45.

Battarbee, R.W., 1980. Diatoms in lake sediments. In *Palaeohydrological Changes in the Temperate Zone in the Last 15,000 Years, Subproject B. Lake and Mire Environments,* ed. B.E. Berglund, I.G.C.P. Project 158, University of Lund, Department of Quaternary Geology.

Bengtsson, L., 1979. Chemical analysis. In *Palaeohydrological Changes in the Temperate Zone in the Last 15,000 Years, Subproject B. Lake and Mire Environments,* ed. B.E. Berglund, I.G.C.P. Project 158, University of Lund, Department of Quaternary Geology.

Birch, P.B., Barnes, R.S. and Spyridakis, D.E., 1980. Recent sedimentation and its relationship with primary productivity in four western Washington lakes. *Limnol. Oceanogr. 25:* 240–7.

Bloemendal, J., Oldfield, F. and Thompson, R., 1979. Magnetic measurements used to assess sediment influx at Llyn Goddionduon. *Nature 280:* 50–3.

Bortleson, G.C. and Lee, G.F., 1974. Phosphorus, iron, and manganese distribution in sediment cores of six Wisconsin lakes. *Limnol. Oceanogr. 19:* 794–801.

Bradbury, J.P., 1978. A paleolimnological comparison of Burntside and Shagawa Lakes, northeastern Minnesota (U.S. Environmental Protection Agency, Ecological Research Report EPA–600/3–78–004).

Bradbury, J.P. and Winter T.C., 1976. Areal distribution and stratigraphy of diatoms in the sediments of Lake Sallie, Minnesota. *Ecology 57:* 1005–14.

Bray, J.T., Bricker, O.P. and Troup, B.N., 1973. Phosphate in interstitial waters of anoxic sediments: oxidation effects during sampling procedure. *Science 180:* 1362–4.

Brugam, R.B., 1978. Human disturbance and the historical development of Linsley Pond. *Ecology 59:* 19–36.

Brugam, R.B. and Speziale, B.J., 1983. Human disturbance and the paleolimnological record of change in the zooplankton community of Lake Harriet, Minnesota. *Ecology 64:* 578–91.

Buckley, D.E. and Cranston, R.E., 1971. Atomic absorption analysis of 18 elements from a single decomposition of aluminosilicate. *Chem. Geol. 7:* 273–84.

Carignan, R. and Flett, R.J., 1981. Postdepositional mobility of phosphorus in lake sediments. *Limnol. Oceanogr. 26:* 361–6.

Carmouze, J.P., Golterman, H.L. and Pedro, G., 1976. The neoformations of sediments in Lake Chad; their influence on the salinity control. In

Interactions Between Sediments and Fresh Water, ed. H.L. Golterman (The Hague).

Chester, M. and Elderfield, H., 1968. The infra-red determination of opal in siliceous deep-sea sediments. *Geochim. cosmochin. Acta. 32*: 1128–40.

Chester, R. and Hughes, M.J., 1967. A chemical technique for the separation of ferro-manganese minerals, carbonate minerals and adsorbed trace elements from pelagic sediments. *Chem. Geol. 2*: 249–62.

Coey, J.M.D., Schindler, D.W. and Weber, F., 1974. Iron compounds in lake sediments. *Can. J. Earth Sci. 11*: 1489–93.

Cooper, L.H.N., 1952. Factors affecting the distribution of silicate in the N. Atlantic Ocean and the formation of N. Atlantic deep water. *J. Mar. biol. Assoc. U.K. 30*: 511–26.

Cowgill, U.M. and Hutchinson, G.E., 1966. La Aguada de Sanda Ana Vieja: the history of a pond in Guatemala. *Arch. Hydrobiol. 62*: 335–72.

Crabtree, K., 1969. Post-glacial diatom zonation of limnic deposits in North Wales. *Mitt. int. Verein. theor. angew. Limnol. 17*: 165–71.

Crocker, R.L. and Major, J., 1955. Soil development in relation to vegetation and surface age at Glacier Bay, Alaska. *J. Ecol. 43*: 427–48.

Cronan, C.S. and Schofield, C.L., 1979. Aluminium leaching response to acid precipitation: effects on high-elevation watersheds in the Northeast. *Science 204*: 304–6.

Davis, M.B. and Ford, M.S., 1982. Sediment focusing in Mirror Lake, New Hampshire. *Limnol. Oceanogr. 27*: 137–50.

Davis, R.B. and Norton, S.A., 1978. Paleolimnologic studies of human impact on lakes in the United States, with emphasis on recent research in New England. *Polskie Archwm Hydrobiol. 25*: 99–115.

Davis, R.B., Thurlow, D.L. and Brewster, F.E., 1975. Effects of burrowing tubificid worms on the exchange of phosphorus between lake sediments and overlying water. *Verh. int. Verein. theor. angew. Limnol. 19*: 382–94.

Davison, W. 1981. Supply of iron and manganese to an anoxic lake basin. *Nature 290*: 241–3.

Dean, W.E., 1981. Carbonate minerals and organic matter in sediments of modern north temperate hard-water lakes. *Spec. Publs. Soc. Econ. Paleontologists Mineralogists 31*: 213–31.

Dearing, J.A., Elner, J.K. and Happey-Wood, C.M., 1981. Recent sediment flux and erosional processes in a Welsh upland lake-catchment based on magnetic susceptibility measurements. *Quaternary Res. 16*: 356–72.

Delfino, J.J., Bortleson, G.C. and Lee, G.F., 1969. Distribution of Mn, Fe, P, Mg, K, Na, and Ca in the surface sediments of Lake Mendota, Wisconsin. *Environ. Sci. Technol. 3*: 1189–92.

Delfino, J.J. and Lee, G.F., 1971. Variation of manganese, dissolved oxygen and related chemical parameters in the bottom waters of Lake Mendota, Wisconsin. *Wat. Res. 5*: 1207–17.

Digerfeldt, G., 1972. The post-glacial development of Lake Trummen. *Folia limnol. Scand. 16*: 1–104.

Digerfeldt, G., 1975. The post-glacial development of Ranviken bay in Lake Immeln. III. Paleolimnology. *Geol. För. Stockh. Förh. 97*: 13–28.

Digerfeldt, G., 1977. The Flandrian development of Lake Flarken, regional vegetation history and paleolimnology (University of Lund, Department of Quaternary Geology Report 13).

Douglas, L.A. and Fiessinger, F., 1971. Degradation of clay minerals by H_2O_2 treatments to oxidize organic matter. Clays Clay Miner. 19: 67–8.

Dudas, M.J. and Harward, M.E., 1971. Effect of dissolution treatment on standard and soil clays. Proc. Soil Sci. Soc. Am. 35: 134–40.

Einsele, W., 1937. Physikalisch-chemische Betrachtung einiger Probleme des limnischen Mangan- und Eisenkreislaufs. Verh. int. Verein. theor. angew. Limnol. 8: 69–84.

Elner, J.K. and Happey-Wood, C.M., 1980. The history of two linked but contrasting lakes in North Wales from a study of pollen, diatoms and chemistry in sediment cores. J. Ecol. 68: 95–121.

Engstrom, D.R., 1983. Chemical stratigraphy of lake sediments as a record of environmental change (Ph.D. thesis, University of Minnesota).

Ericsson, B., 1973. The cation content of Swedish post-glacial sediments as a criterion of paleosalinity. Geol. För Stockh. Förh. 95: 181–220.

Flower, R J , 1980. A study of sediment formation, transport and deposition in Lough Neagh, Northern Ireland, with special reference to diatoms (Ph.D. dissertation, New University of Ulster).

Frink, C.R., 1967. Nutrient budget: rational analysis of eutrophication in a Connecticut lake. Environ. Sci. Technol. 1: 425–8.

Fritz, S.C. and Carlson, R.E., 1982. Stratigraphic diatom and chemical evidence for acid strip-mine lake recovery. Wat. Air Soil Pollut. 17: 151–63.

Gjessing, E.T., 1976. Physical and Chemical Characteristics of Aquatic Humus (Michigan).

Goldberg, E.D., 1954. Marine geochemistry 1. Chemical scavengers of the sea. J. Geol. 62: 249–65.

Goldberg, E.D., 1958. Determination of opal in marine sediments. J. mar. Res. 17: 178–82.

Goulden, C.E., 1964. The history of the cladoceran fauna of the Esthwaite Water (England) and its limnological significance. Arch. Hydrobiol. 60: 1–52.

Guppy, S.F. and Happey-Wood, C.M., 1978. Chemistry of sediments from two linked lakes in North Wales. Freshwater Biol. 8: 401–13.

Håkanson, L., 1977. The influence of wind, fetch and water depth on the distribution of sediments in Lake Vänern, Sweden. Can. J. Earth Sci. 14: 397–412.

Hallberg, R.O., 1972. Sedimentary sulfide mineral formation – an energy circuit system approach. Mineralium Deposita 7: 189–201.

Hecky, R.E. 1976. The use of diatom microfossils in interpreting the distribution of pore water solutes in small Momela Lake (Tanzania). In Interactions Between Sediments and Fresh Water, ed. H.L. Golterman (The Hague).

Hecky, R.E. and Kilham, P., 1973. Diatoms in alkaline, saline lakes: Ecology and geochemical implications. Limnol. Oceanogr. 18: 53–71.

Hesslein, R.H., 1976. An in situ sampler for close interval pore water studies. Limnol. Oceanogr. 21: 912–14.

Holdren, G.C. and Armstrong, D.E., 1980. Factors affecting phosphorus release from intact lake sediment cores. *Environ. Sci. Technol. 14:* 79–87.

Hutchinson, G.E., 1957. *A Treatise on Limnology. I. Geography, Physics and Chemistry* (New York).

Hutchinson, G.E. and Wollack, A., 1940. Studies on Connecticut lake sediments, II. Chemical analyses of a core from Linsley Pond, North Branford. *Am. J. Sci. 238:* 493–517.

Huttunen, P., Meriläinen, J. and Tolonen, K., 1978. The history of a small dystrophied forest lake, southern Finland. *Polskie Archwn Hydrobiol. 25:* 189–202.

Huttunen, P. and Tolonen, K., 1977. Human influence in the history of Lake Lovojärvi, S. Finland. *Suom. Mus. 1975:* 68–105.

Ingmar, T., 1973. Sjöavsnörningar från aktualgeologiska synpunkter. En översikt (University of Lund, Department of Quaternary Geology Report 3).

Iversen, J., 1958. The bearing of glacial and interglacial epochs on the formation and extinction of plant taxa. *Uppsala Univ. Arsskr. 6:* 210–15.

Iversen, J., 1964. Retrogressive vegetational succession in the post-glacial. *J. Ecol. 52:* 59–70.

Jackson, M.L., 1958. *Soil Chemical Analysis – Advanced Course* (published by the author: Madison, WISC).

Jacobson, G.L., Jr and Birks, H.J.B., 1980. Soil development on recent end moraines of the Klutlan Glacier, Yukon Territory, Canada. *Quaternary Res. 14:* 87–100.

Johnson, T.C. and Eisenreich, S.J., 1979. Silica in Lake Superior: mass balance considerations and a model for dynamic response to eutrophication. *Geochim. cosmochin. Acta. 43:* 77–91.

Jones, B.F. and Bowser, C.J., 1978. The mineralogy and related chemistry of lake sediments. In *Lakes, Chemistry, Geology, Physics,* ed. A. Lerman (New York).

Jørgensen, E.G., 1955. Solubility of the silica in diatoms. *Physiologia Pl. 8:* 846–51.

Kamp-Nielsen, L., 1974. Mud–water exchange of phosphate and other ions in undisturbed sediment cores and factors affecting the exchange rates. *Archiv. Hydrobiol. 73:* 218–37.

Kato, K., 1969. Behavior of dissolved silica in connection with oxidation-reduction cycle in lake water. *Geochem. J. 3:* 87–97.

Kelts, K. and Hsü, K.J., 1978. Freshwater carbonate sedimentation. In *Lakes, Chemistry, Geology, Physics,* ed. A. Lerman (New York).

Kemp, A.L.W., Thomas, R.L., Dell, C.I. and Jaquet, J.-M., 1976. Cultural impact on the geochemistry of sediments in Lake Erie. *J. Fish. Res. Bd. Can. 33:* 440–62.

Kjensmo, J., 1968. Late and post-glacial sediments in the small meromictic lake Svinsjøen. *Archiv. Hydrobiol. 65:* 125–41.

Kjensmo, J., 1978. Postglacial sediments in Vilbergtjern, a small meromictic kettle lake. *Polskie Archwm Hydrobiol. 25:* 207–16.

Koljonen, T. and Carlson, L., 1975. Behaviour of the major elements and minerals in sediments of four humic lakes in south-western Finland. *Fennia 137:* 1–47.

Krause, G.L., Schelske, C.L. and Davis, C.O. 1983. Comparison of three wet-alkaline methods of digestion of biogenic silica in water. *Freshwater Biol.* 13: 73–81.

Kukkonen, E., 1973. Sedimentation and typological development in the basin of the lake Lohjanjärvi, south Finland. *Bull. Geol. Surv. Finland.* 261: 1–67.

Lahann, R.W., 1976. The effect of trace metal extraction procedures on clay minerals. *J. Envir. Sci. Hlth A.* 11: 639–62.

Lamb, H.F., 1980. Late Quaternary vegetational history of southeastern Labrador. *Arctic. Alpine Res.* 12: 117–35.

Lee, G.F., Sonzogni, W.C. and Spear, R.D., 1976. Significance of oxic vs anoxic conditions for Lake Mendota sediments. In *Interactions Between Sediments and Fresh Water*, ed. H.L. Golterman (The Hague).

Lehman, J.T., 1975. Reconstructing the rate of accumulation of lake sediments: the effect of sediment focusing. *Quaternary Res.* 5: 541–50.

Leinen, M., 1977. A normative calculation technique for determining opal in deep-sea sediments. *Geochim. cosmochim. Acta.* 41: 671–6.

Lewin, J.C., 1961. The dissolution of silica from diatom walls. *Geochim. cosmochim. Acta.* 21: 182–98.

Lewis, W.M., Jr and Weibezahn, F.H., 1981. Chemistry of a 7.5-m sediment core from Lake Valencia, Venezuela. *Limnol. Oceanogr.* 26: 907–24.

Lijklema, L., 1976. The role of iron in the exchange of phosphate between water and sediments. In *Interactions Between Sediments and Fresh Water*, ed. H.L. Golterman (The Hague).

Likens, G.E. and Davis, M.B., 1975. Post-glacial history of Mirror Lake and its watershed in New Hampshire, U.S.A.: an initial report. *Verh. int. Verein. theor. angew. Limnol.* 19: 982–93.

Livingstone, D.A. and Boykin, J.C., 1962. Vertical distribution of phosphorus in Linsley Pond mud. *Limnol. Oceanogr.* 7: 57–62.

Lynn, D.C. and Bonatti, E., 1965. Mobility of manganese in diagenesis of deep-sea sediments. *Mar. Geol.* 3: 457–74.

Mackereth, F.J.H., 1966. Some chemical observations on post-glacial lake sediments. *Phil. Trans. R. Soc. B.* 250: 165–213.

Malo, B.A., 1977. Partial extraction of metals from aquatic sediments. *Environ. Sci. Technol.* 11: 277–82.

Manheim, F.T., 1966. A hydraulic squeezer for obtaining interstitial water from consolidated and unconsolidated sediments (U.S. Geological Survey Professional Paper 550–C: C256–C261).

Mattson, S., Alvsaker, E., Koutler-Anderson, E., Barkoff, E. and Vahtras, K., 1950. Phosphate relationships of soil and plant. VI. The salt effect on phosphate solubility in pedalfer soils. *Lantbr-Högsk. Annlr.* 17: 141–60.

Mayer, L.M., Liotta, F.P. and Norton, S.A., 1982. Hypolimnetic redox and phosphorus cycling in hypereutrophic Lake Sebasticook, Maine. *Wat. Res.* 16: 1189–96.

Mehra, O.P. and Jackson, M.L., 1960. Iron oxide removal from soils and clays by a dithionite-citrate system buffered with sodium-bicarbonate. *Clays and Clay Minerals, Proceedings of the Seventh National Conference on Clays and Clay Minerals* 7: 317–27.

Meriläinen, J., 1973. The dissolution of diatom frustules and its palaeoecological interpretation (University of Lund, Department of Quaternary Geology Report 3).

Mortimer, C.H., 1941–1942. The exchange of dissolved substances between mud and water in lakes. *J. Ecol. 29:* 208–329; *30:* 147–201.

Mortimer, C.H., 1971. Chemical exchanges between sediments and water in the Great Lakes – speculations regulatory mechanisms. *Limnol. Oceanogr. 16:* 387–404.

Müller, G., and Förstner, U., 1973. Recent iron ore formation in Lake Malawi, Africa. *Mineralium Deposita 8:* 278–90.

Norton, S.A. and Hess, C.T., 1980. Atmospheric deposition in Norway during the last 300 years as recorded in SNSF lake sediments, I. Sediment dating and chemical stratigraphy. In *Ecological Impact of Acid Precipitation*, Proceedings of an International Conference, Sandefjord, Norway, ed. D. Drabløs and A. Tollan (SNSF Project, Oslo).

Norton, S.A. and Sasseville, D.R., 1975. Flux of nutrients by diffusion through the lake sediment-hypolimnion interface for 9 lakes in Maine, U.S.A. *Verh. int. Verein. theor. angew. Limnol. 19:* 372–81.

Nriagu, J.O., 1968. Sulfur metabolism and sedimentary environment: Lake Mendota, Wisconsin. *Limnol. Oceanogr. 13:* 430–9.

Nriagu, J.O., 1978. Dissolved silica in pore waters of Lakes Ontario, Erie, and Superior sediments. *Limnol. Oceanogr. 23:* 53–67.

Nriagu, J.O. and Coker, R.D., 1980. Trace metals in humic and fulvic acids from Lake Ontario sediments. *Environ. Sci. Technol. 14:* 443–6.

Nriagu, J.O. and Dell, C.I., 1974. Diagenetic formation of iron phosphates in recent lake sediments. *Am. Miner. 59:* 934–46.

Otsuki, A. and Wetzel, R.G. 1972. Coprecipitation of phosphate with carbonates in a marl lake. *Limnol. Oceanogr. 17:* 763–7.

Parker, J.I., Conway, H.L. and Yaguchi, E.M., 1977. Dissolution of diatom frustules and recycling of amorphous silicon in Lake Michigan. *J. Fish. Res. Bd. Can. 34:* 545–51.

Parker, J.I. and Edgington, D.N., 1976. Concentration of diatom frustules in Lake Michigan sediment cores. *Limnol. Oceanogr. 21:* 887–93.

Pennington, W., 1981a. Records of a lake's life in time: the sediments. *Hydrobiologia 79:* 197–219.

Pennington, W., 1981b. Sediment composition in relation to the interpretation of pollen data. *Proc. IV int. palynol. Conf., Lucknow (1976–77), 3:* 188–213.

Pennington, W., Haworth, E.Y., Bonny, A.P. and Lishman, J.P., 1972. Lake sediments in northern Scotland. *Phil. Trans. R. Soc. B. 264:* 191–294.

Petr, T., 1976. Bioturbation and exchange of chemicals in the mud–water interface. In *Interactions Between Sediments and Fresh Water*, ed. H.L. Golterman (The Hague).

Quennerstedt, N., 1955. Diatoméerna i Långans sjövegetation. *Acta Phytogeogr. Suec. 36:* 1–208.

Renberg, I., 1976. Paleolimnological investigations in Lake Prästsjön. *Early Norrland 9:* 113–59.

Riedel, W.R., 1959. Siliceous organic remains in pelagic sediments. *Spec. Publs Soc. econ. Paleontologists Mineralogists 7:* 80–91.

Rippey, B., 1977. The behaviour of phosphorus and silicon in undisturbed cores of Lough Neagh sediments. In *Interactions Between Sediments and Freshwater*, ed. H.L. Golterman (The Hague).

Rippey, B., Murphy, J.R. and Stirk, W.K., 1982. Anthropogenically derived changes in the sedimentary flux of Mg, Cr, Ni, Cu, Zn, Hg, Pb, and P in Lough Neagh, Northern Ireland. *Environ. Sci. Technol. 16*: 23–30.

Round, F.E., 1956. The late-glacial and post-glacial diatom succession in the Kentmere Valley deposit. *New Phytol. 56*: 98–126.

Round, F.E., 1961. The diatoms of a core from Esthwaite water. *New Phytol. 60*: 43–59.

Round, F.E., 1964. The diatom sequence in lake deposits: some problems of interpretation. *Verh. int. Verein. theor. angew. Limnol. 15*: 1012–20.

Sasseville, D.R. and Norton, S.A., 1975. Present and historic geochemical relationships in four Maine lakes. *Limnol. Oceanogr. 20*: 699–714.

Sasseville, D.R., Norton, S.A. and Davis, R.B., 1975. Comparative interstitial water and sediment chemistry in oligotrophic and mesotrophic lakes, Maine, U.S.A. *Verh. int. Verein. theor. angew. Limnol. 19*: 367–71

Sasseville, D.R., Takacs, A.P., Norton, S.A. and Davis, R.B., 1974. A large-volume interstitial water sediment squeezer for lake sediments. *Limnol. Oceanogr. 19*: 1001–4.

Saunders, W.M.H., 1965. Phosphate retention by New Zealand soils and its relationship to free sesquioxides, organic matter and other soil properties. *N.Z. Jl agric. Res. 8*: 30–57.

Schindler, D.W., Hesslein, R.H. and Wagemann, R., 1980. Effects of acidification on mobilization of heavy metals and radionuclides from the sediments of a freshwater lake. *Can. J. Fish. Aquat. Sci. 37*: 373–7.

Shapiro, J., 1966. The relation of humic color to iron in natural waters. *Verh. int. Verein. theor. angew. Limnol. 16*: 477–84.

Shapiro, J., Edmondson, W.T. and Allison, D.E., 1971. Changes in the chemical composition of sediments of Lake Washington, 1958–1970. *Limnol. Oceanogr. 17*: 437–52.

Shukla, S.S., Syers, J.K., Williams, J.D.H., Armstrong, D.E. and Harris, R.F., 1971. Sorption of inorganic phosphate by lake sediments. *Proc. Soil. Sci. Soc. Am. 35*: 244–9.

Sly, P.G., 1976. Sedimentary environments in the Great Lakes. In *Interactions Between Sediments and Fresh Water*, ed. H.L. Golterman (The Hague).

Smith, V.H., 1979. Nutrient dependence of primary productivity in lakes. *Limnol. Oceanogr. 24*: 1051–64.

Sommers, L.E., Harris, R.F., Williams, J.D.H., Armstrong, D.E. and Syers, J.K., 1972. Fractionation of organic phosphorus in lake sediments. *Proc. Soil Sci. Soc. Am. 36*: 51–4.

Stevens, R.J. and Gibson, C.E., 1976. Sediment release of phosphorus in Lough Neagh, Northern Ireland. In *Interactions Between Sediments and Fresh Water*, ed. H.L. Golterman (The Hague).

Suhr, N.H. and Ingamells, C.O., 1966. Solution techniques for analysis of silicates. *Analyt. Chem. 38*: 730–4.

Syers, J.K., Harris, R.F. and Armstrong, D.E., 1973. Phosphate chemistry in lake sediments. *J. Environ. Qual. 2*: 1–14.

Tessenow, U., 1966. Untersuchungen über den Kieselsäurehaushalt der Binnengewässer. *Arch. Hydrobiol. suppl. 32:* 1–136.

Tessenow, U., 1972. Losungs- Diffusions- und Sorptionsprozesse in der Oberschicht von See Sedementen I. Ein Langzeitexperiment unter aeroben und anaeroben Bedingungen im Fliessgleichgewicht. *Arch. Hydrobiol. suppl. 38:* 353–98.

Theis, T.L. and McCabe, P.J., 1978. Phosphorus dynamics in hypereutrophic lake sediments. *Wat. Res. 12:* 677–85.

Tolonen, K., 1972. On the paleo-ecology of the Hamptjärn Basin II. bio-and chemostratigraphy. *Early Norrland 1:* 53–77.

Ugolini, F.C., 1966. Soil development and ecological succession in a deglaciated area of Muir Inlet, southeast Alaska (Ohio State University, Institute of Polar Studies Report No. 20).

Ulén, B., 1978. Seston and sediment in Lake Norrviken III. Nutrient release from sediment. *Schweiz. Z. Hydrol. 40:* 286–305.

Vallentyne, J.R., 1963. Isolation of pyrite spherules from recent sediments. *Limnol. Oceanogr. 8:* 16–30.

Vollenweider, R.A., 1968. Scientific fundamentals of the eutrophication of lakes and flowing waters with particular reference to nitrogen and phosphorus as factors in eutrophication. (Organization for Economic Cooperation and Development (OECD), DAS/CSI/68.27, Paris).

Vuorinen, J., 1978. The influence of prior land use on the sediments of a small lake. *Polskie Archwm Hydrobiol. 25:* 443–51.

Walker, R. 1978. Diatom and pollen studies of a sediment profile from Melÿnllyn, a mountain tarn in Snowdonia, North Wales. *New Phytol. 81:* 791–804.

Werner, D., 1966. Die Kieselsäure im Stoffwechsel von *Cyclotella cryptica* Reimann, Lewin and Guillard. *Arch. Mikrobiol. 55:* 278–308.

Williams, J.D.H., Syers, J.K., Shukla, S.S., Harris, R.F. and Armstrong, D.E., 1971a. Levels of inorganic and total phosphorus in lake sediments as related to other sediment parameters. *Environ. Sci. Technol. 5:* 1113–20.

Williams, J.D.H., Syers, J.K., Harris, R.F. and Armstrong, D.E., 1971b. Fractionation of inorganic phosphate in calcareous lake sediments. *Proc. Soil. Sci. Soc. Am. 35:* 250–5.

Williams, J.D.H., Jaquet, J.-M. and Thomas, R.L., 1976a. Forms of phosphorus in the surficial sediments of Lake Erie. *J. Fish. Res. Bd. Can. 33:* 413–29.

Williams, J.D.H., Murphy, T.P. and Mayer, T., 1976b. Rates of accumulation of phosphorus forms in Lake Erie sediments. *J. Fish. Res. Bd. Can. 33:* 430–9.

Winter, T.C., 1978. Ground-water component of lake water and nutrient budgets. *Verh. int. Verein. theor. angew. Limnol. 20:* 438–44.

Wright, R.F., Conroy, N., Dickson, W.T., Harriman, R., Henriksen, A. and Schofield, C.L., 1980. Acidified lake districts of the world: a comparison of water chemistry in southern Norway, southern Sweden, southwestern Scotland, the Adirondack Mountains of New York, and southeastern Ontario. In *Ecological Impact of Acid Precipitation*, Proceedings of an International Conference, Sandefjord, Norway, ed. D. Drabløs and A. Tollan (SNSF Project, Oslo).

2 Organic geochemistry of lacustrine sediments: triterpenoids of higher-plant origin reflecting post-glacial vegetational succession

P. A. Cranwell

Introduction

IN MANY British lakes the sediments contain a complete record of changes occurring within the lake and its environment since the last glaciation (Pennington 1981). A part of this record consists of microfossil assemblages such as pollen and diatom frustules, providing information about the terrestrial vegetational succession and about the aquatic ecology, respectively, together with other morphological residues studied by palaeolimnologists (Frey 1974). Chemical residues, including inorganic species derived by weathering from the lithosphere, and organic compounds derived from terrestrial or aquatic biota, provide a complementary record of changes in erosion rate, organic source material, and environmental factors affecting the degree of preservation.

Organic geochemical studies on lacustrine sediments have focused mainly on the lipid portion of organic matter, which contains a wide range of compound classes useful for correlation with source materials (Barnes and Barnes 1978). Thus, the chain length distributions of straight-chain compounds such as n-alkanes and n-alkanoic acids (Cranwell 1977a, Ishiwatari et al. 1980, Meyers et al. 1980) have been used to distinguish input from allochthonous and autochthonous source organisms. The branched/cyclic (B/C) lipid components contain marker compounds or molecular fossils indicative of organic input derived from micro-organisms, e.g. iso- and anteiso-branched fatty acids (Leo and Parker 1966, Cranwell 1973a) and alkanols (Cranwell 1980), and pentacyclic triterpene derivatives having the hopane carbon skeleton (Ourisson et al. 1979).

Compounds having a polycyclic carbon skeleton are believed to be more resistant to breakdown than acyclic compounds, as inferred from their enhanced abundance in ancient sediments and crude oils (Wakeham and Farrington 1980). Pentacyclic triterpenes containing an oxygen function in the 3-position occur widely in higher plants (Boiteau et al.

1964). These compounds have been detected in deltaic sediments (Corbet *et al.* 1980) and marine sediments (Dastillung 1976, Brassell *et al.* 1980) in which they are regarded as indicators of terrestrial input. Wakeham *et al.* (1980) noted that aromatic hydrocarbons, probably derived from pentacyclic triterpenes by means of reactions initiated either by loss of oxygen at the 3-position or by cleavage of ring A followed by aromatization beginning in ring B, have been found in shales, lignites, and crude petroleum. The recognition of these aromatic hydrocarbons in recent marine and lacustrine sediments (Spyckerelle *et al.* 1977a), together with the presumed precursor compounds (Dastillung 1976), indicates a more rapid formation than was previously thought to occur. Tetracyclic carboxylic acids formed by A-ring cleavage, possible intermediates in the second of the postulated aromatization mechanisms, have been found in deltaic and lacustrine sediments (Corbet *et al.* 1980) and their formation has been attributed to microbial activity.

Assessment of higher-plant input to recent lacustrine sediments using triterpenes therefore requires analysis of both the 3-oxygenated triterpenes, the primary plant products, and the tetracyclic acids and partially-aromatic hydrocarbons, secondary products resulting from early diagenesis. A suite of unidentified triterpenoid carboxylic acids, present in lacustrine sediments and previously attributed to allochthonous input (Cranwell 1977a), has now been shown to correspond with those reported by Corbet *et al.* (1980). Pentacyclic triterpenoid alcohols and ketones have also been detected in the same sediments. In this paper correlation of (1) the distributions of all these derivatives of 3-oxygenated pentacyclic compounds (ketones, alcohols, carboxylic acids, and aromatic hydrocarbons), and (2) changes in relative abundance of the constituents in each class with changes in higher-plant input, as recorded by the pollen content of the sediment, is attempted for 10 sediment samples representing various stages of the post-glacial vegetational succession at four sites.

Methods

Sampling sites

Cam Loch (area 2.6km^2, maximum depth 37m) in Sutherland, north-west Scotland (grid reference NC 210240), lies in a drainage basin of complex geology, containing Torridonian and Cambrian sedimentary rocks, Lewisian gneiss, and outcrops of Durness limestone. Sediment sections 1–4 (see table 2.1) were taken from core 725, obtained with a 6m corer (Mackereth 1958) in 9.5m of overlying water. Since this core did not reach the earliest post-glacial sediments, sections 5 and 6 were taken from core 729, which penetrated the glacial clay and was obtained in 14m water depth. Pollen data for the sediment profile were kindly provided by Professor W. Tutin (personal communication); data for the older sediment are taken from Pennington (1975).

Blelham Tarn (area $0.1km^2$, maximum depth 14m) is now one of the most productive lakes of the English Lake District (Jones 1972). The environmental history of the drainage basin during the post-glacial period has been summarized by Pennington *et al.* (1977). The section analysed for triterpenes was deposited in the lake as a result of soil disturbance resulting from forest clearance AD c. 1100; pollen data for this horizon are taken from Pennington *et al.* (1976).

Loch Clair (area $1.5km^2$, maximum depth 32m), Ross and Cromarty, north-west Scotland (Grid Reference NG 0056), lies in a drainage basin consisting of sedimentary rocks (sandstones and quartzites) and having drift cover on the lower parts of the catchment. Present vegetation consists of pine–birch forest in areas with good drainage and *Sphagnum-Calluna* bog where drainage is impeded. Pollen analysis of a sediment core (Pennington *et al.* 1972) showed that peat soils have developed over the past 5000 years with decreasing percentages of pine and birch during the present regional pollen zone beginning c. 950 BC. The uppermost 20cm layer from the profundal sediment profile was obtained using a Gilson surface sampler and sectioned horizontally at the 10cm level, which was deposited 140 ± 20 years ago from ^{210}Pb dating (Tutin, personal communication). The organic matter in the 20cm horizon is estimated, by interpolation from ^{14}C dating of older deposits, to be 380 ± 50 years old, assuming constant accumulation rate.

Crose Mere (area $0.15km^2$, maximum depth 9.2m) is a productive mere of the Shropshire-Cheshire plain (National Grid Reference SJ 430305), fed mainly by the sub-surface flow of groundwater. The seasonal phytoplankton succession and physical limnology of the mere have been studied by Reynolds (1973a, 1973b). Pollen analysis of the post-glacial sediment has been reported by Beales (1976) and the organic geochemistry of the 12–25cm section used in this study has been studied by Cranwell (1978).

EXTRACTION AND ISOLATION

Two methods of extraction were used, differing in the stage at which the sediment was subjected to acidic hydrolysis. In procedure A acidic hydrolysis preceded solvent extraction, which then gave the total lipids; this method was used for the Cam Loch (Cranwell 1977b) and Blelham Tarn sediments. The total lipids were saponified and separated into neutral and acidic portions by solvent partition. Chromatographic separation of the neutral product on alumina has been reported by Cranwell (1977b). Ketones were isolated from the benzene eluate by thin-layer chromatography (TLC); urea adduction removed n-alkan-2-ones enabling the branched/cyclic (B/C) ketones to be isolated from the non-adducted material. Monohydroxylic compounds were eluted from the alumina column with chloroform; n-alkanols were separated by urea adduction and sterols were precipitated from the non-adduct by treatment with digitonin. B/C alcohols were isolated from the residue by TLC (Cranwell 1980).

In procedure B solvent-extractable lipids were fractionated to give free and esterified components; subsequent acidic hydrolysis of residual sediment gave a bound lipid component, as described for Crose Mere (Cranwell 1978). Extraction procedure B was also used for the sediments from Loch Clair. The compound classes reported here were obtained from the extractable lipids, which were resolved into neutral and acidic portions by chromatography on KOH-impregnated silica gel (McCarthy and Duthie 1962). The neutral portion was separated by chromatography on silica gel, as described previously (Cranwell 1978) giving a weakly-polar fraction containing aromatic hydrocarbons, ketones and alkyl esters, and a polar fraction containing hydroxylic compounds. TLC separation of the weakly-polar fraction (Cranwell and Volkman 1981) gave partially-aromatic hydrocarbons eluting just ahead of the alkyl esters, and ketones from a more polar band. B/C ketones and alcohols were obtained as described above.

The acidic portion from either procedure A or B was esterified with diazomethane and the saturated monocarboxylic esters were isolated by TLC procedures (Cranwell 1973a); B/C esters were obtained after removal of n-alkanoic methyl esters by urea adduction.

PREPARATION OF RING A-OPENED TRITERPENE ACIDS

Lupeol, lupanol, α- and β-amyrin were each oxidized to the corresponding ketones by standard methods (Ives and O'Neill 1958). Photochemical decomposition of the ketones was accomplished by irradiation of a heptane solution with a high pressure mercury lamp for 1.5 hours while passing oxygen through the solution (Aoyagi et al. 1973). The solvent was removed, the residue was dissolved in diethyl ether and acidic products were isolated, and then esterified with diazomethane. The monocarboxylic methyl esters were obtained by TLC on silica gel developing with hexane-diethyl ether (85:15).

ANALYSIS

The ketones, monocarboxylic esters, alcohols, and aromatic hydrocarbons were analysed by gas chromatography (GC) using a fused silica capillary column (25m × 0.25mm inner diameter) coated with SE-30, hydrogen as carrier gas and temperature programming of 150–275°/3°/min. Alcohols were converted into trimethylsilylether or methyl ether derivatives for gas chromatography.

Computerized gas chromatography-mass spectrometry (C-GC-MS) used a fused silica column (25m × 0.2mm inner diameter) coated with methyl silicone fluid or a glass capillary column (20m × 0.25mm inner diameter) coated with OV-1 coupled to a Finnigan 4000 quadrupole filter mass spectrometer. Mass spectra were acquired and edited using an INCOS 2300 data system.

Compound identification

Sedimentary triterpene ketones and alcohols were identified by a combination of co-elution with authentic compounds on GC and comparison of mass spectra with those of authentic compounds where these were available; literature mass spectral data alone were used to assign structures to other constituents. C-GC-MS of the monoesters obtained by photochemical decomposition of the triterpene ketones supported structures analogous to those reported by Aoyagi *et al.* (1973); triterpenoid acids present in sediments were identified by GC co-elution and mass spectral comparison with these photochemical products. Aromatic hydrocarbons were identified by comparison of mass spectra with published data (Spyckerelle 1975, Wakeham *et al.* 1980).

Results

The abundance of the B/C lipid fractions and the proportion of higher-plant derived triterpenes in these lipids are presented in table 2.1, together with the age, carbon content, and major pollen zonation of the sediments. Although the triterpenes of higher-plant origin occurred only in the extractable lipids from Loch Clair and Crose Mere sediments, in which extraction procedure B was used, the abundance figures in table 2.1 include the bound lipids to allow comparison of the proportion of triterpenes with those in the other sediments.

B/C acids

Pentacyclic triterpene acids with an extended hopane skeleton and equivalent chain length (ECL) values (Miwa 1963) greater than 31 on SE-30 GC columns are ubiquitous in these sediments and have been previously reported for the Cam Loch profile (Cranwell 1977a) and Crose

Table 2.1.   Abundance of lipid fractions containing triterpenes in lacustrine sediments.

Source	Age[1] (years)	Organic carbon (%)	Abundance of B/C lipid fractions[2]				Pollen zones[3]
			Ketones	Alkanoic acids	Alcohols	Aromatic hydro-carbons	
Cam Loch 1	1000–1300	12.49	160 (70)	310 (11)	300 (32)		NS VI
2	2200–2700	12.36	90 (60)	260 (15)	220 (30)		NS VI
3	3200–4000	11.91	80 (60)	230 (17)	200 (60)		NS VI
4	5000–6200	11.03	20 (35)	150 (20)	100 (50)		NS Vi + Vii
5	8300–9200	7.94	50 (35)	210 (12)	110 (12)		NS III
6	11900–12700	1.80	NA	340 (5)	NA		B (Godwin II)
Blelham Tarn	c. 1000	16.0	100 (30)	420 (14)	400 (tr)		VIII (Godwin)
Loch Clair 1	<140	8.7	400 (50)	590 (5)	350 (70)	80	NS VI
2	140–380	13.6	250 (60)	180 (8)	410 (70)	60	NS VI
Crose Mere	NA	11.8	190 (40)	330 (tr)	520 (tr)	40	VIII (Godwin)

1. Figures for Cam Loch and Blelham Tarn from ^{14}C dating of 10cm sections of core (W. Tutin, personal communication); date for Loch Clair 1 from ^{210}Pb-dating (W. Tutin, personal communication), maximum age of Loch Clair 2 estimated from ^{14}C-dating of older deposits. Crose Mere not anlysed by ^{210}Pb-dating; hard-water error makes ^{14}C dates too old (Beales 1976).
2. Expressed as parts C per million parts organic C in sediment and calculated assuming following C contents: ketones, 82%; alkanoic esters, 78%; alcohols, 82%; aromatic hydrocarbons, 91%. Figures in parentheses show percentage of triterpene-derived compounds in total for each class; tr = trace. Aromatic hydrocarbons consist of naphthalenes and phenanthrenes only.
3. Northern Scotland regional pollen zones (NS) (Pennington et al. 1972) or Godwin pollen zones.
   NA = not analysed.

Mere (Cranwell 1978). A series of constituents with ECL values in the 27–30 range, shown in fig. 2.1, occur in all these sediments with relative abundances shown in histogram form (fig. 2.2). Mass spectra of the components 2, 5, 6, 7 and 9 from Loch Clair 2 are shown in fig. 2.3. Resolution of constituents 6 and 7 during GC-MS was, however, lower than that shown in fig. 2.1. Corresponding peaks in other sediment samples gave very similar spectra. Peaks 2, 5, and 6 each show a molecular ion (M^+) at m/e 456 ($C_{31}H_{52}O_2$) consistent with a pentacyclic saturated or tetracyclic monoenoic monocarboxylic acid methyl ester. Peaks 7 and 9 show M^+ 458 ($C_{31}H_{54}O_2$) consistent with a tetracyclic saturated triterpene ester. The mass spectra show M-15 and M-87 fragment ions and one or more of the following intense fragment ions, m/e 177, 189, 191, 203, 218, characteristic of various triterpene skeleta (Budzikiewicz et al. 1963).

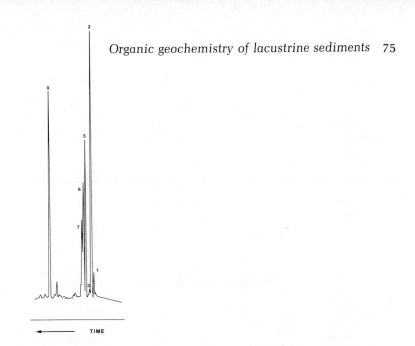

Fig. 2.1 Triterpene region of the GC of saturated branched/cyclic fatty acid methyl esters from Loch Clair 2 sediment. Structural assignments are given in key to fig. 2.2; constituents 4 and 8 absent in this sediment.

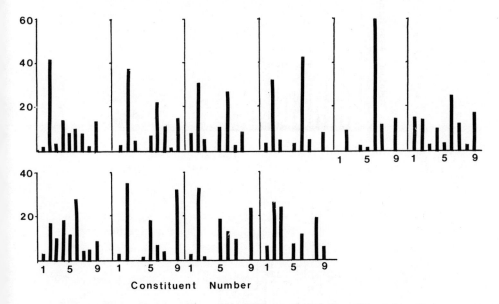

Constituent Number

Fig. 2.2 Percentage composition of the triterpenoid acids in sediments from Cam Loch 1–6 (upper) Blelham Tarn, Loch Clair 1 and 2, and Crose Mere (lower). Constituents, numbered as in fig. 2.1, are 3,4-seco ring A acids derived from taraxerone (peak 2), α-amyrenone (6 and 8), lupanone (7), friedelin (9). Peaks 1, 3, 4, and 5 unidentified.

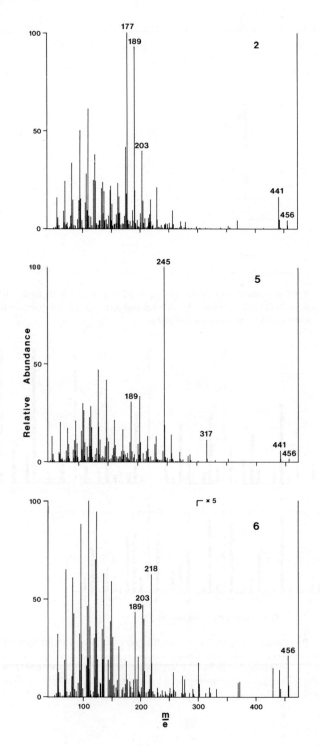

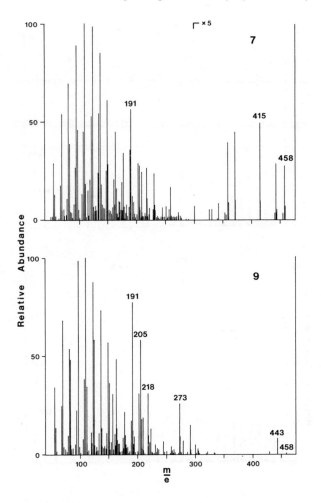

Fig. 2.3 Mass spectra of triterpenoid acid methyl esters from Loch Clair 2 sediment. Numbers refer to peaks in fig. 2.1.

Fatty acids obtained by photolysis of the monoenoic pentacyclic ketones α-amyrenone (I), β-amyrenone (II), and lupenone (III) consisted of

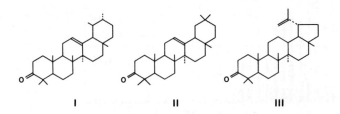

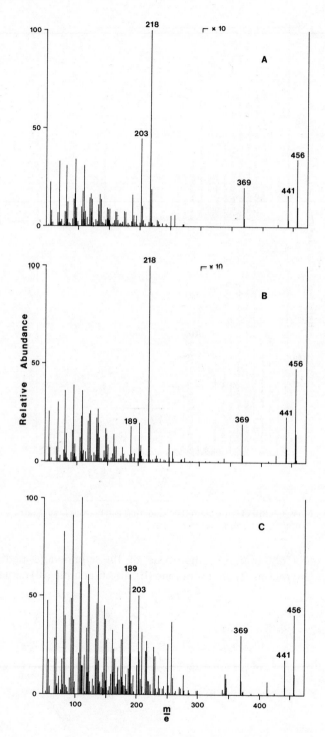

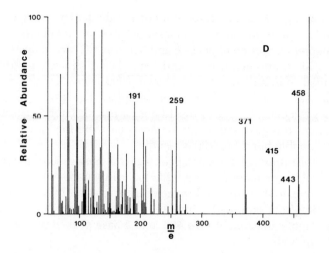

Fig. 2.4   Mass spectra of synthetic acid methyl esters obtained by photolysis of β-amyrenone (A), α-amyrenone (B), lupenone (C), and lupanone (D).

two products from each precursor, showing molecular ions (as methyl esters) at m/e 456 and 454, respectively; the saturated pentacyclic ketone, lupanone (IV) gave products with M⁺ 458 and 456. Mass spectra of all these products showed an M-87 fragment ion due to loss of the ester-containing side chain from partial structures, of the type (V), derived from ring A. Mass spectra of the more-saturated component obtained from each ketone are shown in fig. 2.4.

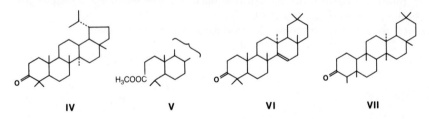

Comparison of these products with the sedimentary esters showed that the β-amyrenone derivative with M⁺ 456 co-eluted with GLC peak 2 isolated from the sediment; the mass spectra, however, differed. The weak m/e 218 fragment ion in component 2 indicates the absence of compounds derived from the oleanane or ursane carbon skeleton (Budzikiewicz et al. 1963). The α-amyrenone and lupanone products with M⁺ 456 and 458, respectively, which are partially resolved on GC, co-eluted with the sedimentary constituents 6 and 7, respectively. Mass spectra of the latter (fig. 2.3) contain many fragment ions characteristic of the co-eluting synthetic acids (figs. 2.4B and 2.4D), supporting the tentative identification based on GC co-elution. Identical mass spectra were not obtained due, in part, to poorer resolution of peaks 6 and 7 during GC-MS analysis. The GC trace of acids from Loch Clair 1 and Cam Loch 1 showed an additional unidentified constituent, eluting between peaks 6 and 7, which may have occurred less prominently in other samples. The presence of acids containing the olean-12-ene or urs-12-ene carbon skeleton in a 'saturated' acid fraction from lake sediments is consistent with the hindered nature of the Δ^{12} double bond (de Mayo 1959).

Among the minor constituents, peak 4 gave a mass spectrum with M^+ at m/e 456 and fragmentation very similar to the lupenone product (fig. 2.4c). The exocyclic double bond in lupene derivatives should result, however, in separation from saturated compounds by argentation TLC. Peak 5, also showing a molecular ion at m/e 456, and a base peak at m/e 245, could not be identified.

KETONES

The triterpene region of the gas chromatogram of a sedimentary B/C ketone fraction is shown in fig. 2.5. Pentacyclic triterpenoid ketones tentatively identified from their mass spectra, in the absence of standards, were peak 2 = taraxerone (VI) (M^+ 424), peak 4 = $\Delta^{13(18)}$-oleanen-3-one (M^+ 424), and peak 8 = friedelin (VII) (M^+ 426). Compounds identified by mass spectral comparison and GC co-elution with authentic samples were α and β-amyrenone, lupenone and lupanone (see key to fig. 2.6). Changes in relative distribution of ketones between the sediment samples are shown as histograms (fig. 2.6).

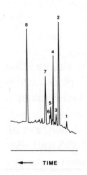

— TIME

Fig. 2.5   Triterpenoid constituents in GC trace of B/C ketones from Loch Clair 2 sediment. Identities of numbered peaks are given in key to fig. 2.6.

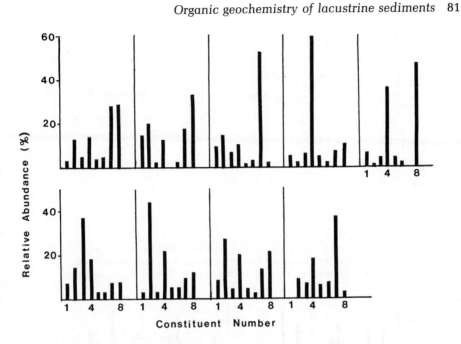

Fig. 2.6   Percentage composition of triterpene ketones in sediments from Cam Loch 1–5 (upper, left to right), Blelham Tarn, Loch Clair 1 and 2, and Crose Mere (lower, left to right). Constituents, numbered as in fig. 2.5, are: 1, unidentified, 2 = taraxerone, 3 = β-amyrenone, 4 = $\Delta^{13(18)}$-oleanen-3-one, 5 = α-amyrenone, 6 = lupenone, 7 = lupanone, 8 = friedelin.

ALCOHOLS

The presence of *iso-* and *anteiso-*branched alkanols in Cam Loch 5, Loch Clair 1, and Crose Mere has been reported (Cranwell 1980); these compounds were also detected in the other sediments by GC and GC-MS analyses. The triterpene region of the gas chromatogram obtained from a typical sediment is shown in fig. 2.7. Mass spectrometry showed seven constituents (peaks 2, 4–8, 10) with molecular ions consistent with pentacyclic monoenoic alcohols (M^+ 440 using methyl ethers, 498 using trimethylsilyl ethers), two di-unsaturated compounds (peaks 1 and 3), and two constituents (peaks 9 and 11) which were saturated pentacyclic triterpenes. GC-coelution studies and comparison with mass spectra of authentic samples enabled β-amyrin (peak 4), α-amyrin (peak 7), and lupeol (peak 8) to be identified, while the presence of lupanol (peak 9) and taraxerol (peak 2) was suggested by the mass spectra. Only trace amounts of triterpene alcohols occurred in the Blelham Tarn and Crose Mere sediments; the relative abundance of constituents in the remaining sediments is shown in fig. 2.8.

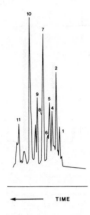

Fig. 2.7   Triterpenoid constituents in GC trace of B/C alcohols (as trimethylsilyl ether derivatives) isolated from Cam Loch 5 sediment. Structural assignments for numbered peaks are given in key to fig. 2.8; Compound 3 absent in this sediment.

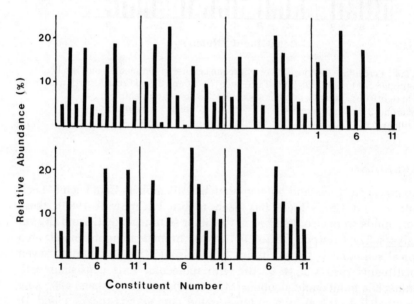

Fig. 2.8   Percentage composition of triterpene alcohols in sediments from Cam Loch 1–4 (upper), Cam Loch 5, Loch Clair 1 and 2 (lower). Peaks, numbered as in fig. 2.7, are: 4 = β-amyrin, 7 = α-amyrin, 8 = lupeol (usually better resolved than in fig. 2.7), 9 = lupanol, 1,3,5,6,10 = unidentified. Tentative assignment for peak 2 is given in text.

AROMATIC HYDROCARBONS

Mass spectrometry showed the presence in both sediments from Loch

Clair of 3 isomers with M^+ at m/e 292 and 4 isomers having M^+ at m/e 274. In the Crose Mere sediment the presence of the former isomers, in similar relative amounts to Loch Clair, was demonstrated by GC co-elution. Proposed structures, shown in fig. 2.9, are based on interpretation of mass spectra which agree well with those obtained by Wakeham *et al.* (1980) for compounds a, b, d, and e. The mass spectra of compounds c and f are in good agreement with those of synthetic specimens (Spyckerelle *et al.* 1977), and the GC elution order of these compounds is the same as that reported by Wakeham *et al.* The mass spectrum of compound g is very similar to that of compound e, only differing in showing a slightly stronger molecular ion relative to the M-15 base peak. The relative abundance of the isomers having geminal and vicinal substitution is shown in table 2.2.

Table 2.2. Relative abundance of geminal to vicinal dimethyl substituted aromatic hydrocarbons in sediments.

Source	Mol. wt of arom series	Relative abundances[1]	
		c:b	f:e
Loch Clair 1	292	0.38	—
	274	—	1.63
Loch Clair 2	292	0.55	—
	274	—	1.70
Crose Mere	292	0.35	—

1. For structures, see fig. 2.9.

Discussion

In a series of sediments varying in age any attempt to correlate organic composition with source material requires some consideration of factors, other than primary input, which may influence the abundance and distribution of the constituents. The term diagenesis is used for changes in organic natural compounds following the death of the source organism. These changes may be microbially-mediated, as in recent sediments, or induced by high temperature or pressure, as in older, deeper deposits (Bluck 1969). Early diagenesis leads to a preferential depletion of lower molecular weight n-alkanes, alkanols, and alkanoic acids during aerobic decay of algae and in initial stages of peat formation (Quirk 1978). Similar features occur in surficial sediments from the oligotrophic Loch Clair (Cranwell 1981), showing that allochthonous material reaching the loch has not attained a stable lipid composition. Although the carbon skeleton of pentacyclic triterpenes is more resistant to microbial breakdown than

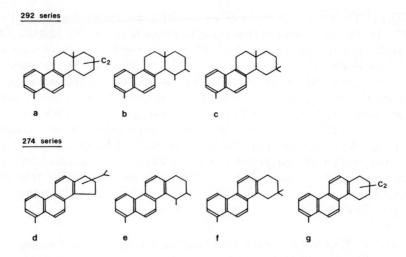

292 series

a   b   c

274 series

d   e   f   g

Fig. 2.9   Triterpene-related aromatic hydrocarbons in lacustrine sediments. GC elution times increase in alphabetical order.

that of the straight chain compounds mentioned above, diagenesis of the 3-oxygenated compounds may lead to changes in (a) oxidation state, (b) composition of the triterpenes, and (c) ring structure.

In relation to the first pathway it is probable that the acids and aromatic hydrocarbons detected in sediments are derived from triterpene alcohols or the corresponding ketones, the primary plant products. The 3-hydroxy derivatives are ubiquitous in green plants, occurring more widely than the ketones, as in the Gramineae (Ohmoto et al. 1970). Ketones, common in the more primitive plant divisions (Boiteau et al. 1964) and in the bark of trees (Sainsbury 1970), could also be formed by oxidation of the 3-alcohols. The acids can be formed by photochemical cleavage of the ketones, a process which may occur in the precursor plant material, since acids formed by 3,4-cleavage of ring A in triterpenes have been detected in higher plants (Devon and Scott 1972). However, the small number of species in which these acids have been reported, contrasted with their widespread distribution in sediments, may indicate that ring A-opened acids are formed by microorganisms through reactions producing 3-ketones in an excited state, analogous to the photochemical mechanism (Corbet et al. 1980). The presence in recent sediments of aromatic hydrocarbons derived from pentacyclic triterpenes by loss of ring A, together with the presumed precursors, 3,4-seco-ring A acids (Dastillung 1976, Corbet et al. 1980), demonstrates the extent and also the incomplete nature of this diagenetic process in sediments. Indeed, sediments from Loch Clair contain compounds derived from α-amyrin in all four compound classes studied, the parent compound being a major component of the alcohols, but the related ketone being a minor constituent of the ketones, possibly reflecting the composition of the source material in these recent sediments. Saturated lupanes do not appear to be widely distributed in plants so that the lupane-derived compounds (acids, ketones, aromatic hydrocarbons, and alcohols) in these sediments may result from reduction of the reactive exocyclic double bond in lupeol, a widespread possible precursor.

Changes in composition of the suite of triterpenes may occur if some ring structures undergo more rapid diagenesis than others. In Loch Clair sediment the abundance ratio of the aromatic hydrocarbon with geminal dimethyl to that with vicinal dimethyl group increases with degree of aromatization (0.38 for M^+ 292 components; 1.63 for M^+ 274 in section 1). A similar result, reported by Wakeham et al. (1980), was attributed to preferential degradation of β-amyrin derivatives compared to α-amyrin derivatives. In Loch Clair section 2 these ratios increase to 0.55 and 1.70 respectively, suggesting that preferential degradation of β-amyrin derivatives continues in the lower horizon. These deductions are supported by the relative abundance of presumed precursors of the aromatic hydrocarbons, thus β-amyrenone is only a minor component of the ketones from Loch Clair sediment (fig. 2.6) and the acid derived from β-amyrenone by 3,4-cleavage was not detectable, whereas α-amyrenone and the corresponding carboxylic acid were more abundant. The parent triterpene alcohols were both present, however, with the α-isomer more abundant, in contrast with the results of Dastillung (1976) who found no β-amyrin in sediments of the Baltic Sea containing geminal dimethyl aromatic hydrocarbons. In Loch Clair sediment both amyrin isomers occurred in part as esters, a form in which stability against diagenetic attack was enhanced, relative to free alcohols (Cranwell and Volkman 1981).

The third diagenetic pathway which may change the original triterpene composition derived from the source organisms, namely changes in ring structure, may be significant as many such changes occur under acidic conditions (de Mayo 1959). Among the triterpenes identified in these sediments, taraxerol derivatives re-arrange to β-amyrin derivatives in acetic acid (p^cK 4.65) (Ives and O'Neill 1958), the lupene skeleton re-arranges to that of olean-13(18)-ene, in which the double bond occupies the most stable position, and friedelane derivatives are interconvertible with those of the oleanane series. The acidic reducing conditions occurring in peats and sediments may be suitable for some of these skeletal transformations to occur. No time scale has been established for these conversions under environmental conditions but re-arranged compounds occur in petroleum source rocks (Hills et al. 1968). Sediments derived from acidic peat (Loch Clair 1 and 2, Cam Loch 1 and 2) all show a prominent component in the cyclic acids (peak 2, fig. 2.2) which co-elutes with, but shows a mass spectrum different from, the photochemical product from β-amyrenone. Taraxerol and taraxerone are dominant triterpenes in *Sphagnum*, a peat-forming species (Ives and O'Neill 1958); also, an acidic constituent showing a mass spectrum identical with that of peak 2 in these sediments has been obtained from fresh *Sphagnum* moss (M.M. Quirk, personal communication), suggesting that constituent 2 in the acids is the 3,4-seco acid formed from taraxerone. The absence of a strong fragment ion at m/e 218, characteristic of acids derived from α- and β-amyrin (fig. 2.4), in the mass spectrum of constituent 2 from Cam Loch sediments up to 8000 years old suggests that the taraxerone → β-amyrenone re-arrangement is not occurring in these sediments. The distribution of ketones in the peat-derived sediments (Loch Clair 1 and 2, Cam Loch 1 and 2, in order of increasing age) also shows no significant increase in relative abundance of β-amyrenone to accompany the fall in relative abundance of taraxerone with increasing age in this sequence of sediments, again suggesting that re-arrangement of the nucleus has not occurred.

Having established that skeletal re-arrangement is unlikely in recent sediments but that selective degradation may have distorted the distribution of marker compounds characteristic of higher-plant input, *via* changes in oxidation state as in β-amyrin derivatives, there follows an evaluation of literature data on the distribution of triterpenes in the main plant input to the sediments, based on their pollen content.

The three major pentacyclic triterpenoid families, structurally related to β-amyrin, α-amyrin, and lupeol, respectively, occur widely in the plant kingdom as alcohols or esterified derivatives, while 3-keto triterpenoids are common in the more primitive divisions (Boiteau et al. 1964). The distribution of triterpenoids in the Gramineae was reviewed recently by Ohmoto et al. (1970). In addition to these families, the sediments also contain derivatives based on two other pentacyclic triterpene skeleta, those of taraxerene and friedelane, which occur extensively in plants (Sainsbury 1970). The pollen composition of the sediments is shown in table 2.3; the principal triterpene alcohols or ketones characteristic of the major plant families are given in table 2.4.

The relationship between the pollen incorporated into a lake sediment and the composition of the terrestrial vegetation from which it is derived is complex; there is a large North American literature on the subject. It has been recently studied in Cumbria (Bonny 1976, 1978, 1980). Variation between species in the magnitude of pollen production, the means and efficiency of pollen dispersal, and differences in sedimentation rate of the pollen within the water column are major factors affecting the relationship.

Table 2.3. Pollen and spore content of sediments.

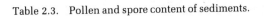

Sediment	Pinus	Quercus	Betula	Alnus	Ulmus	Σ Arboreal	Corylus & Myrica	Salix	Gramineae	Cyperaceae	Calluna	Total herbs	Sphagnum
						Pollen and spore percentages[4]							
Cam Loch 1	2	—	3	11	—	16	16	—	13	7	40	8	3
Cam Loch 2	5	—	14	20	—	39	7	1	11	21	11	10	8
Cam Loch 3	10	—	14	15	—	39	8	+	21	13	12	7	5
Cam Loch 4[1]	58	3	5	3	3	72	7	6	5	5	3	2	12
Cam Loch 5	13	5	25	—	—	43	23	6	9	12	4	3	15
Cam Loch 6[2]	3	—	4	—	—	7	—	5	17	45	—	9	—
Loch Clair 1	11	+	11	6	+	30	14	—	9	5	35	7	5
Loch Clair 2	16	—	8	4	—	28	17	—	6	4	41	4	16
Crose Mere[3]	5					c. 35			36		<2	>20	—
Blelham Tarn	—	11	13	26	—	50	23	—	8	10	2	7	6

1. Figures relate to oldest sediment from this section; uppermost part showed Pinus 28%, Total Arboreal pollen 58%.
2. Empetrum 12, Juniperus 4, in addition to taxa listed.
3. Data from Beales (1976).
4. +, <1%.

Few studies on the geolipids of lacustrine sediments have attempted to correlate the lipid composition with pollen content; correlation of lipid distribution patterns with changes in the ratio of allochthonous to autochthonous sediment input, resulting from a change in trophic status, has received more attention (Barnes and Barnes 1978). A difference between the chain-length distribution of n-alkanes from sediment horizons having a pollen assemblage indicative of acidic peat input and that in horizons derived from forested catchments has been noted in these and other sediments (Cranwell 1973b, 1977b, 1981). As in the n-alkanes, certain broad similarities may be noted in the distribution of triterpenoid ketone, alcohol, and derived acid fractions from sediments having a similar pollen content, also some correlation between the major sediment constituents and the triterpene content of the contributing species is evident.

In the peat-derived sediments (Cam Loch 1 and 2, Loch Clair 1 and 2) similarities in the abundance of structurally-related acidic and ketonic constituents of these sediments (figs. 2.2 and 2.6) were noted above in discussing possible diagenesis. The alcohol fractions of these sediments

Table 2.4.   Pentacyclic triterpenes in plant species.

Triterpenes	Major plant families[1]					
	Pinaceae	Fagaceae	Betulaceae	Gramineae	Muscineae	Ericaceae
β-Amyrin				+		+
α-Amyrin					+	+
Lupeol	+			+		+
Lupenone			+			
Friedelinol		+				
epi-Friedelinol	+	+		+		+
Friedelin	+	+		+		
Taraxerol			+	+	+	+
Taraxerone			+	+	+	
Olean-13(18)-en-3-one			+			

1. +, compound present in some members of family. Data from Boiteau et al. (1964), Ohmoto et al. (1970), and Sainsbury (1970).

show taraxerol and α-amyrin as major constituents (fig. 2.8). The ketone and acid related to taraxerol were also major constituents in their respective compound class in these sediments. These triterpenes occur in typical species forming acidic peat deposits, such as Sphagnum (Ives and O'Neill 1958) and members of the Ericaceae (Boiteau et al. 1964). The relative abundance of the acids derived from these alcohols, constituents 2 and 6, respectively in Cam Loch 3 is intermediate between those in sections 2 and 4, analogous to the n-alkane distribution patterns in these sediments and other evidence (Cranwell 1977b) that section 3 was deposited at a time of increased erosion of organic soils following deforestation and development of peat.

Sediments with a high Betula pollen content, Cam Loch 5 and Blelham Tarn, show some similarities in triterpene composition. The acids show a dominance of the constituent derived from α-amyrin, which itself is a major alcohol in these Cam Loch sediments, although triterpene alcohols 1–11 (fig. 2.8) were virtually absent in Blelham Tarn. Although α-amyrin is widespread in nature (Boiteau et al. 1964), it had not specifically been reported in members of the Betulaceae (table 2.2), so that this correlation may be coincidental. The ketones from these sediments, however, show a major constituent tentatively identified as $\Delta^{13(18)}$-oleanen-3-one, which has only been found in a member of the Betulaceae (Boiteau et al. 1964). As modern analytical techniques are more sensitive and can also separate minor constituents more efficiently, further chemotaxonomic studies are needed to confirm this implied correlation. Analogues of the relationship between triterpene composition and pollen content of lacustrine sediments might be obtained by similar analyses of surfical soils beneath deciduous or coniferous forest canopies, grassland, and peat-forming vegetation.

Conclusions

Correlation of sedimentary distributions of 3-oxygenated pentacyclic triterpenes and derivative compounds with higher-plant input, as inferred from pollen analysis, is reported for sediments varying in age and pollen content. In addition to the non-quantitative relationship between pollen content and composition of the local terrestrial biota, diagenesis of the triterpenes, via microbial transformations producing a change in oxidation state or selective diagenesis of certain structural components, may make any correlation with source organisms difficult. The widespread distribution in plants of the major triterpenes found in sediments also reduces their value for distinguishing differences in source material. However, peat-derived sediments showed a correlation in structure between major alcohol, ketone, and acid constituents which are known to occur in the plant species forming acidic peat deposits. Sediments derived from forest soils also showed similarities in triterpene composition, but these correlated less well with the known distribution of triterpenes in the source organisms. This poorer correlation may result from a lack of chemotaxonomic data for a sufficient number of plant species.

Acknowledgments

I thank Professor W. Tutin, F.R.S., for provision of sediment cores and for useful discussions. Dr E.Y. Haworth, Mr B. Walker, Mr P.V. Allen, and Mr W. Askew assisted in field work. Mr T.I. Furnass kindly prepared many of the figures.

I thank Professor G. Eglinton, F.R.S., and Dr J.R. Maxwell (Bristol University) for the use of GC-MS facilities provided by grants GR3/2951 and GR3/3758 from the Natural Environmental Research Council, and Mrs A.P. Gowar for assistance with GC-MS and data processing.

References

Aoyagi, R., Tsuyuki, T. and Takahashi, T., 1973. One-step synthesis of putranjivic acid from friedelin. *Bull. Chem. Soc. Japan* 46: 692.
Barnes, M.A. and Barnes, W.C., 1978. Organic compounds in lake sediments. In *Lakes, Chemistry, Geology, Physics*, ed. A. Lerman, 127–52 (New York).

Beales, P.W., 1976. Palaeolimnological studies of a Shropshire mere (Ph.D. thesis, University of Cambridge).

Bluck, B.J., 1969. Introduction to sedimentology. In *Organic Geochemistry*, ed. G. Eglinton and M.T.J. Murphy, 245–61 (New York).

Boiteau, P., Pasich, B. and Ratsimamanga, A.R., 1964. *Les triterpénoïdes en physiologie végétale et animale* (Paris).

Bonny, A.P., 1976. Recruitment of pollen to the seston and sediment of some Lake District lakes. *J. Ecol. 64:* 859–87.

Bonny, A.P., 1978. The effect of pollen recruitment processes on pollen distribution over the sediment surface of a small lake in Cumbria. *J. Ecol. 66:* 385–416.

Bonny, A.P., 1980. Seasonal and annual variation over five years in contemporary airborne pollen trapped at a Cumbrian lake. *J. Ecol. 68:* 421–41.

Brassell, S.C., Comet, P.A., Eglinton, G., Isaacson, P.J., McEvoy, J., Maxwell, J.R., Thomson, I.D., Tibbetts, P.J.C. and Volkman, J.K., 1980. Preliminary lipid analyses of sections 440A-7-6, 440B-3-5, 440B-8-4, 440B-68-2, and 436-11-4· legs 56 and 57, Deep Sea Drilling Project. In *Initial Reports of the Deep Sea Drilling Project 56 and 57:* 1367–90.

Budzikiewicz, H., Wilson, J.M. and Djerassi, C., 1963. Mass spectrometry in structural and stereochemical problems. XXXII. Pentacyclic triterpenes. *J. Am. Chem. Soc. 85:* 3688–99.

Corbet, B., Albrecht, P. and Ourisson, G., 1980. Photochemical or photomimetic fossil triterpenoids in sediments and petroleum. *J. Am. chem. Soc. 102:* 1171–3.

Cranwell, P.A., 1973a. Branched-chain and cyclopropanoid acids in a recent sediment. *Chem. Geol. 11:* 307–13.

Cranwell, P.A., 1973b. Chain-length distribution of *n*-alkanes from lake sediments in relation to post-glacial environmental change. *Freshwater Biol. 3:* 259–65.

Cranwell, P.A., 1977a. Organic compounds as indicators of allochthonous and autochthonous input to lake sediments. In *Interactions Between Sediments and Fresh Water, 1976,* ed. H.L. Golterman, 133–40 (The Hague).

Cranwell, P.A., 1977b. Organic geochemistry of Cam Loch (Sutherland) sediments. *Chem. Geol. 20:* 205–21.

Cranwell, P.A., 1978. Extractable and bound lipid components in a freshwater sediment. *Geochim. cosmochim. Acta 42:* 1523–32.

Cranwell, P.A., 1980. Branched/cyclic alkanols in lacustrine sediments (Great Britain): recognition of *iso-* and *anteiso*-branching and stereochemical analysis of homologous alkan-2-ols. *Chem. Geol. 30:* 15–26.

Cranwell, P.A., 1981. Diagenesis of free and bound lipids in terrestrial detritus deposited in a lacustrine sediment. *Org. Geochem. 3:* 79–89.

Cranwell, P.A. and Volkman, J.K., 1981. Alkyl and steryl esters in a recent lacustrine sediment. *Chem. Geol. 32:* 29–43.

Dastillung, M., 1976. Lipides de sédiment récents (Ph.D. thesis, University of Strasbourg).

Devon, T.K. and Scott, A.I., 1972. *Handbook of Naturally Occurring Compounds, Vol. 2.*

Frey, D.G., 1974. Paleolimnology. *Mitt. int. Verein. theor. angew. Limnol. 20:* 95–123.

Hills, I.R., Smith, G.W. and Whitehead, E.V., 1968. Optically active spirotriterpane in petroleum distillates. *Nature 219:* 243–6.

Ishiwatari, R., Ogura, K. and Horie, S., 1980. Organic geochemistry of a lacustrine sediment (Lake Haruna, Japan). *Chem. Geol. 29:* 261–80.

Ives, D.A.J. and O'Neill, A.N., 1958. The chemistry of peat II. The triterpenes of peat moss (Sphagnum). *Canad. J. Chem. 36:* 926–30.

Jones, J.G., 1972. Studies on freshwater micro-organisms: phosphatase activity in lakes of differing degrees of eutrophication. *J. Ecol. 60:* 777–91.

Leo, R.F. and Parker, P.L., 1966. Branched-chain fatty acids in sediments. *Science 152:* 649–50.

Mackereth, F.J.H., 1958. A portable core sampler for lake deposits. *Limnol. Oceanogr. 3:* 181–91.

McCarthy, R.D. and Duthie, A.H., 1962. A rapid quantitative method for the separation of free fatty acids from other lipids. *J. Lipid Res. 3:* 117–19.

Mayo, P. de, 1959. *The Higher Terpenoids.*

Meyers, P.A., Bourbonniere, R.A. and Takeuchi, N., 1980. Hydrocarbons and fatty acids in two cores of Lake Huron sediments. *Geochim. cosmochim. Acta 44:* 1215–21.

Miwa, T.K., 1963. Identification of peaks in gas–liquid chromatography. *J. Am. Oil Chem. Soc. 40:* 309–13.

Ohmoto, T., Ikuse, M. and Natori, S., 1970. Triterpenoids of the Gramineae. *Phytochemistry 9:* 2137–48.

Ourisson, G., Albrecht, P. and Rohmer, M., 1979. The Hopanoids. Paleochemistry and biochemistry of a group of natural products. *Pure Appl. Chem. 51:* 709–29.

Pennington, W., 1975. A chronostratigraphic comparison of late-Weichselian and late-Devensian subdivisions, illustrated by two radiocarbon-dated profiles from western Britain. *Boreas 4:* 157–71.

Pennington, W., 1981. Records of a lake's life in time: the sediments. *Hydrobiologia 79:* 197–219.

Pennington, W., Haworth, E.Y., Bonny, A.P. and Lishman, J.P., 1972. Lake sediments in northern Scotland. *Phil. Trans. R. Soc. B. 264:* 191–294.

Pennington, W., Cambray, R.S., Eakins, J.D. and Harkness, D.D., 1976. Radionuclide dating of the recent sediments of Blelham Tarn. *Freshwater Biol. 6:* 317–31.

Pennington, W., Cranwell, P.A., Haworth, E.Y., Bonny, A.P. and Lishman, J.P., 1977. Interpreting the environmental record in the sediments of Blelham Tarn. *Rep. Freshwat. biol. Assoc. 45:* 37–47.

Quirk, M.M., 1978. Lipids of peat and lake environments (Ph.D. thesis, University of Bristol).

Reynolds, C.S., 1973a. The phytoplankton of Crose Mere, Shropshire. *Br. phycol. J. 8:* 153–52.

Reynolds, C.S., 1973b. Growth and buoyancy of *Microcystis aeruginosa* Kütz. emend. Elenkin in a shallow eutrophic lake. *Proc. R. Soc. B. 184:* 29–50.

Sainsbury, M., 1970. Friedelin and epifriedelinol from the bark of *Prunus turfosa* and a review of their natural distribution. *Phytochemistry 9:* 2209–15.

Spyckerelle, C., 1975. Constituants aromatiques de sédiments (Ph.D. thesis, University of Strasbourg).

Spyckerelle, C., Greiner, A.Ch., Albrecht, P. and Ourisson, G., 1977. Hydrocarbures aromatiques d'origine geologique III. Un tetrahydrochrysene, derive de triterpenes, dans des sediments Récent et Anciens. *J. Chem. Res. (M):* 3746–77.

Wakeham, S.G. and Farrington, J.W., 1980. Hydrocarbons in contemporary aquatic sediments. In *Contaminants and Sediments, Vol. 1,* ed. R.A. Baker, 3–32 (Michigan).

Wakeham, S.G., Schaffner, C. and Giger, W., 1980. Polycyclic aromatic hydrocarbons in Recent lake sediments – II. Compounds derived from biogenic precursors during early diagenesis. *Geochim. cosmochim. Acta* 44: 415–29.

3   Empirical testing of ^{210}Pb-dating models for lake sediments

F. Oldfield and P. G. Appleby

Introduction

FROM palaeoecological studies tracing the history of man's effect on the environment, it is clear that in many places the period of greatest impact and most dramatic transformation lies within the last 150 years. Lake sediments provide a basis for reconstructing many aspects of this impact, for estimating rates of change, for establishing a 'dynamic' baseline in environmental monitoring programmes and for developing a continuum of insight into environmental processes and their effects (Oldfield 1977). Of central importance to all these aims is the establishment of detailed and accurate chronologies of sedimentation. These are required not only for dating events but also for calculating rates of sedimentation, of element flux, and of microfossil deposition.

One of the most promising methods of dating on a time scale of 100–200 years is by means of ^{210}Pb, a natural radioisotope with a half-life of 22.26 years. The development of this technique was initiated by Goldberg (1963), and it was first applied to the dating of lake sediments by Krishnaswamy et al. (1971). Pennington et al. (1976) recorded the first systematic attempt to apply and evaluate the technique in Britain. Recent studies have used ^{210}Pb-dated sediment sequences for reconstructing the history of eutrophication (Battarbee 1978), of catchment erosion (Oldfield et al. in press), and of heavy metal (Edgington and Robbins 1976a) and nutrient fluxes (Osborne and Moss 1977).

Despite the great significance of the technique, the precise nature of the processes by which ^{210}Pb-accumulates in lake sediments is not yet well understood. For lakes with constant rates of sediment accumulation this is not of great consequence, and in these cases the ^{210}Pb-derived dates have generally been validated by independent dating techniques. Where sedimentation rates have varied, however, an understanding of these processes is crucial, and here controversy still surrounds the assumptions used for deriving rates and sedimentation rates from ^{210}Pb-measurements. The present paper outlines ways in which this controversy may be resolved through rigorous empirical testing of results for both internal consistency and external validation by independent techniques. This theme is set within the context of a broader consideration of the behaviour of ^{210}Pb in the environment and of the assumptions involved in alternative dating models, and it is illustrated by a series of case studies arising from published work on the Great Lakes of North America.

Lead-210 in lake sediments

Lead-210 is a member of the ^{238}U decay series. Disintegration of the intermediate isotope ^{226}Ra (half-life 1622 years) yields the inert gas ^{226}Rn. This in turn decays (with a half-life of 3.83 days) through a series of short-lived isotopes to ^{210}Pb. Radium-226 is supplied to lake sediments as part of the particulate erosive input (A. in fig. 3.1). The ^{210}Pb formed by

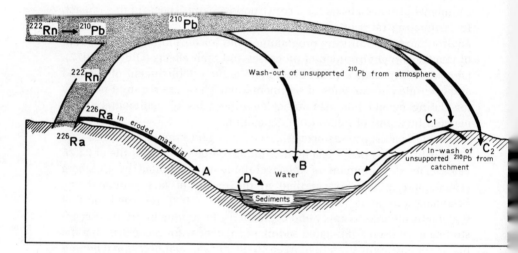

Fig. 3.1   Pathways by which ^{210}Pb reaches lake sediments (see text).

the *in situ* decay of this radium is termed the 'supported ^{210}Pb' and is normally assumed to be in radioactive equilibrium with the radium. In general, however, this equilibrium will be disturbed by a supply of ^{210}Pb from other sources. Fig. 3.1 outlines the main pathways by which excess ^{210}Pb reaches the sediments. Three components are identified:

B. Direct atmospheric fallout. A fraction of the radon atoms formed by ^{226}Ra decay in soils escape into the intersticos and then diffuse through the soil into the atmosphere. The decay of radon in the atmosphere yields ^{210}Pb, which may be removed either by dry deposition or wet fallout.

Lead-210 falling directly into lakes is absorbed onto sediment particles and deposited on the bed of the lake.

C. Indirect atmospheric fallout. Atmospheric ^{210}Pb also reaches the lake indirectly *via* the catchment. Although the distinction may be less clear-cut in practice, it is convenient to separate a C_1 component which is incorporated into the drainage net and flows quickly to the lake without being detained on solid terrestrial particles and a C_2 component which may have a long residence time in the catchment before being delivered to the lake in association with the erosive input of fine surface particulates to which it is attached.

D. Radon decay in the water column. Radon is delivered to the lake waters by diffusion from the underlying sediments, and by the decay of ^{226}Ra in the water column and inflowing streams. A part of the radon is lost by diffusion across the surface of the lake, and the remainder decays in the water column to ^{210}Pb.

Lead-210 activity (components B, C, and D) in excess of the supported activity is called the 'excess' or 'unsupported' ^{210}Pb. The principal source of unsupported ^{210}Pb is generally taken to be direct atmospheric fallout (component B), although the importance of the other sources has not been extensively evaluated. Studies by Benninger *et al.* (1975) on the fate of ^{210}Pb in the Susquehanna River system showed that dissolved ^{210}Pb in the river waters (component C_1) was quickly removed from solution by suspended particles. Further, stream-borne particles (component C_2) carried away no more than 0.8% of the atmospheric flux of ^{210}Pb which reached the catchment soils. Studies by Hammond *et al.* (1975) and Krishnaswamy and Lal (1978) of the production of ^{222}Rn and ^{210}Pb in the lake waters (component D) have indicated that this component too may be negligible, around two orders of magnitude lower than the atmospheric flux. On the other hand, Imboden and Stiller (personal communication) have estimated that both the riverine input and *in situ* production of ^{210}Pb in Lake Kinneret, Israel, are of the same order of magnitude as the direct atmospheric flux. Although, for climatic reasons, the latter case may well be atypical, it nonetheless illustrates the need for further investigation of these components.

In dating by ^{210}Pb it is the unsupported component only which is used since once incorporated in the sediment it decays exponentially with time in accordance with its half-life. The supported ^{210}Pb-activity is estimated by assay of the ^{226}Ra. Although radon diffusion through the sediments may result in a small disequilibrium between the ^{226}Ra and supported ^{210}Pb near the sediment–water interface, provided that the total ^{210}Pb-activity is well in excess of the ^{226}Ra-activity, a correction for this will generally be negligible. Once the supported ^{210}Pb-activity has been estimated, the unsupported ^{210}Pb can be determined by subtraction from

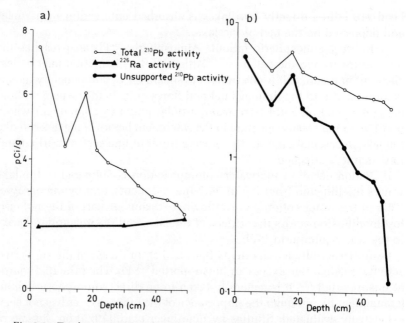

Fig. 3.2   Total, supported, and unsupported ^{210}Pb activity. 3.2(a) plots the total
^{210}Pb activity and the ^{226}Ra activity v. depth for a core from Lake Lovojarvi in
Finland (Appleby et al. 1979), using a natural scale. 3.2(b) plots the total ^{210}Pb
activity and the unsupported ^{210}Pb activity v. depth, using a logarithmic scale.

the total ^{210}Pb-activity (fig. 3.2). Ideally, total ^{210}Pb- and ^{226}Ra-assays
should be carried out on every sample. In practice, total ^{210}Pb-determina-
tions may be scattered down a profile with intervening levels unanalysed,
and the supported component is often estimated from only two or three
^{226}Ra-determinations, from amalgamated samples, or simply from the
total ^{210}Pb-activity of sediments too old to give any significant disintegration from unsupported ^{210}Pb.

In determining a ^{210}Pb-chronology it is assumed that the unsupported
^{210}Pb, once incorporated in a sediment layer, declines with age in
accordance with the ^{210}Pb radioactive-decay law. The validity of the
chronology will then rest on the accuracy with which the ^{210}Pb-dating
model represents the ^{210}Pb delivery mechanisms outlined above. This
assumes that the unsupported ^{210}Pb, once incorporated in the sediments,
does not 'migrate' or diffuse via the pore water of the sediment. Strong
empirical evidence in support of this assumption is derived from the
presence of sharply defined peaks and inflexions in some ^{210}Pb-profiles,
and from profiles considered here and elsewhere in which the unsupported ^{210}Pb-activity at independently dated depths is as expected from
the operation of in situ radioactive decay alone. There is evidence in some

cases for redistribution of ^{210}Pb in association with sediment mixing or sediment resuspension. Where this takes place, the decay law will operate only for sediment layers beneath the zone of mixing.

Alternative assumptions and models

Most studies of ^{210}Pb-dating accept that the supply of unsupported ^{210}Pb to the lake waters is governed primarily by local or regional meteorological factors, and hence is reasonably constant on timescales of the order of 100–200 years. There is some evidence (Krishnaswamy and Lal 1978) that the atmospheric ^{210}Pb-flux (component B, fig. 3.1) may vary significantly on timescales of the order of a year. Since, in practice, most sediment samples span several years of accumulation, they will tend to smooth out this type of short-term variation.

Few authors have set out fully and explicitly the assumptions they have used, though most in calculating age/depth profiles and sedimentation rates from ^{210}Pb-measurements have assumed, at least implicitly, a constant flux of unsupported ^{210}Pb from the lake waters to the sediment. In the majority of cases, the assumption of a constant flux is coupled also with an assumed constant dry-mass sedimentation rate. Where both assumptions are satisfied the unsupported ^{210}Pb-concentration C in the sediments will vary exponentially with the cumulative dry-mass of sediment m in accordance with the formula (Robbins 1978: 341)

$$C = C(o)e^{-km/r}, \tag{1}$$

where $C(o)$ is the unsupported ^{210}Pb-concentration at the sediment–water interface, r is the dry-mass sedimentation rate, and

$$k = \frac{\ln 2}{22.26} \approx 0.03114\,\mathrm{yr}^{-1}$$

is the ^{210}Pb-decay constant. If the unsupported ^{210}Pb-concentration C is plotted on a logarithmic scale against the cumulative dry-mass m, the resulting ^{210}Pb-profile will be linear, with slope $-k/r$. The sedimentation rate r can be determined graphically from the mean slope of the profile, or alternatively by a least-squares fit procedure. This dating model has been called the 'simple' model (Robbins 1978), or constant flux constant sedimentation rate (c.f.:c.s.) model. Linear profiles characterizing an exponential variation in C have been observed in many lakes and have provided a basis for calculating mean sedimentation rates for the past 100–150 years (e.g. Farmer 1978).

When the ^{210}Pb-activity is plotted against depth in the core, as distinct from cumulative dry-mass, non-linearities in the profile may appear near the surface simply as a result of the reduced compaction of near-surface sediments. Many authors have observed more fundamental non-lineari- ties in the ^{210}Pb-profiles, however, ranging from depressed ^{210}Pb-concen- trations in the top few centimetres (Koide *et al.* 1973: 1176) to strongly 'kinked' profiles exhibiting non-monotonic behaviour at depths ranging from 6cm to below 30cm (e.g. Oldfield *et al.* 1978: 339).

A major aim of the present paper is to evaluate alternative ways of treating such profiles. One approach is to interpret the non-linearities by assuming that there is a constant net rate of supply of ^{210}Pb from the lake waters to the sediments, irrespective of changes which may have occurred in the net dry mass sedimentation rate. This dating model we have termed the c.r.s. (constant rate of supply) model. The procedure for calculating ^{210}Pb-dates using this model was first outlined by Goldberg (1963), and is set out in more detail in Appleby and Oldfield (1978), and in Robbins (1978). Briefly, if

$$A = \int_m^\infty C dm = \int_x^\infty \rho C dx \qquad (2)$$

is the cumulative residual unsupported ^{210}Pb beneath sediments of depth x or cumulative dry mass m (and $\rho = dm/dx$ is the dry wt/wet vol ratio), it is shown that the age t of sediments of depth x satisfies the equation

$$A = A(o)e^{-kt}, \qquad (3)$$

where $A(o)$ is the total residual unsupported ^{210}Pb in the sediment column. A and $A(o)$ are calculated by direct numerical integration of the ^{210}Pb profile. The age of sediments of depth x is then given by

$$t = \frac{1}{k} \ln \frac{A(o)}{A}. \qquad (4)$$

The sedimentation rate at each time is calculated from the formula

$$r = \frac{kA}{C}. \qquad (5)$$

The mean ^{210}Pb supply rate is

$$P = kA(o). \qquad (6)$$

Although inflected and non-monotonic profiles may be expected if there is a varying sedimentation rate but a constant supply of ^{210}Pb, as in the c.r.s. model, other possible causes include migration of ^{210}Pb through the interstitial waters near the sediment–water interface (Koide *et al.* 1973), mixing of the near-surface sediments by physical (Petit 1974) or biological (Robbins *et al.* 1977) processes, and post-depositional redis- tribution of sediment either discontinuously through slumping (Edging-

ton and Robbins 1976b), or more or less continuously by sediment
erosion.

The effect of sediment mixing on the c.f.:c.s. model has been considered by Robbins *et al.* (1977). Their model assumes a constant supply
of ^{210}Pb to the sediments, a constant dry-mass sedimentation rate (which
we again denote by r), and steady-state mixing throughout a zone of fixed
thickness s. The movement of ^{210}Pb in the zone of mixing is represented as
a diffusion process, leading to complete homogenization in the case of
rapid steady-state mixing. Beneath the zone of mixing the unsupported
^{210}Pb-profile is shown to be linear, declining exponentially in accordance
with the formula

$$C = C(s)e^{-k(m-W)/r} \qquad (7)$$

(c.f. equation (1)), where $C(s)$ is the unsupported ^{210}Pb-concentration at
the base of the zone of mixing, and W is the dry mass of sediment in the
zone of mixing. The sedimentation rate r is thus given by the mean slope
of this part of the profile, exactly as for the unmixed case.

The c.r.s. model can be similarly modified to take account of sediment
mixing. The assumptions in this case are a constant rate of supply of ^{210}Pb
to the sediments irrespective of variations in the sedimentation rate, and
steady-state mixing throughout a zone of fixed thickness. If the base of the
zone of mixing that existed at time t in the past is currently at a depth ξ,
$A(\xi)$ is the residual unsupported ^{210}Pb currently beneath this depth, and
$A_m(\xi)$ is the residual unsupported ^{210}Pb in the core from this zone,
equation (3) must be modified to read

$$A(\xi) + A_m(\xi) = A(o)e^{-kt}. \qquad (8)$$

$A(\xi)$ and $A(o)$ are again calculated by numerical integration of the
^{210}Pb-profile. Assuming that the mixing processes modify the ^{210}Pb-profile
in the zone of mixing in a roughly similar way at different times, A_m can
be approximated by

$$A_m = \frac{C}{C(s)} A_m(o), \qquad (9)$$

where $A_m(o)$ is the ^{210}Pb-content of the current zone of mixing. From the
values of these quantities equation (8) can be used to calculate the relation
between ξ and t. Given the dry mass of sediment W in the zone of mixing,
this relation can be used to construct the depth/age relation for the
nominal sediment–water interface. When there is rapid steady-state
mixing,

$$A_m = WC, \qquad (10)$$

and the ξ/t relation is given by

$$t = \frac{1}{k} \ln \frac{A(o)}{A + WC}. \tag{11}$$

Several points here require emphasis.

1. Where mixing has occurred, and the ^{210}Pb-flux and sedimentation rate have been constant, the modified c.f.:c.s. and c.r.s models will give the same results. Where, however, there is a non-linear ^{210}Pb-profile beneath the zone of mixing, varying sedimentation rates may be inferred, and the c.f.:c.s. model will not be applicable.

2. Mixing and changes in sedimentation rates are not mutually exclusive, therefore the former cannot be used to dispose of evidence which may be indicative of the latter, the more so since mixing rarely, if ever, implies total homogenization.

3. Human impact has often led to changes in sedimentation rates over the last few decades and centuries as a result of, for example, catchment erosion or eutrophication. It is inherently unsatisfactory to assume constant sedimentation except where there is strong evidence in favour of such an assumption, or there is conflicting evidence and the assumption of constant sedimentation is deductively more conservative.

Lead-210 profiles from laminated cores (Appleby et al. 1979) show quite clearly that depressed near-surface concentrations may be associated with accelerated sedimentation rates. Where mixing is suspected, estimates of the thickness of the zone of mixing cannot therefore be based simply on the shape of the ^{210}Pb-profile. Under favourable circumstances, evidence for the thickness of the zone of mixing and the extent of the mixing process can be derived from visual examination of the sediment, with particular attention to the extent and nature of material derived from singular events (Davis 1967). Clearly defined features in the near-surface sediments, and differential gradients in parameters other than ^{210}Pb, are evidence against substantial mixing. Where the extent of mixing is difficult to assess, it is worth noting that c.r.s.-derived age/depth relations assuming no mixing differ significantly from those derived assuming rapid steady-state mixing only when the thickness of the zone of mixing exceeds ~15% of the depth of the ^{210}Pb-profile.

Where there has been horizontal post-depositional sediment redistribution, the constant ^{210}Pb-supply hypothesis, although not valid for a single core, may be valid for the lake bed as a whole. Where this is the case, and the processes have been relatively smooth, a c.r.s. chronology may still be established by correlating ^{210}Pb-data from a representative set of cores (Oldfield et al. 1980). If ξ is a depth parameter which identifies synchronous levels in different cores, and $B(\xi)$ is the total unsupported ^{210}Pb beneath the entire sediment layer identified by this parameter, the age t of this layer is given by the equation

$$B(\xi) = B(o)e^{-kt} \qquad (12)$$

(c.f. equation (3)). $B(\xi)$ and $B(o)$ are calculated by summing the corresponding ^{210}Pb-residuals for individual cores, weighted by representative areas. The mean ^{210}Pb-flux per unit area to the sediment is

$$P = \frac{kB(o)}{\alpha} \qquad (13)$$

where α is the area of the lake. For ease of calculation these formulae may be expressed in terms of the mean ^{210}Pb-residual per unit area. Core correlations may be established using, for example, magnetic susceptibility profiles, pollen and diatom analyses, or synchronous influx events.

The models discussed so far all assume a constant supply of ^{210}Pb to the sediments, modified possibly by sediment mixing or sediment redistribution. The principal alternative hypothesis used for dating non-linear profiles assumes that sediments have a constant initial unsupported ^{210}Pb-concentration, irrespective of changes which may have occurred in the net dry-mass sedimentation rate. This model has been termed the constant initial concentration (c.i.c.), model, or constant specific activity model (Robbins 1978). Papers by Pennington et al. (1976) and Battarbee (1978) exemplify this approach. Since all sediments have the same initial unsupported ^{210}Pb-concentration, the present unsupported ^{210}Pb-concentration C in a sediment column will vary exponentially with age t in accordance with the formula

$$C = C(o)e^{-kt} \qquad (14)$$

where $C(o)$ is the unsupported ^{210}Pb-concentration at the sediment–water interface. C and $C(o)$ are obtained directly from the ^{210}Pb-measurements. The age t of each sediment layer is then calculated using the formula

$$t = \frac{1}{k} \ln \frac{C(o)}{C} . \qquad (15)$$

There are a number of hypotheses under which c.i.c. calculations are feasible:

1. By assuming, in lakes where sedimentation is dominated by allochthonous particle input, that the main delivery pathway for unsupported ^{210}Pb is route C_2 (fig. 3.1), whereby the ^{210}Pb reaches the sediment in particle-associated form, and that the initial unsupported ^{210}Pb-concentration of the sediment is constant for all depths.
2. By assuming, especially in lakes where sedimentation is dominated by autochthonous material derived from primary production within the lake waters, that increased flux of sedimentary particulates from

the water column will remove proportionally increased amounts of ^{210}Pb from the water to the sediment surface.

3. By assuming a constant sedimentation rate and constant ^{210}Pb-supply to the lake whereby all sediments have the same initial concentration, and that varying sedimentation rates in different parts of the lake and at different times simply reflect varying patterns of sediment deposition.

In a variant of the c.i.c. model, some authors (e.g. Brugam 1978) have divided non-linear profiles into linear segments within each of which constant sedimentation is assumed, with instantaneous shifts in rate at inflexions. The age/depth curve derived from this model is a piecewise linear approximation to the c.i.c.-derived age/depth curve. In the c.r.s. model, it should be noted that instantaneous shifts in the sedimentation rate would also be associated with jumps in the value of the ^{210}Pb-activity (cf. Battarbee et al. in press: fig. 2).

From the above, it will be clear that there are at the outset no automatically and universally valid reasons for choosing particular models or variants for use at specific sites. The construction of a ^{210}Pb-chronology for a given lake will not therefore be simply a matter of calculating dates using one or other of the above models. It will demand in addition an element of model testing in order to determine whether a ^{210}Pb-chronology is feasible, and if so, on what basis the calculations should be made.

When the dry-mass sedimentation rate in a lake has, in reality, been constant or nearly so, the c.i.c., c.f.:c.s., and c.r.s. ^{210}Pb-chronologies will be the same, and the precise nature of the ^{210}Pb-supply mechanism is immaterial. As we have seen, however, in situations where sedimentation rates may have varied, choice of model will be critical. This emphasizes the need for rigorous empirical tests designed to evaluate the consistency and validity of the alternatives. In the following sections we will consider procedures for evaluating the hypotheses outlined above in situations where sedimentation rates may have varied. In this connection, the following questions are crucial:

1. Does the evidence available support the assumption of a constant initial concentration, or of a constant supply of ^{210}Pb as the primary component of a ^{210}Pb-based dating model?

2. Where the evidence supports the assumption of a constant supply of unsupported ^{210}Pb to the sediment, but where sedimentation rates may have varied, are dates and rates using the c.r.s. model consistent with independent evidence? Or does the evidence support the contention that only mean rates can be inferred, and that departures from a linear decline in concentration versus accumulated dry-mass

are best interpreted as 'statistical' errors, or as evidence of mixing or disturbed sedimentation?

Empirical testing

Evidence on the mechanism by which ^{210}Pb is supplied to the sediments can be obtained from the sedimentary record. If the ^{210}Pb-supply is governed by the atmospheric flux to the lake waters (c.r.s. model) then:

1a. Non-monotonic profiles may be expected in response to major changes in sedimentation rate since faster net sediment accumulation will tend to depress initial unsupported ^{210}Pb-concentrations and *vice versa*.

1b. Different cores from the same lake, or from the same depositional zone within a very large lake, or from different lakes in the same general area will normally have comparable total residual unsupported ^{210}Pb-values despite differences in the rates of net-sediment accumulation over the time span of unsupported ^{210}Pb-activity.

On the other hand, if the ^{210}Pb is governed by the sedimentation rate (c.i.c. model):

2a. The plot of unsupported ^{210}Pb-concentration versus depth must show a monotonic decline without significant kinks;

2b. The total cumulative residual unsupported ^{210}Pb (standing crop) in sediment cores from the same lake should vary roughly in proportion to the total dry mass of sediment accumulated over the depth of unsupported ^{210}Pb activity.

In practice, sharpest disagreement has centred on the interpretation of profiles in which the plot of unsupported ^{210}Pb-concentration versus accumulated dry mass of sediment is non-linear and in some cases non-monotonic. In the extreme case of strongly non-monotonic profiles only the c.r.s. model offers a basis for full interpretation, and it is essential to develop criteria which will help to establish whether a given profile can in fact be interpreted in this way, or should be discarded. In general, we have found the value of the total unsupported ^{210}Pb-residual an indispensible parameter in evaluating such data. Table 3.1 summarizes results from a wide variety of sites with non-linear profiles and confirms that in many cases, the total ^{210}Pb-residuals in cores from the same locality fulfil condition 1b of the c.r.s. model, despite up to eight-fold variations in the dry-mass sedimentation rate. Table 3.1 also gives the unsupported ^{210}Pb-flux rates consistent with the ^{210}Pb-residuals. The average value of the ^{210}Pb-flux rate from all the data that we have

Table 3.1. ^{210}Pb-parameters and sedimentation rates for cores from a variety of sites satisfying the c.r.s. criteria.

Coring site		Total residual unsupported ^{210}Pb-content A(o) (pCi cm^{-2})	Unsupported ^{210}Pb-conc. at surface C(o) (pCi g^{-1})	Mean sedimentation rate during		^{210}Pb-flux equivalent to ^{210}Pb-residual P (pCi cm^{-2}yr^{-1})
				(a) past 30 yrs r(o) (g cm^{-2}yr^{-1})	(b) past 100 yrs r̄ (g cm^{-2}yr^{-1})	
Ireland (Oldfield et al. 1978)						
Lower L. Erne	Core SM1	20.7	7.13	0.08	0.037	0.64
	Core FM1	19.2	1.82	0.31	0.12	0.60
Upper L. Erne	Core FM2	14.9	1.26	0.35	0.13	0.46
L. Augher (Battarbee, pers. comm.)		12.7	1.81	0.19	0.11	0.40
Wales (Elner and Wood 1980)						
Llyn Peris	Core A	38.8	1.08	1.07	0.27	1.21
	Core E	42.8	1.99	0.67	0.28	1.33
England						
Rostherne Mere	Core RMII	6.5	2.24	0.084	0.055	0.20
	Core N79	5.4	1.87	0.088	0.052	0.17
Newton Mere		5.3	4.43	0.032	0.029	0.17
Belgium (Oldfield et al. 1980)						
L. Mirwart	Core 1	10.5	3.29	0.092	0.14	0.33
	Core 2	10.0	2.21	0.13	0.055	0.31
	Core 3	10.7	2.34	0.14	0.12	0.33
Finland (Appleby et al. 1979)						
Laukunlampi		20.5	36.7	0.011	0.0072	0.64
Lovojärvi		20.1	5.55	0.078	0.059	0.63
Pääjärvi		24.5	14.2	0.044	0.059	0.76

assembled is 0.55 pCi cm^{-2} yr^{-1}, and is close to estimates of the mean annual deposition rate of ^{210}Pb by direct fallout from the atmosphere (Krishnaswamy and Lal 1978: 158–9). Although these results cannot be taken as proof of the general validity of the c.r.s. model, in practice we have found that in sites where condition 1b is fulfilled, the ^{210}Pb-derived dates have generally been consistent with independent dating evidence from annual laminae (Appleby et al. 1979), from ^{137}Cs data (Oldfield et al. 1978), from pollen-analytical representation of dated vegetation and land-use changes (e.g. the Ambrosia horizon in many north American pollen-diagrams), from palaeomagnetic measurements, from diatom records, and from statigraphically identifiable dated events such as inwash layers from periods of catchment disturbance.

Robbins (1978: 371) suggests that varying ^{210}Pb-residuals (up to an order of magnitude) over the bottom of a lake may arise from variations in the net efficiency of transfer of ^{210}Pb from the atmosphere to the sediments at different locations, and that locally high ^{210}Pb-flux rates to the sediments reflect selective deposition of particles scavenging ^{210}Pb from the water column. Despite this possibility, calculations show that condition 1b does hold in a large number of cases, and does provide a relatively simple initial criterion for assessing the reliability of c.r.s.-derived dates and sedimentation rates.

Table 3.2. ^{210}Pb-parameters for cores from a variety of sites not satisfying the c.r.s. criteria.

Coring site		Total residual unsupported ^{210}Pb-content A(o) (pCi cm^{-2})	Unsupported ^{210}Pb-conc. at surface C(o) (pCi g^{-1})	^{210}Pb-flux equivalent to ^{210}Pb residual (pCicm^{-2}yr-1)	Mean ^{210}Pb-flux for related sites (pCi cm^{-2}yr^{-1})
Ireland (Battarbee 1978)					
Lough Neagh	Core B43	10.3	3.51	0.32	0.40
	Core B46	>21.5	4.0	0.79	
Wales (Wood, pers. comm.)					
Llyn Padarn	Core A	5.8	0.31	0.18	1.27
	Core E	16.1	0.75	0.50	
Finland (K. Tolonen, pers. comm.)					
Takakillo		7.02	23.1	0.22	0.68
L. Erie, U.S.A. (Robbins *et al.* 1978)					
	Core U42	14.6	3.76	0.46	
	Core G16	29.3	11.14	0.91	0.45–0.60
	Core M32	72.7	7.16	2.26	

Table 3.2 gives results from a number of cores which have anomalous or variable total residuals, and hence do not satisfy the requirements of the c.r.s. model. In these cases, at least where the concentration profiles are strongly non-linear, c.i.c. and c.f.:c.s. models are also difficult to sustain. There does not appear to be any evidence that the variations in total residual can be explained by supposing that ^{210}Pb is scavenged by sediment particles from the water column in direct proportion to the sedimentation rate. Although high ^{210}Pb-residuals are sometimes associated with high sedimentation rates, the correlation between these parameters is not strong. Even where the correlation is sustained, there are in many cases inconsistencies between the ^{210}Pb-dates and independent dating evidence regardless of the model used. Mechanisms that could lead to varying ^{210}Pb-residuals include delivery of ^{210}Pb from the catchment in a particle-associated form, or post-depositional redistribution of sediment. In both cases, varying erosion rates suggested by non-linear

[210]Pb-profiles are likely to alter the pathways by which ^{210}Pb is delivered to the sediment, and hence the initial ^{210}Pb-concentrations in the sediments. An understanding of these processes depends in the first place on the construction of a reliable independent chronology. In the case of Lake Takakillo, for example, dating by counting annual laminae (K. Tolonen personal communication) reveals that the ^{210}Pb-deficiency (by comparison with neighbouring sites, see table 3.1) derives not from a loss of sediment but from a decline in the ^{210}Pb-supply to the sediments at the end of the nineteenth century. The reasons for this decline are not clear, and are the subject of further investigation. Where anomalies in the ^{210}Pb-supply are localized, core correlations may be used to transfer reliable ^{210}Pb-chronologies from dated cores outside this zone to the anomalous cores.

When the ^{210}Pb-profiles satisfy the assumptions of the c.r.s. model, errors in the ^{210}Pb-derived dates and sedimentation rates may still arise as a result of vertical mixing processes in the near-surface sediments. The extent of these processes can often be explored by associated stratigraphic, microfossil, or mineral magnetic studies. In many cases, ordered and repeatable variations in stratigraphic units (e.g. laminae), microfossil assemblages (e.g. diatoms and pollen), or magnetic parameters have shown that zones of 'flat' ^{210}Pb-concentrations cannot be interpreted by inferring mixing (cf. Oldfield et al. 1978, Appleby et al. 1979). Where there is evidence of mixing, the modified c.r.s. model (given by equations (8)–(11)) can be used.

In cores where the sedimentation rate has been constant, or nearly so, the mean sedimentation rate can be calculated using the c.f.:c.s. model. Where there are significant variations in the sedimentation rate, however, the slope of the ^{210}Pb-profile does not measure the sedimentation rate (Appleby and Oldfield 1978), and the c.f.:c.s. model is not applicable. If desired, the mean sedimentation rate can in this case be calculated by applying the c.f.:c.s. model to the integrated profile. The piecewise linear variant of the c.f.:c.s. model (Brugam 1978) is not a constant ^{210}Pb-supply model, and is not applicable to cores satisfying the c.r.s. hypothesis. Care is required in interpreting rapid variations in the sedimentation rate given by c.r.s. calculations, especially in older sediments where their significance is in doubt for statistical reasons. In several cases where independent evidence of age and catchment history is available, the validity of the 'anomalous' ^{210}Pb-determinations associated with such features has been confirmed by the demonstration that they are the expression of varying material input from the catchment. Confidence in interpretation is only possible through rigorous independent validation, error analysis (Battarbee et al. 1980) or the repeatability of variations in several cores (Elner and Wood 1980). Techniques for improving confidence in the chronology of older sediments include the amalgamation of data from correlated cores (see equations (12) and (13)), and the use of independently dated horizons. If a sediment layer near the base of the ^{210}Pb-profile has age t_0, and ΔA is the integrated total ^{210}Pb-residual in the core above this horizon, assuming that the ^{210}Pb-flux since the date of this horizon is similar to the ^{210}Pb-flux prior to this date the ^{210}Pb-residual A_0 beneath this layer can be estimated using the formula

$$A_0 = \frac{\Delta A}{e^{kt_0} - 1}.$$

(16)

Using A_0 as a starting point for calculating the dating parameter, the c.r.s.-derived age/depth curve is constrained to pass through the age/depth point of this layer. The purpose of the ^{210}Pb-calculations in this case is then simply to determine variations in the sedimentation rate. In the next section we illustrate the application of ^{210}Pb-dating by outlining some significant case studies based on recalculations of published data from the Great Lakes of North America.

Case studies

LAKE MICHIGAN

Robbins and Edgington (1975) used measurements from eight ^{210}Pb-profiles to calculate mean sedimentation rates in Lake Michigan over the last 150 years. In considering the source of unsupported ^{210}Pb to the sediment Edgington and Robbins (1976a) state that the 'runoff of lead from streams would not be expected to contribute a significant concentration of unsupported ^{210}Pb to the sediments'. This, coupled with Lake Michigan's long residence time and high ratio of lake surface area to drainage basin surface area, suggests that it should come close to fulfilling the assumption of a constant net rate of supply of unsupported ^{210}Pb through time, most of it directly from the atmosphere. However, their assumption of constant dry-mass sedimentation rates would appear to be questionable. Dramatic nineteenth- and twentieth-century changes in sediment yield from the catchment as a result of forest clearance and cultivation can be confidently inferred from both local (Davis 1976) and general (Leopold 1956) studies. It would be contrary to experience elsewhere if these were to have had no impact in changing recent accumulation rates in the zone receiving sediment directly from cleared and farmed land especially alongside the south-east quadrant of the lake.

By applying the c.r.s. model to their published data we have calculated age/depth curves and sedimentation rate histories for Lake Michigan (fig. 3.3).

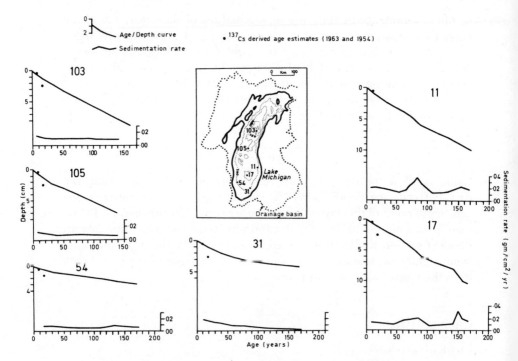

Fig. 3.3   Lake Michigan: age/depth curves and sedimentation rates for six cores, derived from ^{210}Pb measurements assuming a constant net rate of supply of unsupported ^{210}Pb to the sediment (c.r.s. model).

The following points arise from the recalculation of the published data:

1. The ^{210}Pb-data for all cores save numbers 29 and 100A satisfy the c.r.s. model requirement that the total residual unsupported ^{210}Pb-content ($A(o)$) be comparable from core to core within each broad area of the lake. Variations in $A(o)$ for these cores (see table 3.3) do not appear to correlate with variations in the sedimentation rate. There does, however, appear to be a general increase from the south-west corner towards the outfall at the north end. The high unsupported ^{210}Pb-contents of cores 29 and 100A, coupled with surface concentrations comparable to those found in neighbouring cores prompt an interpretation in terms of sediment resuspension and focusing.

2. The mean rate of supply of unsupported ^{210}Pb to the sediments derived directly from the c.r.s. model calculations is 0.19pCi cm^{-2} yr^{-1}. This is in close agreement with Jaworski's estimate of 0.2pCi cm^{-2} yr^{-1} for the atmospheric flux of ^{210}Pb over Lake Michigan quoted in Robbins and Edgington (1975).

Table 3.3. ^{210}Pb-parameters and sedimentation rates for Lake Michigan.

	Total residual unsupported ^{210}Pb-content A(o) (pCi cm^{-2})	Unsupported ^{210}Pb conc. at surface C(o) (pCi cm^{-2})	^{210}Pb-flux equivalent to ^{210}Pb residual P(pCi cm^{-2} yr^{-1})	Estimated sedimentation rates r (g cm^{-2}yr^{-1})	
				(a) 1930	(b) 1965
Core 54	3.93	10.1	0.12	0.0047	0.0076
31	5.02	6.73	0.16	0.0110	0.0180
29	25.50[1]	7.4	0.79		
11	5.42	7.1	0.17	0.0154	0.0200
17	6.06	11.4	0.19	0.0101	0.0132
105	7.85	14.1	0.24	0.0087	0.0143
103	8.33	15.2	0.26	0.0100	0.0140
100A	17.8[1]	18.3	0.55		
			Mean	.0100	.0145

1. No ^{210}Pb derived sedimentation rates are given for these cores (see text).

3. The c.r.s.-derived dry-mass sedimentation rates for cores 11 and 17
    suggest episodes of accelerated sedimentation in the early
    nineteenth century and during the period 1860–1920 (fig. 3.3). The
    latter episode is indicated by the non-monotonic features on both
    ^{210}Pb-profiles at similar depths. The uniformity of the ^{210}Pb-residuals
    precludes an explanation for this feature in terms of sediment
    disturbance due to, for example, storm-induced slumping. The very
    low values of unsupported ^{210}Pb-activity allow only very tentative
    estimates of early nineteenth-century dates and sedimentation rates
    using ^{210}Pb alone. In core 11, however, the *Ambrosia* pollen level has
    been identified and dated by ^{210}Pb to c. 1800–15. This horizon can be
    securely dated to c. 1830 (cf. Davis *et al.* 1971) and if the c.r.s.-
    derived age–depth curve is constrained to pass through this point
    (using equation (16)) a more reliable nineteenth-century chronology
    can be established. Calculations by this method for core 11 confirm
    the pattern of sedimentation suggested above. Cores 11 and 17
    receive direct input of sediment from the large south-east segment of
    the drainage basin (fig. 3.3). Their sedimentation histories can be
    compared with those calculated for Frains Lake by Davis (1976).
    Although the latter is a small lake just beyond the south-east edge of
    the Lake Michigan catchment, it is within an area which has
    experienced a directly comparable history of forest clearance and
    subsequent land use change since 1830 (see, e.g. Paullin 1932). The
    correspondence between Frains Lake sedimentation rate changes
    and those in cores 11 and 17 is, therefore, thought to be significant.

Fig. 3.4 graphs a direct comparison between Davis's estimates and the results from core 11.

Cores 54 and 31 adjacent to the sediment-starved southern end of the lake, and cores 105 and 103 in the central region of the lake, do not appear to have been significantly affected by nineteenth-century events. The monotonically declining slope of the ^{210}Pb-profile from

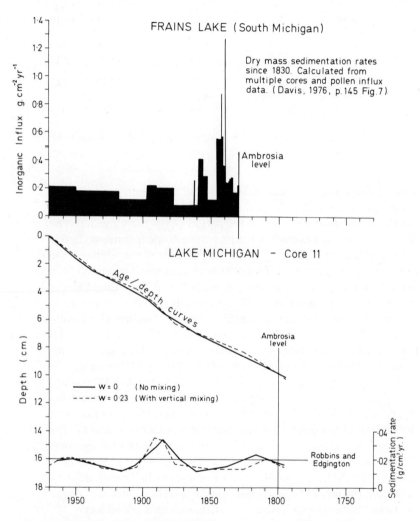

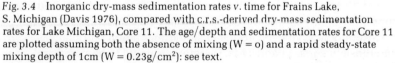

Fig. 3.4   Inorganic dry-mass sedimentation rates *v.* time for Frains Lake, S. Michigan (Davis 1976), compared with c.r.s.-derived dry-mass sedimentation rates for Lake Michigan, Core 11. The age/depth and sedimentation rates for Core 11 are plotted assuming both the absence of mixing (W = o) and a rapid steady-state mixing depth of 1cm (W = 0.23g/cm^2): see text.

core 31, the nearest of these cores to the south-east corner, does, however, indicate a steadily increasing sedimentation rate at this site from ~0.0050g cm^{-2} yr^{-1} in the late nineteenth century to 0.011g cm^{-2} yr^{-1} by 1930.

4. All six cores dated by the c.r.s. calculations appear to show an acceleration in the dry-mass sedimentation during the last 40 years, and this may reflect the effect of cultural eutrophication. The average value of 0.010g cm^{-2} yr^{-1} in 1930 has increased to a present-day value of 0.0145g cm^{-2} yr^{-1}. Error analyses indicate a standard error in the mean value not greater than $\pm 10\%$.

5. The above conclusions are sustained whether or not the effect of possible mixing (as suggested by Robbins and Edgington) is included in the calculations. The use of a mixing model tends to emphasize variations and sharpen inflexions without significantly distorting the original profiles (figs. 3.4 and 3.5).

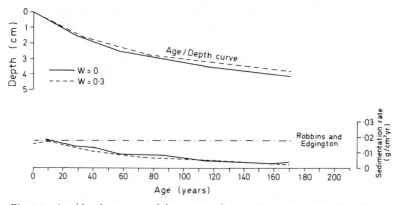

Fig. 3.5 Age/depth curves and dry-mass sedimentation rates *v.* time for Lake Michigan, Core 31. The results assuming no vertical mixing (W = o) are compared with the results assuming rapid steady-state mixing to a depth of 1cm (W = 0.30g/cm^2).

6. Diagrams showing the distribution and thickness of the Waukegan member (Leland *et al.* 1973: 97) indicate that core 29 lies within a relatively localized zone of sediment focusing. Correlation of core 29 with dated cores outside this zone will yield information about the nature of this process. The proximity of cores 29 and 31 to each other, and the similarity of their ^{210}Pb (Robbins and Edgington 1975) and stable-lead (Edgington and Robbins 1976a) concentrations near the sediment–water interface, suggests that sediment focused at core 29 is typical of the material deposited at core 31, and hence that sediments at cores 29 and 31 with similar ^{210}Pb and stable lead

concentrations are contemporaneous. Fig. 3.6a shows the correlation
between sediment depths in cores 29 and 31 when they are matched
on the basis of ^{210}Pb-concentrations, the ^{137}Cs peak (Robbins and
Edgington 1975) and the stable-lead concentrations (Edgington and

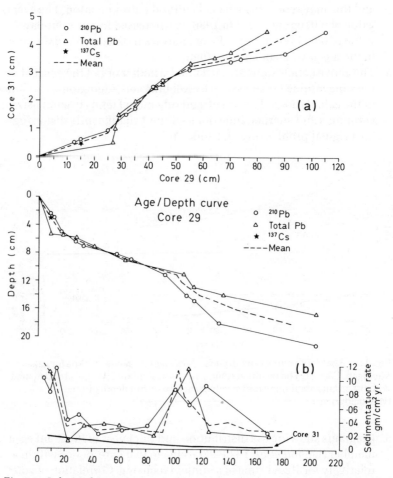

Fig. 3.6   Lake Michigan, Core 29: sediment resuspension and focusing. 3.6(a) plots
the correlation between cores 31 and 29 assuming comparable concentrations of
unsupported ^{210}Pb, stable Pb and ^{137}Cs in each core for sediments of the same age.
Note the close correspondence between the independently derived correlations.
3.6(b) plots age/depth curves and sedimentation rates for core 29 using the
correlations obtained from 3.6(a), and the c.r.s.-derived age/depth curve for core 31.
The sedimentation rate for core 31 has been added for comparison (see text). In all
graphs, the pecked line plots the mean values derived from ^{210}Pb and stable Pb
derived correlations.

Robbins 1976a). The consistency between the three independent sets of data is good, and thus supports the above hypothesis. Fig. 3.6b shows the age/depth curves and sedimentation rates for core 29 derived from these correlations using the c.r.s. age/depth relation for core 31. They suggest that focusing leads to a background rate of sediment deposition at core 29 of three to four times the value at core 31, a figure roughly in agreement with that indicated by the relative thicknesses of the Waukegan member at the two sites. Superimposed on this steady process there appears to have been a period of accelerated sediment deposition during the latter part of the nineteenth century, and during the past 20–30 years. The mean sedimentation rate at core 29 during the past 100 years by this method is $\approx 0.05 \mathrm{g\,cm^{-2}\,yr^{-1}}$, whereas the values given by independent ^{210}Pb calculations all lie in the range 0.09–0.10$\mathrm{g\,cm^{-2}\,yr^{-1}}$.

LAKE ONTARIO

Table 3.4 summarizes ^{210}Pb-parameters and sedimentation rates for five cores from Lake Ontario, derived from data published by Farmer (1978) and Robbins *et al.* (1978). In the case of the pairs of cores from the Niagara Basin and the Rochester Basin there is reasonable comparability in the values of the total ^{210}Pb-residual, and the associated mean ^{210}Pb-flux estimates lie in the expected range. The two cores from the Rochester Basin (Farmer 1978) both have linear ^{210}Pb-profiles, indicating constant

Table 3.4. ^{210}Pb-parameters and sedimentation rates for Lake Ontario.

Coring site	Total residual unsupported ^{210}Pb-content A(o) (pCi cm^{-2})	Unsupported ^{210}Pb- conc. at surface C(o) (pCi g^{-1})	Mean sedimentation rate during the period (a) 1850–1935 (b) 1935–70 (g cm^{-2}yr^{-1})		^{210}Pb-flux equivalent to ^{210}Pb-residual P (pCi cm^{-2}yr^{-1})
Niagara Basin					
Core 72–4 (Farmer 1978)	15.2	9.5	0.0605	0.0542	0.47
Core WB (Robbins *et al.* 1978)	18.2	11.6	0.0194	0.0449	0.57
			0.0242[1]	0.0409[1]	
Rochester Basin					
Core 73–13 (Farmer 1978)	9.9	9.9	0.0272	0.0282	0.31
Core 74–14	13.6	8.9	0.0444	0.0469	0.42
Cape Vincent					
Core KB (Robbins *et al.* 1978)	29.7	14.1	0.044[1]	0.110[1]	0.92

1. Mean sedimentation rates from pollen dates.

sedimentation rates. This is consistent with results obtained for a nearby core by pollen dating (Kemp *et al.* 1974) using the *Ambrosia-* and *Castenea*-horizons. For cores with linear ^{210}Pb-profiles all dating models give similar results. Table 3.4 shows that the c.r.s.-derived sedimentation rates for the periods 1850–1935 and 1935–70 are in close agreement with the values obtained by Farmer using the constant flux/constant sedimentation rate model.

The ^{210}Pb-profiles from the Niagara Basin, in contrast, both have non-linear features, and this may indicate variations in the sedimentation rate associated with the post-1850 colonial development. Fig. 3.7 summarizes the results from core WB, and illustrates the good comparison

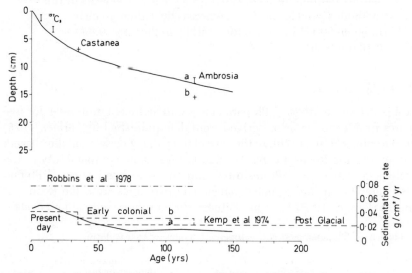

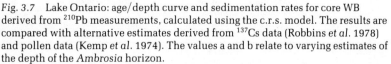

Fig. 3.7   Lake Ontario: age/depth curve and sedimentation rates for core WB derived from ^{210}Pb measurements, calculated using the c.r.s. model. The results are compared with alternative estimates derived from ^{137}Cs data (Robbins *et al.* 1978) and pollen data (Kemp *et al.* 1974). The values a and b relate to varying estimates of the depth of the *Ambrosia* horizon.

between c.r.s.-derived age/depth and sedimentation rate estimates, the previously published pollen study by Kemp *et al.* (1974) and the ^{137}Cs profile. The sedimentation rate appears to increase steadily after c. 1900 from a pre-colonial value of 0.016g cm^{-2} yr^{-1} to a present-day value of 0.050g cm^{-2} yr^{-1}. The mean sedimentation rate of 0.078g cm^{-2} yr^{-1} derived by Robbins *et al.* (1978) using the constant flux/constant sedimentation rate model is significantly higher than this evidence suggests. Where there is an increasing rate of sedimentation the slope of the ^{210}Pb-profile does not measure the sedimentation rate (Appleby and Oldfield 1978), and the constant flux/constant sedimentation rate model will over-estimate the mean sedimentation rate. Robbins *et al.* (1978) suggest that there may have been a loss of sediment from core WB in 1940 due to storm damage. The agreement in the ^{210}Pb-residuals of cores WB and 72–4 indicates that this is unlikely to have occurred. Calculations for core 72–4, directly out from the mouth of the Niagara river, indicate a higher sedimentation rate during the early colonial period, followed by a steady decline during the past 50 years to a present-day value of 0.046g cm^{-2} yr^{-1}.

The ^{210}Pb-content of core KB (Robbins *et al.* 1978) located off Cape Vincent is significantly higher than that of cores in the off-shore basins. This gives rise to doubts about the constancy of the supply of ^{210}Pb to this site, and hence about the validity of ^{210}Pb-dates based on this hypothesis. Indeed, while there is not a great difference between the various ^{210}Pb-chronologies for this site, they all have a relatively poor match with the chronology based on ^{137}Cs and pollen data. The *Ambrosia* and *Castanea* pollen horizons (Kemp *et al.* 1974) suggest a sedimentation rate during the period 1850–1935 similar to that in the off-shore basins. Since 1935, however, the sedimentation rate appears to have increased by a factor of 2–3. Since the ^{210}Pb-content of core KB (table 3.4) is 2–3 times that of the other cores, it would seem that the most likely mechanism for this is sediment focusing.

LAKE HURON

Table 3.5 summarizes ^{210}Pb-parameters and c.r.s.-derived sedimentation rates for a set of cores from Lake Huron, calculated from data published by

Table 3.5.　^{210}Pb-parameters and sedimentation rates for Lake Huron.

ɔring site	Total residual unsupported ^{210}Pb-content A(o) (pCi cm^{-2})	Unsupported ^{210}Pb- conc. at surface C(o) (pCi g^{-1})	Mean sedimentation rate during		^{210}Pb-flux equivalent to ^{210}Pb-residual P (pCi cm^{-2}yr^{-1})
			(a) past 30 yrs r(o) (g cm^{-2}yr^{-1})	(b) past 100 yrs ῑ (g cm^{-2}yr^{-1})	
ᴋe Huron					
ɔre 18[2] (1975)	21.3	8.0	0.063	0.051	0.66
18[2] (1974)	22.0	6.3	0.073	0.061	0.69
14[2] (1975)	21.7	10.3	0.039	0.028	0.67
14[2] (1974)	22.5	8.1	0.042	0.029	0.70
1[3]	24.4	14.7	0.034	0.033[1]	0.76
2[3]	5.4	22.9		0.0065[1]	0.17
3[3]	8.5	19.0		0.0405[1]	0.26
4[3]	15.7	19.8		0.0495[1]	0.49

1. Mean sedimentation rates given by *Ambrosia* pollen data.
2. Robbins, Krezoski and Mozley (1977).
3. Durham and Joshi (1980).

Robbins *et al.* (1977) and Durham and Joshi (1980). The first five cores are all located in the south-east of the lake (see fig. 3.8), and all have comparable residual ^{210}Pb-contents. The data from Robbins *et al.* relate to

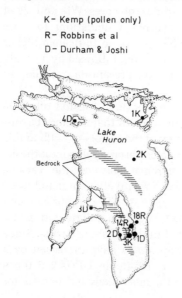

Fig. 3.8   Lake Huron, showing the location of the sediment cores referred to in Kemp *et al.* 1974, Robbins *et al.* 1977, and Durham and Joshi 1980.

measurements in successive years from the same locations. In the case of core 1D the ^{210}Pb-data did not extend down to the equilibrium level, and an allowance has been made for the unrecorded ^{210}Pb using the *Ambrosia* pollen horizon. The measured ^{210}Pb-content was 19.0pCi cm^{-2}. The similarity of the ^{210}Pb-profiles of cores 1R and 1D (fig. 3.9) suggests that they have similar accumulation rates, and this is supported by the c.r.s.-derived chronologies shown in fig. 3.10. Because the authors publishing the original results adopted opposing views of the near-surface trends (Robbins *et al.* inferring sediment mixing, and Durham and Joshi rapid deposition), they obtained quite different accumulation rates, neither of which is compatible with the *Ambrosia* pollen horizon. This occurs at a depth of 15cm in core 1D, and 16.5cm in a nearby core analysed by Kemp *et al.* (1974). While some sediment mixing may well occur, the conformity of these results suggests that homogenization of sediments to a depth of 6cm is unlikely. In core 18R, where there is flattening of the ^{210}Pb-profile only in the top 3cm, the c.r.s.-derived chronology does not differ significantly from that obtained by Robbins *et al.*

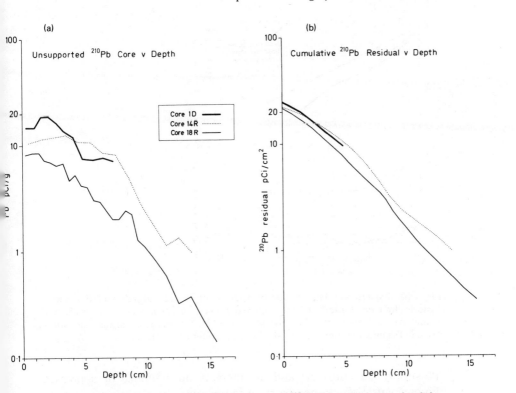

Fig. 3.9 Lake Huron. (a) shows the unsupported ^{210}Pb concentrations *v*. depth for three cores from the south-east corner of the lake. Data from Robbins *et al.* 1979 and Durham and J̈oshi 1980. Robbins *et al.* assume that the flattened part of the profile is due to sediment mixing. Durham and Joshi assume that it is due to rapid sedimentation. (b) shows the cumulative residual unsupported ^{210}Pb in these cores *v*. depth.

The remaining three cores analysed by Durham and Joshi have varying ^{210}Pb-residuals, and this would, in our experience lead us to question the reliability of the associated ^{210}Pb-chronologies. The depth of the *Ambrosia* horizon in core 2D, at 3.4cm, is roughly a quarter of the depth at core 1D, and the ^{210}Pb-content (table 3.5) is lower by a similar factor. A possible explanation for this deficiency is sediment scouring. Since the ^{210}Pb-derived sedimentation rates are similar to those given by the *Ambrosia* pollen data, the process would, however, appear to be reasonably uniform. In cores 3D and 4D the sedimentation rates suggested by the *Ambrosia* horizon, although similar to those for cores 14R and 1D, are considerably greater than the values indicated by ^{210}Pb-calculations. This suggests that even linear ^{210}Pb-profiles may sometimes be unreliable when their ^{210}Pb-contents are not consistent with the assumptions of a uniform

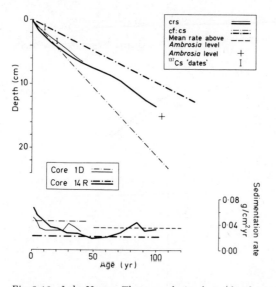

Fig. 3.10   Lake Huron. The c.r.s.-derived age/depth curves and sedimentation rate histories for cores 1D and 14R are compared with original estimates using the cf : cs model. Also shown are the ^{137}Cs 'dates' for core 14R, the *Ambrosia* horizon for core 1D, and the mean sedimentation rate above the *Ambrosia* horizon.

^{210}Pb-supply. In cores 2D and 3D there is no evidence of depressed ^{210}Pb-concentrations in near-surface sediments.

Limitations in the applicability of ^{210}Pb-dating

Although the results summarized in tables 3.1–3.5 confirm that at many sites the total residual unsupported ^{210}Pb-content of sediment cores is consistent with the concept of a uniform flux of ^{210}Pb, they also reveal a significant number of cores with anomalous ^{210}Pb-residuals. At such sites there is no *a priori* reason for supposing that the supply of ^{210}Pb is constant and ^{210}Pb-chronologies based on this premise (c.f.:c.s. and c.r.s.) are thus open to doubts about their reliability. This view is supported by the results from Lakes Ontario and Huron discussed above, and directly confirmed by the data from Lake Takakillo in Finland. In none of these cases does the c.i.c. model provide a better alternative ^{210}Pb-chronology.

One of the aims of ^{210}Pb-chronology must be to identify criteria and site characteristics which allow evaluation of lakes and their sediments as contexts within which reliable ^{210}Pb-based results may be expected. From the above accounts, the following factors may be identified as potential sources of error.

1. Low and/or variable mean residence time of the lake water. In terms of the model in fig. 3.1, this reduces the likelihood of a reasonably large and constant flux of unsupported ^{210}Pb reaching the sediment through the deposition of scavenging fine particulates, and it increases greatly the likelihood that changes in catchment variables such as surface water hydrology and surface soil erosion will significantly disturb the assumptions of a constant ^{210}Pb-flux.
2. Steep underwater slopes. Removal of sediment through slumping has been inferred at some sites though this needs to be tested by more critical stratigraphic comparison wherever it is suspected.
3. Sediment resuspension and focusing. This process will lead to reduced ^{210}Pb-residuals at sites where sediment erosion takes place, and enhanced ^{210}Pb-residuals at sites where focusing occurs. Where sediment redistribution has been reasonably smooth, the constant ^{210}Pb-supply hypothesis, although not valid for a single core, may be valid for the lake bed as a whole. In these cases a c.r.s. chronology may still be established by correlating ^{210}Pb-data from a representative set of cores (see pp. 100–101).
4. Inwash of catchment-derived unsupported ^{210}Pb. Identified in fig. 3.1 as the C pathway, this remains unproven as a significant source of error for lake-sediment dating. It may be an important factor in arable areas of significant topsoil erosion. When it is a major and variable component of the ^{210}Pb supply it is difficult to see how any adjustments to dating models can be made to compensate.
5. Changing rates of autochthonous sedimentation. Although in theory and perhaps in practice for certain lakes, increased rates of autochthonous sedimentation may give rise to increases in the flux of unsupported ^{210}Pb to the sediment, the overwhelming indications of our evidence so far are that this is not a significant source of error in c.r.s. calculations. Case studies have included lakes affected by rapid eutrophication, the development of hypolimnetic oxygen depletion, and a shift to autochthonous sapropel accumulation. So far, this has been demonstrably associated with sharp *declines* in initial concentration, and c.r.s.-derived dates and sedimentation rates have been closely compatible with independent evidence.

Recommended procedures

The main purpose of the present review is not to advocate a particular model but to advance the concept of empirical testing in all ^{210}Pb-dating projects. All projects involving ^{210}Pb-dating should be designed in a way that will give the greatest possible opportunity to answer the following questions:

1. Does the mean flux of unsupported ^{210}Pb to the sediment fall within the range normally expected from measured atmospheric fluxes $(0.2–1.0 pCi\,cm^{-2}\,yr^{-1})$? Although a direct relation between the input and the atmosphere fallout has not been conclusively proved, in practice this has proved a useful first test of likely validity. Where values are much less than the atmospheric range, results have often been poor and problems arising from low residence time or sediment focusing may be indicated.

2. Do cores from nearby sites with different accumulation rates have comparable total residual unsupported ^{210}Pb-values? If so, the prima facie case for a constant rate of supply is strong. If not, sediment displacement, resuspension, and focusing may have occurred.

3. Has mixing had a major effect on any of the profiles studied? It is not sufficient to refer to the presence of benthic organisms to a given depth in the core, and then use this to justify inferences of homogenization. Even when important, active bioturbation rarely has this effect. The significance of mixing is best tested by detailed stratigraphic study.

4. Are the dates and rates derived from the preferred ^{210}Pb-dating model compatible with independent evidence of age from, for example, palaeomagnetic, stratigraphic, or microfossil studies?

   Only when the results of ^{210}Pb-analysis have been evaluated in the above terms can full confidence be placed on their implications for palaeoenvironmental reconstruction. The relative ease with which plausible dates and rates can often be inferred from unevaluated analyses, and the ready confirmation that even in apparently simple situations these can be significantly in error reinforce the point.

Evaluation by means of the types of empirical test described above can be best effected by adopting the following rules:

1. Never rely on a single profile. Even if only one profile is subject to full analysis others can be measured for total residual unsupported ^{210}Pb by means of a single additional assay.

2. Always determine the unsupported ^{210}Pb-concentration of the uppermost sediment and continue analysis to the depth at which unsupported ^{210}Pb ceases to be measurable. Although it will often not be necessary to assay all contiguous samples above this in a profile, the greater the proportion of the core assayed, the more accurate will be the calculation of the residual unsupported ^{210}Pb at each stage in integration and for the whole profile.

3. In the core to be analysed, always determine directly and accurately the dry weight/wet volume ratio for every slice, from the surface down to the level at which unsupported ^{210}Pb ceases to be

detectable. An accurate relation between depth and cumulative dry-mass in the core is essential for calculating the residual ^{210}Pb.

4. Wherever possible, associate ^{210}Pb-analyses directly with independent methods of age determination for as many points in the profile as possible and with methods of detailed stratigraphic correlation between cores. By these means external validation may be established. In addition, detailed stratigraphic correlation of dated profiles allows close evaluation of internal consistency, correlation between dated and undated profiles allows transfer of ^{210}Pb-chronologies to undated cores, and close stratigraphic study of possible mixing zones allows some check on the likelihood of this having homogenized the sedimentary record. In these last connections, use of mineral magnetic characteristics are attractive by virtue of the ease and speed with which they can be measured non-destructively on whole cores and single samples, before further analyses are begun (Thompson *et al.* 1975, 1980; Dearing 1983; Oldfield *et al.* in press).

Future research

The above propositions point to the need not only for further studies but for some reorientation of their scope and focus.

First, we believe there is an overriding case for combining empirical studies of ^{210}Pb-sources, fluxes, and inventories in lakes and watersheds with closely and independently controlled dating studies at the same sites. The two types of study tend to have been undertaken independently with consequent loss of valuable insights. Second, where only dating is envisaged, empirical testing of the type outlined above should, wherever possible, be built into the research strategy from the outset. Finally, the role of particular processes in relation to c.r.s. calculations calls for special attention. We would identify:

1. The role of scavenging particulates and the relationship between the flux of these and the flux of unsupported ^{210}Pb to the sediment surface,
2. The possible input of unsupported ^{210}Pb associated with fine particulates from the catchment surface,
3. The impact of sediment focusing on total residual unsupported ^{210}Pb-values and on the consistency and validity of ^{210}Pb-based dates.

All the above may introduce second-order variables into c.r.s. calculations and in some cases at least they are potentially capable of integration into a modified model.

References

Appleby, P.G. and Oldfield, F., 1978. The calculation of lead-210 dates assuming a constant rate of supply of unsupported ^{210}Pb to the sediment. *Catena. 5:* 1–8.

Appleby, P.G., Oldfield, F., Thompson, R., Huttunen, P. and Tolonen, K., 1979. ^{210}Pb dating of annually laminated lake sediments from Finland. *Nature 280:* 53–5.

Battarbee, R.W., 1978. Observations on the recent history of Lough Neagh and its drainage basin. *Phil. Trans. R. Soc. B. 281:* 303–45.

Battarbee, R.W., Appleby, P.G., Odell, K. and Flower, R.J., in press. ^{210}Pb dating of Scottish lake sediments, afforestation and accelerated soil erosion.

Battarbee, R.W., Digerfeldt, G., Appleby, P.G. and Oldfield, F., 1980. Palaeoecological studies of the recent development of Lake Växjösjön. III Reassessment of recent chronology on the basis of modified ^{210}Pb dates. *Arch. Hydrobiol. 89:* 440–6.

Benninger, L.K., Lewis, D.M. and Turekian, K.K., 1975. The use of natural ^{210}Pb as a heavy metal tracer in the river–estuarine system. In *Marine Chemistry in the Coastal Environment*, ed. T.M. Church, *Symp. Ser. Am. chem. Soc. 18:* 202–10.

Brugam, R.B., 1978. Pollen indicators of land-use change in Southern Connecticut. *Quaternary Res. 9:* 349–62.

Davis, M.B., 1976. Erosion rates and land use history in Southern Michigan. *Environ. Conserv. 3:* 139–48.

Davis, M.B., Brubaker, L.B. and Beiswenger, J.M., 1971. Pollen grains in lake sediments. Pollen percentages in surface sediments from Southern Michigan. *Quaternary Res. 1:* 450–67.

Davis, R.B., 1967. Pollen studies of near surface sediments in Maine lakes. In *Quaternary Palaeoecology*, ed. E.J. Cushing and H.E. Wright, 153–7 (Yale).

Dearing, J.A., 1983. Changing patterns of sedimentation in a small lake in Scania, S. Sweden. *Hydrobiologia 103:* 59–64.

Durham, R.W. and Joshi, S.R., 1980. Recent sedimentation rates, ^{210}Pb fluxes, and particle settling velocities in Lake Huron, Laurentian Great Lakes. *Chem. Geol. 31:* 53–66.

Edgington, D.N. and Robbins, J.A., 1976a. Records of lead deposition in Lake Michigan sediments since 1800. *Environ. Sci. Technol. 10:* 266–74.

Edgington, D.N. and Robbins, J.A., 1976b. Pattern of deposition of natural and fall-out radionuclides in the sediments of Lake Michigan and their relation to limnological processes. In *Environmental Biogeochemistry 2*, ed. J.O. Nriagu, 705–29 (Ann Arbor).

Elner, S. and Wood, C., 1980. The history of two linked but contrasting lakes in N. Wales from a study of pollen, diatoms and chemistry in sediment cores. *J. Ecol. 68:* 95–121.

Farmer, J.G., 1978. The determination of sedimentation rates in Lake Ontario using the ^{210}Pb dating method. *Can. J. Earth Sci. 15:* 431–7.

Goldberg, E.D., 1963. Geochronology with ^{210}Pb. In *Radioactive Dating*, 121–31, International Atomic Energy Agency, Vienna.

Hammond, D.E., Simpson, E.J. and Mathieu, G., 1975. Methane and Radon-222 as tracers for mechanisms of exchange across the sediment–water interface in the Hudson River Estuary. *Mar. Chem.* 7: 119–232.

Kemp, A.L.W., Anderson, T.W., Thomas, R.L. and Mudrochova, A., 1974. Sedimentation rates and recent sediment history of Lakes Ontario, Erie and Huron. *J. Sedim. Petrol.* 44: 207–18.

Koide, M., Bruland, K.W., and Goldberg, E.D., 1973. Th-288/Th-232 and Pb-210 geochronologies in marine and lake sediments. *Geochim. cosmochim. Acta* 37: 1171–87.

Krishnaswamy, S., Lal, D., Martin, J.M. and Meybeck, M., 1971. Geochronology of lake sediments. *Earth Planet. Sci. Lett.* 11: 407–14.

Krishnaswamy, S., and Lal, D., 1978. Radionuclide limnochronology. In *Lakes, Chemistry, Geology and Physics*, ed. A. Lerman, 153–77 (New York).

Leland, H.V., Shukla, S.S. and Shimp, N.F., 1973. Factors affecting distribution of lead and other trace elements in sediments of Southern Lake Michigan. In *Trace Metals and Metal–Organic Interactions in Natural Waters*, ed. P.C. Singer, 89–129 (Ann Arbor).

Leopold, L.B., 1956. Land use and sediment yield. In *Man's Role in Changing the Face of the Earth*, ed. W.L. Thomas, 639–47 (Chicago).

Oldfield, F., 1977. Lakes and their drainage basins as units of sediment-based ecological study. *Prog. Phys. Geogr.* 1: 460–504.

Oldfield, F., Appleby, P.G. and Battarbee, R.W., 1978. Alternative ^{210}Pb dating; results from the New Guinea Highlands and Lough Erne. *Nature* 271: 339–42.

Oldfield, F., Appleby, P.G. and Worsley, A.T., in press. *Evidence from Lake Sediments for Recent Erosion Rates in the Highlands of Papua New Guinea*.

Oldfield, F., Appleby, P.G. and Thompson, R., 1980. Palaeoecological studies of lakes in the Highlands of Papua New Guinea: I. Chronology of sedimentation. *J. Ecol.* 68: 457–77.

Osborne, P.L. and Moss, B., 1977. Palaeolimnology and trends in phosphorus and iron budgets of an old man-made lake, Barton Broad, Norfolk. *Freshwater Biol.* 7: 213–33.

Paullin, C., 1982. *Atlas of the Historical Geography of the United States* (Washington).

Pennington, W., Cambray, R.S., Eakins, J.D. and Harkness, D.D., 1976. Radionuclide dating of the recent sediments of Blelham Tarn. *Freshwater Biol.* 6: 317–31.

Petit, D., 1974. Pb-210 et isotopes stable du plomb dans des sediments lacustres. *Earth Planet Sci. Lett.* 23: 199–205.

Robbins, J.A., 1978. Geochemical and geophysical applications of radioactive lead. In *Biogeochemistry of Lead in the Environment*, ed. J.O. Nriagu, 285–393 (Elsevier: Holland).

Robbins, J.A. and Edgington, D.N., 1975. Determination of recent sedimentation rates in Lakes Michigan using Pb-210 and Cs-137. *Geochim. cosmochim. Acta* 39: 285–304.

Robbins, J.A., Edgington, D.N. and Kemp, A.L.W., 1978. Comparative Pb-210, Cs-137, and pollen geochronologies of sediments from Lakes Ontario and Erie. *Quaternary Res.* 10: 256–78.

Robbins, J.A., Krezoski, J.R. and Mosley, S.C., 1977. Radioactivity in sediments of the Great Lakes: Post-depositional redistribution by deposit feeding organisms. *Earth Planet Sci. Lett. 36:* 325–33.

Thompson, R., Battarbee, R.W., O'Sullivan, P.E. and Oldfield, F., 1975. Magnetic susceptibility of lake sediments. *Limnol. Oceanogr. 20:* 687–98.

Thompson, R., Stober, J.C., Turner, G.J., Bloemendel, J., Dearing, J., Oldfield, F. and Rummery, T.A., 1980. Environmental applications of magnetic measurements. *Science 207:* 481–6.

4 The transfer of natural and artificial radionuclides to Brotherswater from its catchment

J. D. Eakins, R. S. Cambray, K. C. Chambers and A. E. Lally

Introduction

THE SURFACE of a lake and its catchment area are exposed to a relatively constant flux of natural radionuclides. These are produced either by cosmic ray interactions with stable elements in the upper atmosphere, or by decay of radioelements in the earth's crust. The lake and its catchment are also exposed to a variable flux of artificial nuclides from three main sources:

1. The atmospheric testing of distant nuclear weapons (weapon fallout).
2. Discharges from local nuclear installations.
3. The return to the atmosphere and subsequent disintegration of satellites containing nuclear power sources.

Radionuclides deposited on the catchment may be retained by soil and vegetation or transferred to the lake *via* the input water courses. The purpose of this study, which took place from 1976 to 1978, was to estimate the transfer of certain radionuclides from a water catchment to its lake. The nuclides chosen for the investigation were ^{210}Pb and its granddaughter ^{210}Po, $^{239+240}$Pu and ^{238}Pu, and ^{137}Cs.

^{210}Pb has a half-life of 22.26 years and decays by beta emission to ^{210}Bi. It is a naturally occurring radionuclide in the ^{238}U decay series (part of which is shown in fig. 4.1) and is present in the atmosphere as a result of the following series of events. ^{226}Ra in the earth's crust decays to the rare gas ^{222}Rn, which diffuses into the atmosphere at an average rate of 42 atoms min cm^{-2} of land surface (Israel 1951). ^{222}Rn has a half-life of 3.8 days and decays *via* a series of short-lived daughters to ^{210}Pb.

^{210}Po has a half-life of 138 days, and is present in the atmosphere as a result of decay of ^{210}Pb *via* ^{210}Bi. The concentration of ^{210}Po in air is dependent, among other factors, on the residence time of ^{210}Pb in the atmosphere. This has been variously reported between 9.6 days and a few weeks (Burton and Stewart 1960, Peirson *et al.* 1966, Francis *et al.* 1970).

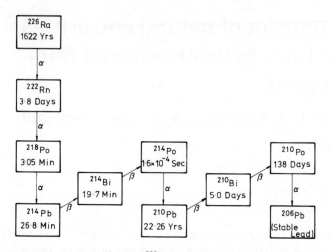

Fig. 4.1   Decay products of ^{226}Ra.

$^{239+240}$Pu, ^{238}Pu, and ^{137}Cs are present in the atmosphere as a result of the testing of nuclear weapons. The concentration of these nuclides in air reached a maximum in the northern hemisphere in 1963 but at the time of the present studies their concentrations had declined to about 1% of the peak values. These nuclides are also present locally in air near fuel reprocessing sites as a result of discharges both to the sea and the atmosphere (Pattenden et al. 1980). ^{238}Pu was also released into the atmosphere by the disintegration of a satellite power source (SNAP-9A) upon re-entry into the atmosphere in 1964 (Harley 1964).

Brotherswater was chosen for the study primarily because it is a simple system, with single major inlet and outlet streams. The catchment is also large relative to the lake area and hence small percentage transfers of nuclides from the catchment to the lake are more readily detected. Soil, water, sediment samples, and vegetation were collected on several occasions and analysed for their radionuclide contents. Using these data an attempt has been made to produce a budget of nuclides on the catchment, in the lake, and in its sediment. From this inventory it is possible to deduce the extent of transfer of radionuclides from the catchment to the lake.

Brotherswater, the lake and its catchment

Brotherswater is situated in the eastern half of the Lake District in Cumbria, at the foot of the Kirkstone Pass and adjacent to the A592 road. It is a small lake with a surface area of $2.44 \times 10^5 m^2$, a maximum depth of 15m, and a normal volume of about $1.6 \times 10^6 m^3$.

The lake is fed primarily by four becks, Kirkstone, Caiston, Caudale, and Hartsop (fig. 4.2) which merge to form a common input stream at the

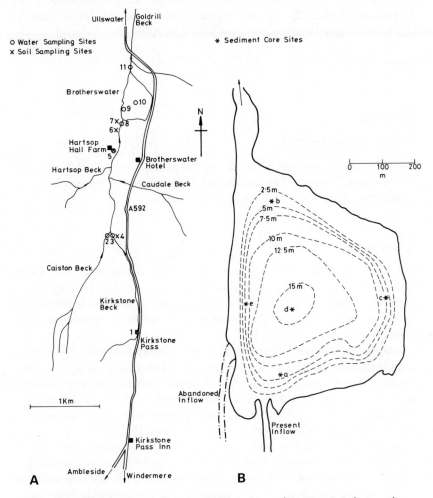

Fig. 4.2   Brotherswater sampling sites. A. Stream sampling sites. B. Lake sampling sites.

south-west corner of the lake. The outlet stream, Goldrill Beck, leaves the lake at the north-west corner. Flow measurements (Chambers 1978) taken over the period 1975–7 indicate that the common inlet stream carries $4.4 \times 10^7 m^3 y^{-1}$ of water which accounts for 90% of that leaving the lake *via* Goldrill Beck. The balance is presumably accounted for by the water entering the lake in local drainage from the catchment and direct input by rain. The mean residence time of water entering the lake is about seven days. The area of the catchment is approximately $1.15 \times 10^7 m^2$, giving a catchment to lake ratio of 47. Fig. 4.2A shows the main water courses and sampling sites; fig. 4.2B shows the lake in more detail with the results of a bathymetric survey by Chambers and P.V. Allen. Also shown are the locations from which sediment cores were taken for analysis as part of the present study.

Methods

PHYSICAL PARAMETERS

The methods of measurement of the major lake and catchment parameters, water flows and sediment loads have been described elsewhere (Chambers 1978).

SAMPLING

Sediment cores for total radionuclide content and radiometric dating were taken using a short pneumatically driven corer (Mackereth 1969) of diameter 6.3cm. Sediment was extracted from the corer, in sections 1cm thick, by means of a free piston driven hydraulically up the core tube (Mackereth 1969).

Soil samples were obtained from undisturbed sites of open aspect on fairly level ground. The samples were taken with a large percussion corer (7.8cm internal diameter) to a depth of about 30cm. Cores were normally divided horizontally into 5cm sections to determine the distribution of radioactivity with depth. At least two cores were taken from each site and corresponding sections were combined for analysis.

Samples of stream and lake water were generally collected in a bucket by wading, either to the middle of a stream or as far from the lake shore as possible in about 60cm of water. Samples were also taken at different depths in the lake from a boat, using a weighted sample tube and a peristaltic pump to draw the water to the surface. Samples were normally analysed without filtration, although large items of detritus, such as leaves, were removed.

Rainwater samples were collected monthly from Hartsop Hall Farm over a period of a year, using a 25cm diameter funnel and polythene bottle. The rainfall at this site was also measured each month using a 5-inch (12.7cm) Bradford rain gauge.

ANALYSIS

For plutonium analysis, soil and sediment samples were dried, ground, and ashed at 450°C. The determination was carried out on aliquots of the ash by acid leaching and anion exchange using ^{236}Pu as an internal tracer followed by alpha spectrometry (Lally and Eakins 1978). The activities of ^{239}Pu and ^{240}Pu are expressed as the sum of the two nuclides as their alpha energies, at 5.15 and 5.16MeV, are normally indistinguishable by alpha spectrometry. Plutonium in water was determined by a simplification of the above procedure.

^{137}Cs was determined in samples of dried and ground soil and sediment by gamma-ray spectrometry using a lithium drifted germanium detector. Waters were analysed similarly after concentration by evaporation. The spectra were interpreted by least-squares fitting using automatic data processing (Salmon and Creevy 1971).

^{210}Pb was determined in sediments by dry distillation of ^{210}Po, using the method described by Eakins and Morrison (1978). ^{210}Pb and ^{210}Po in vegetation, soils, and waters were determined by the methods described by Lally and Eakins (1978), using stable lead and ^{208}Po to determine the radiochemical recovery of ^{210}Pb, and ^{210}Po respectively.

Results

SOIL SAMPLES

Samples were taken on three separate occasions from two sites on the upper catchment, near the Kirkstone beck at site 1 (Kirkstone Upper) and at site 4 at the junction of the Kirkstone and Caiston becks. Three sites were sampled similarly on the lower catchment, at site 5 (Hartsop Hall) and sites 6 and 7 on the delta where the main input stream runs into the lake (Delta A and Delta B). The locations of these sites are shown in fig. 4.2. The mean ^{137}Cs and plutonium contents of samples, expressed as pCi cm^{-2} of deposit, are presented in table 4.1.

The mean ratio of 57 for ^{137}Cs/$^{239+240}$Pu may be compared with an average of 56 reported by Cawse (1978) for soil samples collected over the whole of the U.K. The observed ^{238}Pu/$^{239+240}$Pu ratio of 0.04 is the same as that reported by Hardy *et al.* (1973) for soil in the northern hemisphere.

The deposition of fallout from nuclear weapons at any site is proportional to rainfall (Peirson and Salmon 1959) and the long-term average rainfall at Brotherswater is 2400mm y^{-1} (Meteorological Office 1981). The mean deposit of ^{137}Cs per 1000mm of rain at this time is therefore 10pCi cm^{-2} and the corresponding figure for $^{239+240}$Pu is 0.18pCi cm^{-2}. Cawse (1978) also found mean values of 10pCi cm^{-2} and 0.18pCi cm^{-2} for ^{137}Cs

Table 4.1. ^{137}Cs and plutonium in soil.

Site	^{137}Cs (pCi cm^{-2})	^{238}Pu (pCi cm^{-2})	$239 + 240_{Pu}$ (pCi cm^{-2})	$\dfrac{^{137}Cs}{239 + 240_{Pu}}$	$\dfrac{^{238}Pu}{239 + 240_{Pu}}$
1. Kirkstone Upper	22	0.022	0.45	49	0.049
4. Kirkstone/Caiston	26	0.018	0.40	64	0.045
5. Hartsop Hall	24	0.018	0.44	54	0.041
6. Brotherswater Delta A	26	0.016	0.38	67	0.042
7. Brotherswater Delta B	21	0.016	0.41	50	0.039
Mean	24	0.018	0.42	57	0.043

and $^{239+240}$Pu, respectively. The agreement between these figures and between the ratios in the preceding paragraph confirms that ^{137}Cs and plutonium in soil at Brotherswater can be accounted for by fallout from nuclear weapons and debris from the SNAP 9A satellite.

At each of the five sampling sites, at least 70% of the ^{137}Cs and plutonium was found in the top 5cm of soil with 90% in the top 10cm. Cawse (1978) reported that the retention of ^{137}Cs by the top 15cm of soil averaged 75% of the amount present to 30cm depth.

^{210}Pb and ^{226}Ra were determined in cores from site 1 on the upper catchment and site 7 on the lower catchment. The results of these measurements are presented in table 4.2. The unsupported ^{210}Pb content

Table 4.2. ^{210}Pb and ^{226}Ra in soil.

Depth (cm)	Site 1		Site 7	
	^{210}Pb(pCi cm^{-2})	^{226}Ra(pCi cm^{-2})	^{210}Pb(pCi cm^{-2})	^{226}Ra(pCi cm^{-2})
0–5	10.64	1.86	14.09	2.01
5–10	7.80	3.05	3.85	2.66
10–15	2.50	2.60	3.22	2.49
15–20	3.05	3.30	3.10	2.64
20–25	3.17	2.71	2.40	2.33
25–30	3.48	3.29	2.36	2.57
Total	30.64	16.81	29.00	14.70

of each core (the difference between the total ^{210}Pb and ^{226}Ra content) is 13.8pCi cm^{-2} for site 1 and 14.3pCi cm^{-2} for site 7 with a mean value of 14.1pCi cm^{-2}. This would require an annual deposit of about 0.4pCi cm^{-2} of ^{210}Pb to maintain equilibrium, which is in agreement with the value reported by Nozaki et al. (1978) for the eastern United States, although the sites are not necessarily comparable.

LAKE SEDIMENT CORES

Sediment cores were taken at each of the sites marked a–e in fig. 4.2. The organic contents of cores a, b, and d (from the inlet, outlet, and centre of the lake) were determined by loss in weight on ignition at 450°C. The results of these measurements are presented in fig. 4.3 as percent organic

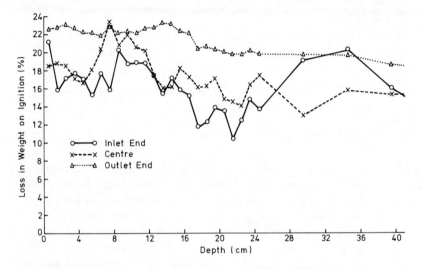

Fig. 4.3   Loss in weight on ignition of sediment samples.

matter plotted against depth. The inlet core generally has the lowest organic content although there is a peak between 20–40cm. The outlet core has the largest organic content with the centre core values lying, in general, between the two.

The bulk density of these three cores (expressed as g cm^{-3} wet volume) was also measured to a depth of 12cm. The results are shown in fig. 4.4, from which it can be seen that the density decreases from the inlet to the outlet as the organic content increases.

The distribution of ^{137}Cs with depth for each of the five cores is shown in fig. 4.5. The total ^{137}Cs was calculated to a depth of 25cm for each core (see fig. 4.5) and may be compared with the mean value of 24pCi cm^{-2} found on the catchment (table 4.1). The sediment-accumulation rates are calculated using the 1963 maximum of ^{137}Cs as a reference point (Pennington et al. 1973). Although the profiles are rather broad, estimates can be made with reasonable certainty for cores a–d. However, the maximum in core e is very ill-defined and the result quoted must be treated with caution.

The distribution of the total ^{210}Pb content with depth and that unsup-

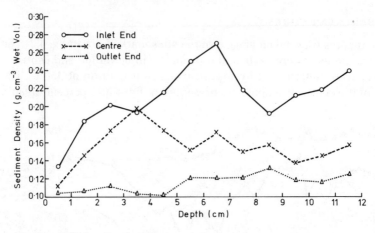

Fig. 4.4   Bulk density of sediment samples.

ported by ^{226}Ra is shown for each core in fig. 4.6. The integrated unsupported ^{210}Pb in each core, expressed as deposit per unit area, is also given. Sediment horizons can be dated and accumulation rates derived from the decrease in unsupported ^{210}Pb with depth, using either the constant initial concentration (c.i.c.) model (Pennington et al. 1976), or the constant rate of supply (c.r.s.) approach (Appleby and Oldfield 1978). In table 4.3 sediment-accumulation rates for the top 20cm of each core (derived using the c.i.c. model) are compared with those obtained from the ^{137}Cs measurements and there is good agreement. Also shown are accumulation rates prior to about 1940.

In each of the five ^{210}Pb profiles (fig. 4.6) there is a discontinuity in the curve about 35 years ago. This occurs at different depths in the sediment due to varying rates of accumulation in different parts of the lake. A possible explanation for the discontinuity is that the lower catchment may have been ploughed during the Second World War and subsequently allowed to revert to grassland. This could have resulted in increased

Table 4.3.   Sediment-accumulation rates in Brotherswater.

Core	Recent accumulation rate (mm y^{-1})		Accumulation rate before ~1940 by^{210}Pb
	^{137}Cs	^{210}Pb	mm y^{-1}
Inlet	5	4	4
Outlet	9	10	3
East	3	4	3
Centre	5	6	2
West	4	3	3

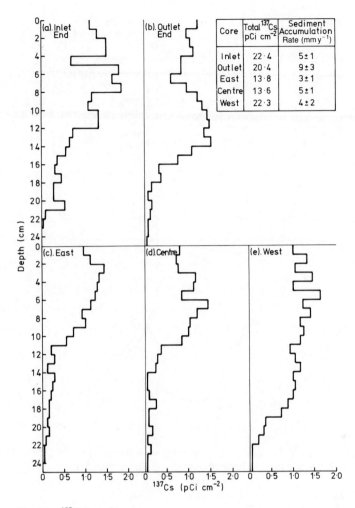

Core	Total ^{137}Cs pCi cm^{-2}	Sediment Accumulation Rate (mm y^{-1})
Inlet	22·4	5±1
Outlet	20·4	9±3
East	13·8	3±1
Centre	13·6	5±1
West	22·3	4±2

Fig. 4.5 ^{137}Cs in sediment cores.

erosion of surface soil, rich in ^{210}Pb, and account for the observed phenomenon.

There is a much larger perturbation in the ^{210}Pb profile at the inlet end of the lake between 50–100 years ago. This may be accounted for by a diversion of the main inlet stream at about the turn of the century, for which there is some cartographic evidence although the exact date of the diversion is not known. The increase in ^{210}Pb in this core between 50–100 years ago can be related to an increase in the organic content of sediment which occurs between 20–40cm (fig. 4.3). The high accumulation rate of sediment at the outlet end of the lake, as shown by both dating

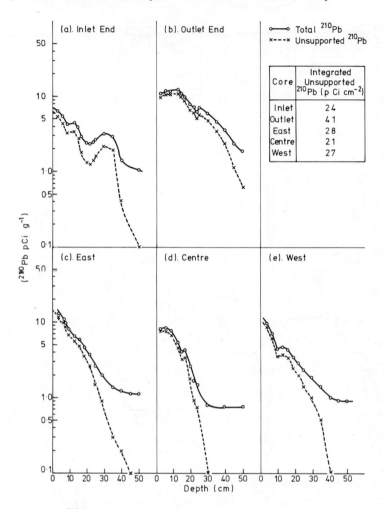

Fig. 4.6 ^{210}Pb in sediment cores.

techniques, is also associated with a high organic content. This is about 25% greater than that found at the centre of the lake, and results in a lower density (fig. 4.4). It appears that the less dense organic material is transported preferentially to the outlet end of the lake before being deposited. An additional factor is that the dominant wind funnels the waves to the shallow waters at the outlet and resuspension of sediments probably contributes to the higher accumulation rate.

$^{239+240}$Pu was measured in the inlet, centre, and outlet cores and the profiles obtained were broadly similar to those for ^{137}Cs. The accumulated deposits of $^{239+240}$Pu in sediment are shown in table 4.4 and compared

Table 4.4. ^{137}Cs and $^{239+240}$Pu deposits in Brotherswater sediment.

Core	^{137}Cs (pCi cm^{-2})	$^{239+240}$Pu (pCi cm^{-2})	$\dfrac{^{137}\text{Cs}}{^{239+240}\text{Pu}}$
Inlet	22.4	0.40	56
Centre	13.6	0.38	36
Outlet	20.4	0.68	30

with those obtained for ^{137}Cs. The ^{137}Cs/$^{239+240}$Pu ratio in the inlet core is similar to that found in soil (table 4.1), but in the centre and outlet cores the ratio is reduced, implying either a deficiency of ^{137}Cs or an excess of $^{239+240}$Pu.

WATER SAMPLES

Rainwater was collected monthly at Hartsop Hall (site 5) from November 1976 to October 1977 and analysed for ^{137}Cs and ^{210}Pb. The results are given in table 4.5 together with the monthly rainfall. The weighted mean ^{137}Cs concentration was 0.82pCi 1^{-1}, which is in reasonable agreement with the value of 0.68pCi 1^{-1} found at Milford Haven over the same period (Cambray *et al.* 1978). The mean ^{210}Pb concentration was 1.9pCi 1^{-1}, suggesting an annual deposit of 0.40pCi cm^{-2}. However, the total annual rainfall was only about 88% of the long-term average value of 2400mm y^{-1} (Meteorological Office 1981). Using the long-term average

Table 4.5. ^{137}Cs and ^{210}Pb in rain at Hartsop Hall.

Month	Rainfall (mm)	^{137}Cs(pCi 1^{-1})	^{210}Pb(pCi 1^{-1})
1976			
November	215	0.58	2.0
December	164	0.27	2.5
1977			
January	218	0.17	2.3
February	112	1.25	1.4
March	227	0.39	2.6
April	166	0.82	2.5
May June	190	0.83	2.2
July	101	1.88	1.1
August	216	1.38	1.3
September	194	1.28	1.5
October	305	0.86	1.3
Weighted mean		0.82	1.9

Table 4.6. ^{210}Pb, ^{210}Po, and ^{137}Cs in stream and lake waters. (Values for ^{137}Cs in parentheses.)

Date	Upper Kirkstone beck Site 1	Middle Kirkstone beck Site 3	Caiston beck Site 2
	[^{210}Pb](fCi l^{-1})	[^{210}Pb](fCi l^{-1})	[^{210}Pb](fCi l^{-1})
1976			
26 July	23 ± 15	<20	14 ± 12
	(<40)	(60 ± 40)	(210 ± 80)
22 September	740 ± 30	64 ± 22	24 ± 14
	(190 ± 80)	(140 ± 60)	(120 ± 50)
2 November	120 ± 20	47 ± 16	55 ± 17
	(310 ± 30)	(70 ± 40)	(40 ± 40)
1977			
2 May	—	—	—
14 November	—	—	—

value the annual deposition of ^{210}Pb on the catchment is increased to 0.46pCi cm^{-2}.

Stream and lake water samples were collected on several occasions and analysed for ^{210}Pb, ^{210}Po, and ^{137}Cs. The results are given in table 4.6 progressing down the catchment from the left (Upper Kirkstone) to the lake outlet stream on the right. The values for ^{137}Cs are shown in parentheses as fCi l^{-1} below those for ^{210}Pb.

The results in table 4.6 all refer to unfiltered water. However, a sample of lake water collected on 24 January 1978 was filtered through a Whatman GFA glass-fibre filter paper, which removes particles greater than 1.6μm in diameter, and the filtrate and suspended particulate were analysed separately for ^{210}Pb. The particulate was found to contribute 18fCi l^{-1} and the filtrate <5fCi l^{-1}. Over 75% of the ^{210}Pb in the water sample was therefore associated with the suspended particulate. The concentration of ^{210}Pb in the particulate was 10.3pCi g^{-1}.

Water samples were also taken on 22 September 1976 from the centre of the lake (site 10) at the surface and at depths of 7.5m and 15m. These samples were analysed for ^{137}Cs and ^{210}Pb and the results are given in table 4.7.

On 26 July 1976 water levels were low and both the ^{210}Pb and ^{137}Cs concentrations were at or near the limit of detection. However, sampling in September and November 1976 followed periods of heavy rain and concentrations of both nuclides increased. A comparison of the ^{210}Pb data for these samples with the average value of about 2pCi l^{-1} in rainwater

Lake inlet stream, Site 8			Lake water	Lake outlet stream, Site 11		
^{210}Pb	^{210}Po	$\dfrac{^{210}Po}{^{210}Pb}$	^{210}Pb	^{210}Pb	^{210}Po	$\dfrac{^{210}Po}{^{210}Pb}$
(fCi l^{-1})	(fCi l^{-1})		(fCi l^{-1})	(fCi l^{-1})	(fCi l^{-1})	
14 ± 6 (<50)	37 ± 7	2.6	12 ± 6 (70 ± 40)	23 ± 15 (50 ± 40)	78 ± 4	3.4
24 ± 17 (150 ± 60)	27 ± 2	1.1	47 ± 15 (110 ± 50)	34 ± 12 (120 ± 50)	82 ± 3	2.4
70 ± 30	19 ± 4	0.24	64 ± 20	53 ± 12	29 ± 5	0.55
38 ± 12 (140 ± 60)	28 ± 6	0.74	32 ± 12 (160 ± 50)	46 ± 15 (130 ± 50)	55 ± 6	1.2
60 ± 19	21 ± 4	0.35	61 ± 25	87 ± 12	37 ± 6	0.43

indicates a considerable and progressive loss of ^{210}Pb from the water in its passage down the catchment. The ^{137}Cs is also much reduced in stream and lake water compared with its average value in rain of about 1pCi l^{-1}. The concentrations of ^{210}Pb and ^{137}Cs in water entering the lake are similar to those leaving it and also to water taken from the bottom at the centre of the lake. However, the ^{210}Pb in surface water at the centre of the lake is much greater than that of bottom water. This increase may be due either to slow mixing of rainwater falling directly into the lake or to fine organic debris in the surface water. The uniformity of the ^{137}Cs content with depth appears to exclude the first possibility and the evidence that the bulk of the ^{210}Pb is associated with particulate material supports the second.

A noteworthy feature is the behaviour of the ^{210}Po to ^{210}Pb ratio in the input and output water of the lake. The ratio in rainwater is normally about 0.2 which is consistent with a mean residence time in the atmosphere of about one month (Peirson et al. 1966). However, in July

Table 4.7. ^{137}Cs and ^{210}Pb in unfiltered waters from Brotherswater centre (Site 10).

Depth	^{137}Cs (fCi l^{-1})	^{210}Pb (fCi l^{-1})
Surface	170 ± 70	140 ± 22
7.5 m	150 ± 60	sample lost
15 m	130 ± 50	20 ± 7

1976, the ratio in the input stream was 2.6 and that in the output water was 3.4. These ratios are much greater than the equilibrium value of 1.02 and indicate preferential removal of ^{210}Pb from the water and/or preferential removal of ^{210}Po from the catchment. In September, after rain, the ratios were 1.1 at the input and 2.4 at the output. In November, after exceptionally heavy rain, the ^{210}Po to ^{210}Pb ratio in the input water was 0.26 and for the output stream 0.55. In 1977 water levels were much higher in November than in May and the results followed the same trend. On each occasion the ^{210}Po concentration in water leaving the lake was higher than that entering, indicating some removal of ^{210}Po from the ^{210}Pb deposits in the lake. The reduction in the ^{210}Po to ^{210}Pb ratio in stream water with increasing rainfall is associated with a corresponding increase in ^{210}Pb content and implies either recruitment of ^{210}Pb from the catchment or a reduction in the loss to the catchment from the faster-moving water.

A further sample of the lake water collected on 24 January 1978 was filtered through a Whatman GFA filter and the filtrate and insoluble material analysed separately for $^{239+240}$Pu. The filtered water contained 0.6 ± 0.2fCi 1^{-1} and the insoluble fraction was equivalent to 2.2 ± 0.7fCi 1^{-1}. It is evident therefore that $^{239+240}$Pu is similar to ^{210}Pb in that in lake water it is associated mainly with the suspended particulate fraction.

VEGETATION

In order to identify possible sources of ^{210}Pb, samples of dead leaves entering the lake in the autumn by the input stream, leaf mould from the steep and wooded western shore, and decaying vegetation from a transient feeder stream were collected and analysed. The results of these measurements are presented in table 4.8.

The concentrations of ^{210}Pb found may be compared with the value of 10.3pCi g^{-1} in the suspended material in lake water and it is evident they are of the same order. There is particularly good agreement between the lake-water suspended matter and the decaying vegetation from the feeder stream. Furthermore an examination of fig. 4.6 shows that the average

Table 4.8. ^{210}Pb in vegetation samples.

Sample	Date	^{210}Pb (pCi g^{-1} dry wt)
Dead leaves from input stream	22 Sept. 1976	1.68 ± 0.07
	10 Oct. 1976	3.07 ± 0.11
	11 Oct. 1976	$2.94 + 0.10$
Decaying vegetation from feeder stream	5 Oct. 1977	9.58 ± 0.67
Leaf mould from western shore	5 Oct. 1977	4.19 ± 0.32

^{210}Pb concentration in surface sediment was about 11pCi g^{-1} which is in close agreement with these two values. Although the concentration of ^{210}Pb in leaf mould on the western slope is quite high, there is no evidence from the ^{210}Pb content of the western sediment core that much of this material is entering the lake.

An inventory of radionuclides in Brotherswater and its catchment

It is possible from the analytical results to derive an inventory of radionuclides in Brotherswater and its catchment. This is presented in table 4.9 where the contents of the three compartments of the system and the percentage of the total in each are calculated for ^{137}Cs, unsupported ^{210}Pb, and $^{239+240}$Pu.

Table 4.9.　^{137}Cs, ^{210}Pb, and $^{239+240}$Pu in Brotherswater and its catchment.

Catchment			Lake sediment			Lake water		
^{137}Cs (mCi)	^{210}Pb (mCi)	$^{239+240}$Pu (mCi)	^{137}Cs (mCi)	^{210}Pb (mCi)	$^{239+240}$Pu (mCi)	^{137}Cs (mCi)	^{210}Pb (mCi)	$^{239+240}$Pu (mCi)
2800	1600	48	43	69	1.2	0.2	0.07	0.004
(98.5%)	(95.9%)	(97.6%)	(1.5%)	(4.1%)	(2.4%)	(0.007%)	(0.004%)	(0.008%)

There is considerable confidence in the catchment and lake-sediment values, which are of cumulative deposits and readily measurable. However, the lake-water results are derived from a limited number of samples (only one in the case of $^{239+240}$Pu) in which the concentrations of radionuclides were often near the limit of detection. In consequence, the lake-water inventory is subject to considerable uncertainty. Furthermore, as the concentrations are likely to change both seasonally and, in the case of ^{137}Cs and $^{239+240}$Pu, with the incidence of nuclear weapons tests, the lake-water contents are only representative of the periods when the samples were taken. Nevertheless, it can be reasonably concluded that of the ^{137}Cs, ^{210}Pb, and $^{239+240}$Pu present in the lake, probably less than 1% is present in the water.

It is instructive to calculate the annual input of ^{210}Pb to Brotherswater. The major route of ^{210}Pb from the catchment is *via* the main input stream. The mean concentration observed in input stream water was 40fCi l^{-1} of which perhaps 25% was in solution (see above). The total volume of water entering the lake by this stream is about 4.4×10^7m^{3}y^{-1} which suggests that about 0.4mCi y^{-1} enters the lake in solution. The annual suspended sediment load entering the lake is $\approx 1.3 \times 10^9$g, over 90% of which is transported during brief rising flood conditions. Unfortunately these conditions did not coincide with any of the visits made in this study. However, if the concentration of ^{210}Pb in the annual sediment load is 10pCi g^{-1} as observed in January 1978 (in agreement with the concentration of ^{210}Pb in the top sections of the sediment core and with decaying vegetation on the catchment) it contributes about 13mCi y^{-1} to the lake. A further source of ^{210}Pb is dead leaves. It has been estimated that about 400kg are transported in the autumn to the lake by the inlet stream. From the data in table 4.8 these could contain a further 1μCi of ^{210}Pb, which is insignificant compared with other sources. The annual supply of ^{210}Pb to the lake, including direct input by rain, is therefore about 15mCi. The ^{210}Pb content of the lake sediment is 69mCi which, because of radioactive decay, requires an annual addition of 2.1mCi to maintain equilibrium. Thus only about 14% of the ^{210}Pb entering the lake is incorporated in sediment, the remainder leaving by the outlet. This explains the apparently anomalous results for ^{210}Pb in table 4.6, which show no significant difference between the contents of the inlet stream, the lake water, and the outlet stream. The concentrations are so low that the uncertainties would mask any small differences. None of the samples were collected in rising flood conditions when high sediment loads, and hence high ^{210}Pb concentrations, would be expected. Although water levels were fairly high on some occasions, they were on falling floods when sediment loads are much lower (Gregory and Walling 1973).

A similar explanation may apply to ^{137}Cs which, from the data in table 4.6 also appears to have similar concentrations in inlet, lake, and outlet waters. ^{137}Cs is believed to be retained on the catchment by ion exchange on clay minerals (Francis and Brinkley 1976) and hence should be at higher concentrations in water when erosion occurs during rising floods. Similarly $^{239+240}$Pu present in the lake water is associated mainly with the suspended particulate and it would be expected that levels would be at a maximum when sediment loads are highest.

Discussion

The deposits of radionuclides present in the sediment of Brotherswater are approximately equal to those expected from direct input to the lake by rain, but it is believed that this is fortuitous. Carlsson (1978) has estimated that 1.9% of the freshly deposited ^{137}Cs and 0.56% per year of the accumulated ^{137}Cs on the catchment is transported to a lake. This suggests a transport of 16mCi (97% of which is derived from the accumulated deposit) to Brotherswater from its catchment in 1976–7. From the Hartsop

Hall rain data, 0.4mCi of ^{137}Cs fell directly into the lake with rain during this period, about 2% of the calculated annual supply to the lake water. The average annual increment of ^{137}Cs to the sediment at this time was about 1.2mCi and hence only 34% at most of the ^{137}Cs deposited in sediment during this period could have been derived from direct input *via* rain. However, the retention by sediment was only 8% of the calculated input and there is no reason to believe that ^{137}Cs deposited directly into the lake will be accumulated in sediment at a faster rate than that entering from the catchment; in fact the converse is probably true.

Carlsson (1978) estimated that for Lake Ulkesjön in Sweden, 38% of the ^{137}Cs entering the lake was transferred to sediment. However, this lake is shallower and has a larger mean residence time (0.7 years) than Brotherswater (7 days). These factors no doubt account for the differences in transfer.

Work by Khristianova *et al.* (1973) on Mozhayskoye Reservoir in 1968 showed that 15% of the ^{137}Cs present in the lake water came from direct input by rain. This is not inconsistent with our value of 2%, because at that time the atmospheric levels were about five times those in 1976 (Cambray *et al.* 1969).

It is interesting to apply Carlsson's factors for transfer of ^{137}Cs from a catchment to the case of ^{210}Pb in Brotherswater. They give an annual input to the lake of 10mCi, in reasonable agreement with our value of 15mCi. This implies a similar mechanism for transport of both ^{137}Cs and ^{210}Pb, namely transfer by suspended sediment.

Nozaki *et al.* (1978) found a linear relationship between excess ^{210}Pb and loss in weight on ignition in soil cores from three different sites. This, together with the correlation between ^{210}Pb and organic contents in the inlet and outlet sediment cores, is taken as confirmation that the bulk of the ^{210}Pb present in lake sediment is transferred from the catchment with decaying organic matter.

The mean residence time of water in Brotherswater is about seven days. However, about 40% of the water enters the lake in relatively infrequent flood events when the residence time will be considerably less. The density of the sediment is approximately $1.5\,g\,cm^{-3}$ and the majority of the particles are $<2\mu m$ in diameter, although they contribute only 20–30% of the total weight of the sediment. Because of their greater specific surface area, the smaller particles are likely to transport the majority of the radionuclides, which are probably attached by mechanisms that are surface dependent.

The average depth of Brotherswater is 6.6m and a $2\mu m$ particle of density $1.5\,g\,cm^{-3}$ takes about 20 days to fall this distance in water. It is not surprising therefore that less than 15% of the ^{210}Pb, and probably other nuclides, entering the lake are deposited in the bottom sediment.

If the input stream carried an annual suspended sediment load of 1.3×10^6kg, this implies an erosion rate on the catchment of about 0.11kg$\,$m$^2\,$y^{-1}. The majority of the Brotherswater catchment is of steep relief, for which Young (1969) quotes a natural erosion rate of $\approx$0.05kg$\,$m$^{-2}\,$y^{-1}. However, part of the catchment is agricultural land for which erosion rates are normally much higher. This study therefore suggests that erosion in the Brotherswater catchment is within the normal range.

Conclusions

Brotherswater and its catchment have been studied to investigate the movement and fate of naturally occurring ^{210}Pb and artificially produced ^{137}Cs, $^{239+240}$Pu, and ^{238}Pu. These nuclides were measured in soil, lake sediment, lake water, stream water, rainwater, and vegetation. The results have shown that ^{137}Cs and plutonium in soil on the Brotherswater catchment can be entirely accounted for by fallout from nuclear weapons and debris from the SNAP-9A satellite.

The study has demonstrated that the transfer of ^{137}Cs, ^{210}Pb, and plutonium from the catchment to the lake is small and that accumulations in the lake sediment represent 1.5, 4.1, and 2.4% respectively of that retained on the catchment. The radionuclide content of the lake water is generally much less than 1% of that retained in the lake sediment. However, the lake-water content may increase in periods of rising floods.

The results of the investigation suggest that the bulk of the ^{210}Pb present in the lake sediment is transferred from the catchment in association with decaying organic matter. ^{137}Cs is probably transferred with clay minerals and $^{239+240}$Pu has also been shown to be associated with suspended particulate.

It has been shown that it is possible to obtain valuable historical data on the history of the lake by sediment-dating using radiometric techniques. Good agreement on sediment-accumulation rates was obtained using the widely different techniques of ^{137}Cs and ^{210}Pb dating and a clear indication was observed of the diversion of the main inlet stream around AD 1900.

Acknowledgments

The authors would like to thank Professor W. Tutin, F.R.S., Dr Elizabeth Y. Haworth, and Mr P.V. Allen of the Freshwater Biological Association at Windermere for assistance in taking the lake-sediment cores and general advice and encouragement throughout the whole of this study. We would also like to thank Mr and Mrs J. Allen of Hartsop Hall Farm for collecting the rainwater samples and Miss E.M.R. Fisher, Mr D. Kilworth, and Mr P. Flavel for valuable technical assistance.

References

Appleby, P.G. and Oldfield, F., 1978. The calculation of lead-210 dates assuming a constant rate of supply of unsupported ^{210}Pb to the sediment. *Catena 7*: 1–8.

Burton, W.M. and Stewart, N.G., 1960. Use of long lived natural radioactivity as an atmospheric tracer. *Nature 186*: 584.

Cambray, R.S., Fisher, E.M.R., Brooks, W.L. and Peirson, D.H., 1969. Radioactive fallout in air and rain: results to the middle of 1969. (AERE-R 6212 H.M.S.O.)

Cambray, R.S., Fisher, E.M.R., Playford, K. and Peirson, D.H., 1978. Radioactive fallout in air and rain: results to the end of 1977. (AERE-R 9016 H.M.S.O.)

Carlsson, G., 1978. A model for the movement and loss of ^{137}Cs in a small watershed. *Hlth. Phys. 34*: 33–7.

Cawse, P.A., 1978. Environmental and Medical Sciences Division Progress Report, 1977, ed. W.M. Hainge (AERE-PR-EMS 5, 80. H.M.S.O.).

Cawse, P.A., 1978, ed. W.M. Hainge (AERE-PR-EMS 6 (1979) 128. H.M.S.O.).

Chambers, K.C., 1978. Source-sediment relationship in the Cumbrian Lakes. (Ph.D. thesis, University of Reading.)

Eakins, J.D. and Morrison, R.T., 1978. A new procedure for the determination of lead-210 in lake and marine sediments. *Int. J. Appl. Radiat. Isotopes 29*: 531–6.

Francis, C.W., Chesters, G. and Haskin, L.A., 1970. The determination of Pb-210 mean residence time in the atmosphere. *Environ. Sci. Technol. 4*: 586.

Francis, C.W. and Brinkley, F.S., 1976. Preferential adsorption of ^{137}Cs to micaceous minerals in contaminated freshwater sediment. *Nature 260*: 511–13.

Gregory, K.J. and Walling, D.E., 1973. Drainage Basin Form and Process. A Geomorphological Approach.

Hardy, E.P., Krey, P.W. and Volchok, H.L., 1973. Global inventory and distribution of fallout plutonium. *Nature 241*: 444–5.

Harley, J.H., 1964. Possible ^{238}Pu distribution from a satellite failure. (HASL-149, USAEC 138–41.)

Israel, H., 1951. Radioactivity in the atmosphere. In *Compendium of Meteorology*, ed. T.F. Malone (American Meteorological Society: Boston).

Khristianova, L.A., Anikiyev, V.V. and Vinogradova, N.N., 1973. Distribution of radioactive isotopes in a reservoir system. U.S.S.R. State Committee on the Use of Atomic Energy. Translated by Health & Safety Laboratory (USAEC as UNSCEAR Document A/AC-8/G/L. 1474).

Lally, A.E. and Eakins, J.D., 1978. Some recent advances in environmental analysis at AERE Harwell. In *Symposium on the Determination of Radionuclides in Environmental and Biological Materials* (Central Electricity Generating Board, Sudbury House, London).

Mackereth, F.J.H., 1969. A short core sampler for sub-aqueous deposits. *Limnol. Oceanogr. 14:* 145–51.

Meteorological office computer index of rainfall stations data (1981).

Nozaki, Y., DeMaster, D.H., Lewis, D.M. and Turekian, K.K., 1978. Atmospheric ^{210}Pb fluxes determined from soil profiles. *J. Geophys. 83:* 4047–51.

Pattenden, N.J., Cambray, R.S., Playford, K., Eakins, J.D. and Fisher, E.M.R., 1980. Studies of environmental radioactivity in Cumbria, Part 3. Measurements of radionuclides in airborne and deposited material. (AERE-R 9857 H.M.S.O.)

Peirson, D.H. and Salmon, L., 1959. Gamma-radiation from deposited fallout. *Nature 184:* 1678–9.

Peirson, D.H., Cambray, R.S. and Spicer, G.S., 1966. Lead-210 and polonium-210 in the atmosphere. *Tellus 18:* 428.

Pennington, W., Cambray, R.S. and Fisher, E.M.R., 1973. Observations on lake sediments using fallout ^{137}Cs as a tracer. *Nature 242:* 324–6.

Pennington, W., Cambray, R.S., Eakins, J.D. and Harkness, D.D., 1976. Radionuclide dating of the recent sediments of Blelham Tarn. *Freshwater Biol. 6:* 317–31.

Salmon, L. and Creevy, M.G., 1971. Nuclear techniques in environmental pollution. *Proceedings of the Salzburg Conference Symposium,* 47–61 (IAEA, Vienna).

Young, A., 1969. Present rate of land erosion. *Nature 224:* 851–2.

5   A global review of palaeomagnetic results from wet lake sediments

R. Thompson

Preamble

PALAEOMAGNETIC investigations of lake sediments were pioneered at the Freshwater Biological Association's Laboratory beside Windermere by the late John Mackereth. Magnetic analyses of recent lake sediments from all continents of the world have now been carried out by over a dozen laboratories due to the stimulation of Mackereth's original work (Mackereth 1971). I became involved in palaeolimnological studies on visiting the Freshwater Biological Association in vacation periods during my postgraduate research at Newcastle. I happened to have been born within 100 yards of Mackereth's home in Bowness-on-Windermere. Following John Mackereth's untimely death in 1972 I was able to pursue palaeolimnological studies further as a Mackereth Memorial Fellow. This came about largely through Winifred Tutin's wish that palaeomagnetic work should continue at the Freshwater Biological Association. By this time palaeomagnetic results from dry lakes in Western America had been published and interest in the magnetism of recent sediments was growing elsewhere in the world.

Key high-quality palaeomagnetic records representative of six localities in the world have been selected from the scores of lakes which have been investigated in the last decade. The six records are presented as regional master curves. They can be used, in conjunction with a table of the calibrated ^{14}C ages of magnetic features in the curves, in magneto-stratigraphic investigations. The six records have also been mathematically combined in order to determine the global properties of the Holocene (post-glacial) geomagnetic field.

Introduction

The earth's magnetic field is continuously changing. Its most spectacular changes involve complete reversals of the field. These so-called polarity reversals have not occurred in the last 100,000 years but can be recognized by the signature they have left in the remanent magnetization of many rocks and sediments. Although the geomagnetic field is rather weak at the earth's surface, it is powerful enough to influence the direction of magnetization of iron oxide crystals. If the magnetization of such crystals becomes locked into a rock by a natural event such as the freezing of an igneous lava flow or the cementation of a fine-grained sediment then a record of the magnetic field can be preserved.

Ancient magnetic field changes can be investigated by sampling geological materials and measuring their weak but highly stable magnetic remanence using very sensitive instruments called magnetometers. Unfortunately the majority of natural materials do not hold a record of past geomagnetic field changes. Some materials simply contain no iron oxide crystals, some have crystals the wrong size, some have lost their earlier magnetic records due to geological events such as diagenesis or heating, and others may have had their records complicated by the superimposition of secondary magnetizations on their primary magnetic component.

Fortunately, many lake sediments hold much simpler magnetic records than those of older rocks. This means that the range of laboratory and data analysis techniques which have been devised by palaeomagnetists in order to decipher the complex magnetizations of older rocks can be applied to lake sediments to recover remarkably detailed records of the Holocene geomagnetic field. Indeed, fine-grained organic lake sediments have proved to be particularly amenable to palaeomagnetic investigations. Nevertheless, a wide range in quality of lake sediment magnetic records is still found. Some lake sediments have been excellent at monitoring the ancient geomagnetic field while others have been complete failures. Such variability opens up a number of pitfalls into which the unwary investigator may tumble and provokes lively debates about the interpretation of palaeolimnomagnetic data. A number of practical guidelines are tabulated below which can be used to assess the quality of palaeomagnetic results from lake sediments in order to help the investigator avoid the worst of the pitfalls. These guidelines have been employed in selecting one high-quality record from each of six localities around the world. These six records have been used to form regional master curves for magnetostratigraphic work. They have also been processed mathematically in order to explore and appraise some of the global properties of the Holocene geomagnetic field.

Regional master curves

The geomagnetic field exhibits variations on all time scales from millions of years to fractions of a second. Palaeomagnetic measurements of lake sediments can be used to investigate field changes on time scales of hundreds and thousands of years. Holocene field changes from six different regions of the world are summarized as curves of declination and inclination plotted against ^{14}C age in fig. 5.1. These direction changes have all been computed from individual lake sediment records chosen on the quality of their palaeomagnetic measurements and their ^{14}C chronologies.

QUALITY CONTROL

As the majority of geological materials do not hold a reliable record of ancient field changes, it is extremely important to demonstrate some credibility for palaeomagnetic data before accepting them for geophysical or magnetostratigraphic studies. The normal approach to this palaeomagnetic assessment problem is to employ reliability criteria. Such practical reliability criteria should include (1) reproducibility among samples, cores, and lakes, (2) geological tests such as independence from lithology, (3) statistical tests such as serial correlation, (4) comparisons with historically documented geomagnetic field behaviour, and (5) physical laboratory magnetic stability tests. The main collection and reliability criteria to be aimed for in palaeomagnetic studies of recent sediments are summarized in table 5.1. Probably the most important criteria are those concerned with reproducibility and with sedimentological disturbances. A particularly simple quantitative reliability measure

Table 5.1. Reliability criteria.

1	Localities	$\geqslant$2 per region
2	Coring sites	Flat; $\geqslant$2 sites per lake
3	Cores	Length $\geqslant$3m or overlapping
4	Sediments	Fresh; gyttja/fine silt/clay; undisturbed; no marked lithological changes
5	Sample density	$\geqslant$1 sample every 5cm
6	Stability	MDF $\geqslant$10mT or 1 week zero field storage
7	Continuity	Average solid angular difference between paired or contiguous samples $\leqslant$10°
8	Surface cores	Match to historical or archaeomagnetic secular variation records
9	Reproducibility	Demonstrable directional features in stacked records based on correlations independent of the directional data
10	Dating	Chronostratigraphy *and* litho- or bio-stratigraphy; sufficient dates to characterize sediment accumulation rate well and to assess possibility of unconformities. [For example, cores spanning last 10,000 years could have ^{14}C and pollen stratigraphy with at least 1 date every 2000 years]

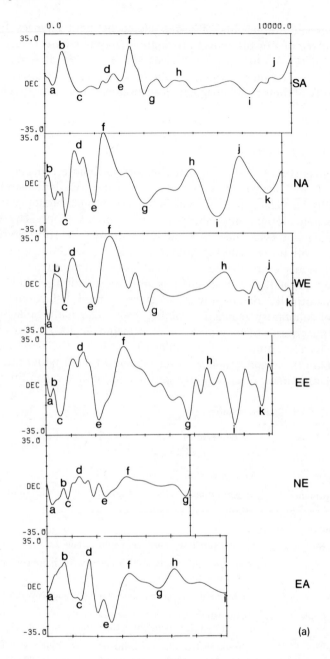

Fig. 5.1(a)   Regional declination master curves. Tree ring calibrated time scale in calendar years BP (0 BP = AD 1950). SA South Australia; NA North America; WE Western Europe; EE Eastern Europe; NE Near East; EA East Asia.

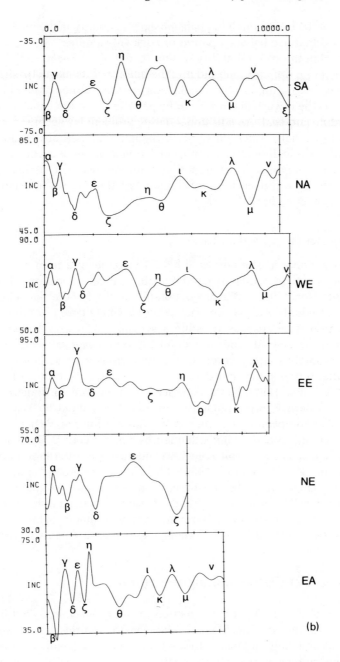

Fig. 5.1(b)   Regional inclination master curves. Time scale and localities as in fig. 5.1(a).

which can be used in judging palaeomagnetic quality is the median solid angular difference between paired or adjacent samples.

Assessing the true reliability of the ^{14}C dating of a lake sediment can often be more difficult than judging the quality of its magnetic signature. A major ^{14}C dating problem concerns natural carbon contamination, caused by the inwash of old soils or by particles of bedrock materials such as graphite and coal. An additional dating problem is concerned with the well-known hard-water effect of carbonate-rich bedrock regions. On account of these difficulties, duplicate dating techniques involving tephra or pollen zonations in conjunction with radiocarbon dating of lake sediments have proved to be of the greatest importance in establishing palaeomagnetic chronologies.

GEOMAGNETIC MASTER CURVES

The six regional master curves of fig. 5.1 were derived from high-quality palaeomagnetic records using the mathematical procedure described by Thompson (1983). Whenever possible a single core was chosen to cover the whole of the Holocene. Where a single core did not cover as complete a time span as possible a second core was selected from a neighbouring lake in order to extend the time coverage. The lakes and cores selected as type records for the six regions were as follows: South Australia, Bullenmerri core BC extended by Keilambete core KF (Barton and McElhinny 1982); North America, St Croix core 75 (Banerjee et al. 1979); Western Europe, Lomond core LLRPI extended by Windermere core W3 (Thompson and Turner 1979); Eastern Europe, Pääjärvi core P4 (Huttunen and Stober 1980) extended by Lojärvi core D (Tolonen et al. 1975); Near East, Kinneret core K8 (Thompson et al. 1984); East Asia, Kizaki core K3 (Horie et al. 1980). The ages of the broad features of the secular changes have been summarized for magnetostratigraphic use in table 5.2. The most recent part of each regional master curve has been based on a spherical harmonic analysis model (Thompson and Barraclough 1981) of historical geomagnetic measurements (fig. 5.2).

Historical and archaeomagnetic data

Accurate observations of the direction of the geomagnetic field became common in the seventeenth and eighteenth centuries while measurements of the field intensity followed in the nineteenth century. An excellent summary of such observations of the historical geomagnetic field is the catalogue of Veinberg and Shibaev (1969). The observations listed in their catalogue have been combined with archaeomagnetic

Table 5.2.   Ages of magnetostratigraphic features.[1]

Declination

	SA	NA	WE	EE	NE	EA
a	300	—	140	160	220	0
b	680	100	450	300	700	700
c	1300	750	600	600	850	1200
d	2000	1200	1000	1400	1300	1650
e	2800	2000	2000	2200	1900	2200
f	3500	2400	2600	3100	2100	3100
g	4500	4000	4900	5700	2400	4400
h	5500	5900	7100	6500	3200	5100
i	8300	7000	8300	7600	5600	7300
j	9000	7900	9100	8000	—	—
k	—	9000	10000	8700	—	—
l	—	—	—	9000	—	—

Inclination

	SA	NA	WE	EE	NE	EA
α	—	50	240	300	300	—
β	—	420	650	600	550	400
γ	400	750	1150	1300	700	760
δ	900	1200	1650	1900	900	1000
ε	1900	2300	3100	2600	1400	1300
ξ	2600	2900	3800	4600	2000	1550
η	3200	3700	4300	5500	3600	1750
θ	3600	4400	5000	6400	5300	2800
ι	4600	5300	6000	7200	—	4100
κ	6000	6600	7100	7800	—	4600
λ	6800	7700	8300	8600	—	5100
μ	7900	8400	8800	—	—	5600
ν	8600	9600	9700	—	—	6600
ζ	10000	—	—	—	—	—

SA	South Australia	(35°S 140°E)
NA	North America	(45°N 90°W)
WE	Western Europe	(55°N 05°W)
EE	Eastern Europe	(60°N 30°E)
NE	Near East	(30°N 35°E)
EA	Eastern Asia	(35°N 140°E)

---

1. a to l declination turning points. α to ζ inclination turning points. Ages tabulated in calibrated ^{14}C years BP. The pre-2000 BP EA magnetostratigraphic features are taken from Horie et al (1980). The EA ages are rather poorly known, based here on a linear interpolation between the basal tephra layer and the archaeomagnetic features in the upper sediments.

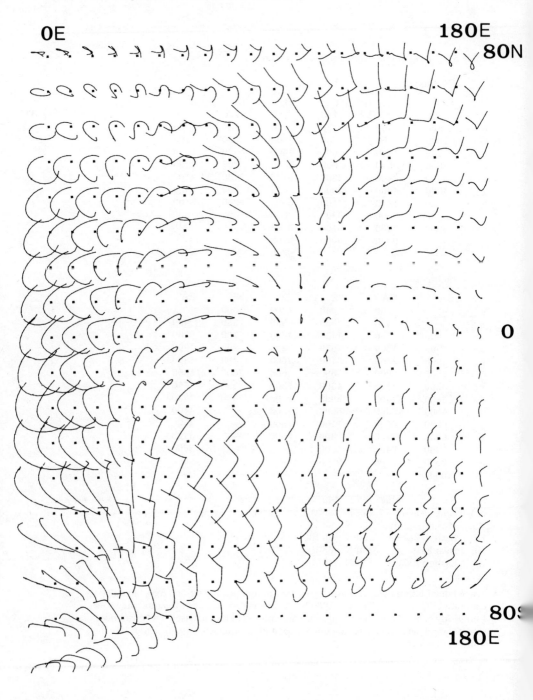

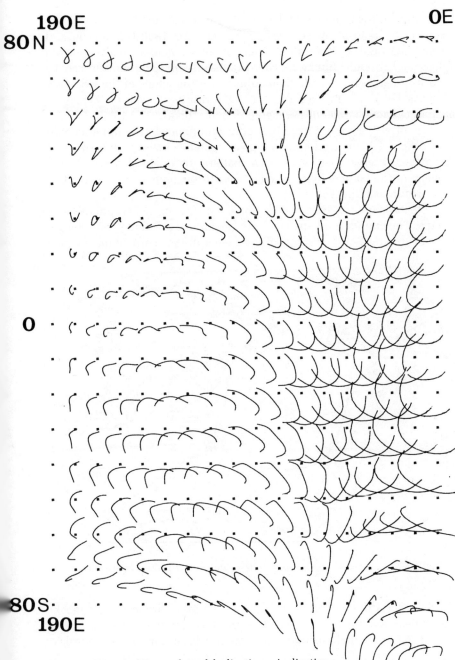

190E   **0E**
80N

0

80S
190E

Fig. 5.2   Historical Bauer plots of declination v. inclination
on a 10° by 10° latitude and longitude grid
for AD 1600 to AD 1975. The well-known clockwise looping observed at London
and Paris (Bauer 1896) can be seen in the upper central part of the diagram.

inclination measurements and used to construct global models of the geomagnetic field at 50 year intervals since AD 1600 (Thompson and Barraclough 1981). The calculations involve potential analyses using spherical harmonic functions (Barraclough 1978). The models can be used in order to interpolate field values over the whole of the world for the past 350 years. Fig. 5.2 illustrates the results of such mathematical analyses as a collection of stereographic plots, on a 10° by 10° grid, of the movement of the local magnetic vector. Prominent in the upper central part of fig. 5.2 is the well-known type of open clockwise loop of Western European observatories. A similar pattern of open clockwise polar movement can be seen to have taken place throughout a region extending as far south as Cape Town.

A cluster analysis of the directional data of fig. 5.2 produced five groups. The boundaries between these five groups are plotted on a map of the world in fig. 5.3. The shape of the cluster in the centre of the diagram

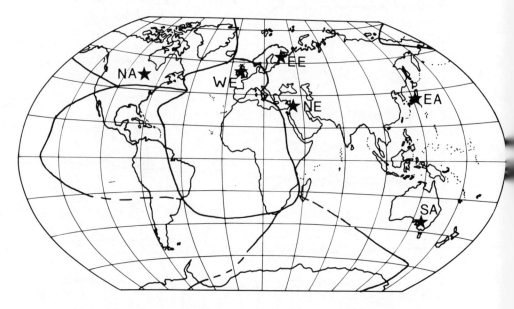

Fig. 5.3 Localities of regional master curves and cluster analysis of historical secular variation records.

clearly reflects the regional extent of the Western European open clock-wise type of looping. The five clusters provide an indication of the size of regions within which secular variation has been similar over the last 300–400 years.

Archaeomagnetic data, although not as accurate as historical observa-tions, are particularly useful in older secular variation studies. Measurements of the remanent magnetization of bricks, tiles, pieces of pottery, and other earthenware have allowed remarkably detailed secular variation records to be produced for Europe (Thellier 1981) and for Japan (Hirooka 1971) over the last 2000 years. Both these archaeomagnetic studies have been used as an aid in dating the lake sediment records of fig. 5.1 and table 5.2.

Global geomagnetic field

One of the longest-standing problems in geophysics is the origin of the earth's magnetic field. Although a number of explanations have been suggested, the only likely mechanisms seem to involve some form of self sustaining electromagnetic dynamo action in the earth's electrically conducting fluid core. It is suggested that this immense natural dynamo operates through the electromagnetic interaction of swirling currents of molten metal driven by thermal convection.

THEORETICAL DYNAMO MODELS

The full dynamo problem, starting from electrodynamic and hydrodynamic equations such as those of Maxwell, Ohm, and Navier-Stokes and incorporating geophysical boundary conditions such as the size and conductivity of the earth's core, is highly complicated. Not surprisingly theoretical models do not as yet duplicate details of observed field changes and so unfortunately they cannot be used for collating and examining our palaeolimnomagnetic data. Instead we have to be content with geometrical models or with simple mathematical formulations. Such models have a minimal physical basis but they can mimic the geomagnetic field changes very well.

DIPOLE MOVEMENT

The simplest of the geometrical models just involves movement of the dipole axis of the earth's magnetic field (fig. 5.4a). In this model secular changes of declination and inclination vary from place to place over the earth's surface. However, they follow a regular pattern such that changes in position of virtual poles are all the same. The virtual polar paths all trace out the movement of the dipole axis. Kawai and Hirooka (1967) have suggested that the broad trends in the Japanese and European archaeomagnetic records for the last 2000 years fit such a model in which dipole movement has dominated secular changes. They proposed that the pattern of dipole movement has been a somewhat complicated 'quasi-

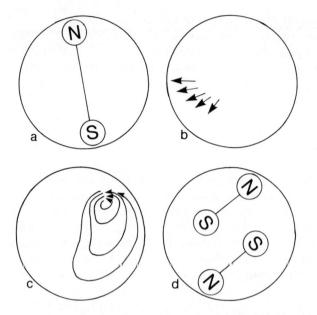

Fig. 5.4   Geomagnetic models: (a) dipole wobble; (b) westward drift; (c) growth and decay of non dipole foci; (d) quadrupole sources.

hypotrochoidal' path which consists of an anti-clockwise precession around the earth's rotation axis on which is superimposed a smaller and more rapid clockwise precession. The dipole movement model does not, however, explain field changes on time scales of tens and hundreds of years, nor does it fit the lake sediment records of fig. 5.1.

NON-DIPOLE CHANGES

Global analyses of historical secular variation indicate that a very different secular change model to that of dipole movement has operated over the last 350 years. The analyses show that the dipole axis of the earth's magnetic field has remained relatively stationary and that secular variation has resulted mainly from changes in the non-dipole field.

1. *Drift of non-dipole sources.* Halley (1692) first noticed that elements of the magnetic field were moving westwards with time. He discussed the 23° movement to the west of the position of zero declination in the South Atlantic during a 90-year period. Analyses by Bullard *et al.* (1950) give the global average westward drift rate of the non-dipole field to be 0.18° of longitude per year. This general drift to the west is taken to be an indication that the outer part of the earth's fluid core is rotating more slowly than the solid mantle. Vestine *et al.* (1947) analysed changes in the

earth's magnetic field between AD 1905 and 1945 and showed that the secular variation field is also drifting westwards. Bullard *et al.* (1950) computed the worldwide average westward drift rate of secular change to be 0.32° per year. The compilations of Vestine *et al.* (1947) clearly showed historical secular variation to have been a regional rather than planetary phenomenon and emphasized the importance of non-dipole field changes.

McNish (1940) modelled the magnetic field in terms of 14 radial dipoles in addition to its main geocentric dipole. By moving a few radial dipoles westward with time a simple secular variation model can be produced. This rather appealing, simple type of model can be employed very successfully to imitate past magnetic field changes (fig. 5.4b). However, Alldredge and Stearns (1969) found that they needed 34 radial dipoles to form a global model and that the positions of the satellite dipoles are quite different when different numbers of dipoles are used in the models. Furthermore, Lowes (1955) has suggested that westward movements account for only about one-third of the total global secular variation changes. So we see that a more complicated model is necessary to explain historical secular changes.

2. *Evolution of non-dipole sources.* Variations in strength of local centres of non-dipole activity can account for the secular variation not explained by westward drift and main dipole change. Such variations can be built into radial dipole models by allowing the strengths of the satellite dipoles to fluctuate with time. Changes in intensity and position of the radial dipoles can then be thought of as a mathematical representation of the growth, decay, and drift of current loops flowing near the core/mantle interface (fig. 5.4c). Turbulence of the core's fluid flow, near its boundary with the solid mantle, is thought to cause such current loops.

Unfortunately, in order to build up a global model which takes all these extra phenomena into account a large number of parameters have to be determined. There are not enough data available to allow the parameters to be calculated mathematically. So any particular satellite dipole model which is chosen to explain Holocene secular variation is just one out of hundreds which could be employed.

3. *Quadrupole and higher order multipoles.* Another mathematical method of representing a planetary magnetic field is in terms of spherical harmonic coefficients. Secular variations are modelled as changes in the spherical harmonic coefficients with time. A centred dipole field is specified by three coefficients of the first order. A non-dipole field is specified by quadrupole, octupole, and higher-order coefficients. A great advantage of the spherical harmonic modelling method is that a unique model can be calculated from observations of the magnetic field. A fourth order model with 24 coefficients was used in the historical field analyses which produced fig. 5.2. Eighth order models with 80 coefficients are

routinely employed to produce the definitive International Geomagnetic Reference Fields which are used on magnetic charts for navigators and surveyors. Abundant, precise, well-distributed data are needed for such spherical harmonic calculations. Unfortunately, the available lake sediment palaeomagnetic data do not satisfy any of these conditions.

COMPARISON OF LAKE SEDIMENT PALAEOMAGNETIC RECORDS

Visual inspection of the master curves of fig. 5.1 reveals that most of the magnetic variations are dominated by long period changes. Spectral analyses show that the power content of all the records peaks at periods of a few thousand years. It is also possible to see (fig. 5.1) that the dominant periodicity at any particular locality changes with time. This is a clear demonstration that geomagnetic changes have been aperiodic through the Holocene. Furthermore by comparing the records of fig. 5.1 it can be seen that the details of the directional changes have been quite different in the six regions. These differences demonstrate the complexity of the global pattern of Holocene geomagnetic field changes and emphasize the importance of non-dipole field components.

At first sight, some short sections of the secular variation records appear to correlate tolerably well, e.g. the declination swings NAb to NAf are similar to the swings WEb to WEf and EEb to EEf (fig. 5.1). However, closer examination of the combined declination and inclination records indicates that these types of correlations are probably spurious alignments which result from the general similarity of the frequency content of the records. Of course, it is always possible to produce some correlation scheme, with an associated geomagnetic model, by squeezing and stretching the time scales, perhaps by invoking variable longitudinal drift rates, and by explaining the lack of quantitative correlations in terms of the geometrical effects of local non-dipole field sources. However, introducing so many parameters or degrees of freedom allows almost any geomagnetic model to be invoked. So an alternative empirical approach is now discussed in an attempt to sidestep these 'over-parameterization' problems.

EMPIRICAL MODEL

Various aspects of the regional lake sediment geomagnetic master curves can be compared and combined in order to build up an empirical model of global field changes over the last ten thousand years. In this empirical approach towards estimating the past behaviour of the geomagnetic field the regional lake sediment palaeomagnetic records have all been subjected to the same closely defined mathematical procedures. Robust averaging of the local records was employed in searching for global geomagnetic features. Virtual pole paths rather than declination and inclination fluctuations were used in the analyses, as polar paths present a convenient method of taking into account certain geometrical effects of site longitude and latitude.

The following discussion concentrates on three aspects of geomagnetic field behaviour. First, longitudinal drift has been assessed by examining the sense of looping of the field vector. Runcorn (1959) showed that westward drift is generally associated with clockwise looping of virtual poles and eastward drift with an anti-clockwise looping. Although exceptions to this relationship may exist (Runcorn 1959, Dodson 1979) the exceptions tend either to be limited geographically or to be caused by rather unusual magnetic sources. Second, quiet periods of secular variation, as opposed to disturbed periods, have been investigated by examining the rate of change of position of the virtual poles in the lake-sediment records. Cox (1975) has drawn attention to the old secular variation records preserved in the remanence of sequences of ancient lava flows. He noted that noisy records occurred which, in addition to their rapid polar motions, tended to have low inclinations. He suggested that the noisy records resulted from the longitudinal drift of giant anomalies of the vertical non-dipole field and that the quiet records were produced at times when the dipole field was free from distortions caused by non-dipole field anomalies. Third, movement of the main dipole axis has been assessed by calculating the average position of the virtual poles of the lake-sediment records. The relative importance of Holocene non-dipole fields can be gauged in this third analysis in terms of the variance of the regional virtual pole positions. The results of these mathematical calculations have been plotted in fig. 5.5 and are summarized in fig. 5.6. The global average intensity of the geomagnetic dipole moment, as derived from archaeomagnetic intensity studies (McElhinny and Senanayake 1982), has also been plotted in fig. 5.5 for comparison.

The mathematical procedures used in these analyses have been described in Thompson (1982). Briefly the procedures involve (1) choosing type magnetic records using the reliability criteria of table 5.1; (2) detrending the paired declination and inclination measurements in order to take into account any twisting or warping of the core tube during coring; (3) allowing for random noise in the palaeomagnetic record by smoothing the data using robust weighted least-squares cubic spline functions in conjunction with a cross validation technique; (4) transforming the depth scales into a tree-ring calibrated ^{14}C time scale (Clark 1975); and finally (5) converting the resulting declination and inclination information (of fig. 5.1) into virtual pole positions at 50 year intervals.

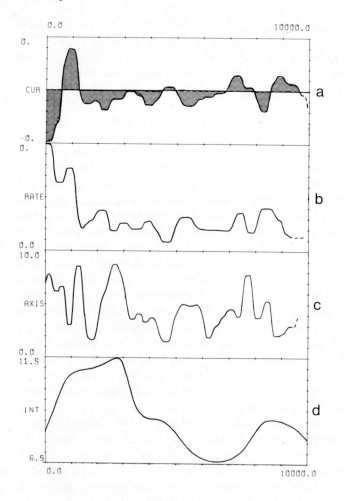

Fig. 5.5   Holocene global averages:
(a) Global average of rate of movement of virtual pole position where positive rates
    correspond to anticlockwise curvature and negative rates correspond to
    clockwise curvature. For most movements of physically plausible geomagnetic
    sources these curvatures correspond to eastward and westward drift of the
    geomagnetic field respectively. Averaging as in fig. 5.5(b).
(b) Global average of magnitude of rate of movement of virtual pole positions over
    the last 10,000 years. Rates calculated from master curves of fig. 5.1. Averaging
    by taking medians every 50 years followed by Tukey's (1977) 7RSSH 7RSSH7,
    twice median smoothing procedure.
(c) Global average of deviation of virtual pole positions from their mean direction.
    N.B. Lake sediment cores are not orientated so the mean direction only
    approximates to the geographic pole position. Averaging as in fig. 5.5(b).
(d) Dipole intensity. Global average. Smooth fit based on McElhinny and
    Senanayake's (1982) world archaeomagnetic intensity mean values.

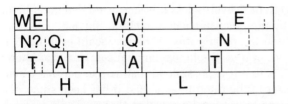

Fig. 5.6   Summary of Holocene global averages of fig. 5.5: (a) E Eastward, W Westward drift; (b) Q quiet, N noisy secular variation; (c) A axial, T tilted dipole orientation; (d) H high, L low dipole intensity.

Averaging of the virtual pole position records was carried out using Tukey's (1977) robust running median method. The specific method employed was, in Tukey's distinctive notation, '7RSSH7RSSH7, twice'.

The average curves of figs. 5.5a and 5.5b are based on the intrinsic equations of the polar paths. Fig. 5.5b uses the rate of change of the polar path as calculated using the gradient of the spline functions fitted to the polar paths. Fig. 5.5a also uses the gradient of the spline functions but in addition it takes into account the sign of the curvature of the polar paths. Curvature can be derived directly from polar paths by combining the gradient and acceleration of their spline functions according to the usual formula (e.g. Thomas 1960). A positive curvature corresponds to anti-clockwise looping of the virtual geomagnetic pole. The curve of fig. 5.5c illustrating axial tilt is based on the solid angle departure of the virtual poles from their average value. The summary dipole intensity curve of fig. 5.5d is based on 1167 palaeointensity results from different parts of the world. The palaeointensity results have been combined into 1000-year averages for the period between 10,000 BP and 3000 BP and into 500-year averages between 3000 BP and the present (McElhinny and Senanayake 1982). A smooth curve has been fitted in fig. 5.5d to their disjoint means.

The most important feature of fig. 5.5a is the change in sign of the average curvature. It is suggested in fig. 5.6 that the clear changes in sign of curvature correspond to reversals in the sense of longitudinal drift. The change from eastward to westward drift during the last 1000 years is particularly interesting as it corresponds well with the change in sense of looping in European archaeomagnetic records (Aitken *et al.* 1964). The complete pattern of fig. 5.5a over the last 10,000 years indicates that there has been no particular preference for either westward or eastward drift.

In fig. 5.5b the most noticeable feature is the rapid increase in rate of polar movement during the last 2000 years. A large part of this increase undoubtedly reflects poor memory effects in the palaeomagnetic record-ing process rather than the behaviour of the geomagnetic field. Some of the lake sediment palaeomagnetic records have been smoothed or degraded in their older sections. Nevertheless it is suggested that quiet field times can be detected by the polar movement method (fig. 5.6).

The differences between the polar paths of the six localities of fig. 5.3 emphasize the importance of non-dipole secular changes. However, there are indications in fig. 5.5c that it is possible to distinguish periods when the dipole axis was closer to the rotation axis than it has been since AD 1600. Also it is interesting to note in fig. 5.6 that these periods of axial alignment of the field occur at similar times to those of quiet-field behaviour.

The empirical model, although lacking the appeal of a simple, all-embracing geometrical model, exhibits a number of interesting features which can be quantitatively derived. Furthermore, the empirical model can be extended and improved as new palaeomagnetic data from lake sediments become available.

Conclusions

Holocene palaeomagnetic secular variations are becoming reasonably well-established in many regions of the world making secular variation magnetostratigraphy a practical chronological tool. The magnetostratigraphy can be used in the same way as the well-established palaeomagnetic reversal stratigraphy has been applied in dating deep-ocean sediments. There are a number of pitfalls in palaeomagnetic work. They can be most easily avoided by paying close attention to the quality of the magnetic data and in particular to their reproducibility.

Holocene field changes have followed distinctive but complicated patterns. Centres of activity of the non-dipole field have been very important in producing the variations in declination and inclination. These centres of activity have been continuously changing. They have disappeared, reappeared, grown, drifted, expanded, and contracted on time scales of tens to thousands of years. They have produced secular variation records of regional extent. Dipole variations, which in contrast produce planetary wide changes, have tended to take place at a slower rate than non-dipole changes but to have been of particular importance in determining intensity changes. The palaeomagnetic results indicate that the Holocene non-dipole field has drifted both eastwards and westwards. They also suggest that the non-dipole field has, for short periods, been less active than at present.

Acknowledgments

Funding of fieldwork by the Natural Environment Research Council and the Royal Society is gratefully acknowledged. I thank M. Hyodo and C. Barton for kindly supplying listings of their palaeomagnetic data.

References

Aitken, M.J., Harold, M.R. and Weaver, G.H., 1964. Some archaeomagnetic evidence concerning the secular variation in Britain. *Nature 201:* 659–60.

Alldredge, L.R. and Stearns, C.O., 1969. Dipole model of the sources of the earth's magnetic field and secular change. *J. Geophys. 74:* 6583–93.

Banerjee, S.K., Lund, S.P. and Levi, S., 1979. Geomagnetic record in Minnesota lake sediments – Absence of the Gothenburg and Eriean excursions. *Geology 7:* 588–91.

Barraclough, D.R., 1978. Spherical harmonic models of the geomagnetic field. *Inst. Geol. Sci. Geomag. Bull. 8.*

Barton, C.E. and McElhinny, M.W., 1982. A 10,000 year geomagnetic secular variation record from three Australian Maars. *Geophys. J.R. astr. Soc. 67:* 465–86.

Bauer, L.A., 1896. On the secular motion of a free magnetic needle II. *Phys. Rev. 3:* 34–48.

Bullard, E.C., Freedman, C., Gellman, H. and Nixon, J., 1950. The westward drift of the earth's magnetic field. *Phil. Trans. R. Soc. A. 243:* 67–92.

Clark, R.M., 1975. A calibration curve for radiocarbon dates. *Antiquity XLIX:* 251–66.

Cox, A., 1975. The frequency of geomagnetic reversals and the symmetry of the non dipole field. *Rev. Geophys. Space Phys. 13:* 35–51.

Dodson, R.E., 1979. Counterclockwise precession of the geomagnetic field vector and westward drift of the non dipole field. *J. Geophys. 84:* 637–44.

Halley, E., 1692. An account of the cause of the change of the variation of the magnetic needle; with an hypothesis of the structure of the internal part of the earth. *Phil. Trans. R. Soc. A. 17:* 563.

Hirooka, K., 1971. Archaeomagnetic study for the past 2,000 years in Southwest Japan. *Mem. Sci. Fac. Kyoto Univ. Ser. Geol. & Min. 38:* 167–207.

Horie, S., Yaskawa, K., Yamamoto, A., Yokoyama, T. and Hyodo, M., 1980. Paleolimnology of Lake Kizaki. *Arch. Hydrobiol. 89:* 407–15.

Huttunen, P. and Stober, J., 1980. Dating of palaeomagnetic records from Finnish lake sediment cores using pollen analysis. *Boreas 9:* 193–202.

Kawai, N. and Hirooka, K., 1967. Wobbling motion of the geomagnetic dipole field in historic time during these 2000 years. *J. Geomagn. Geoelect. 19:* 217–27.

Lowes, F.J., 1955. Secular variation and the non dipole field. *Annls. Géophys. 11:* 91–112.

Mackereth, F.J.H., 1971. On the variations in direction of the horizontal component of remanent magnetization in lake sediments. *Earth Planet. Sci. Lett. 12:* 332-8.

McElhinny, M.W. and Senanayake, W.E., 1982. Variations in the geomagnetic dipole 1: The last 50000 years. *J. Geomagn. Geoelect. 34:* 39–51.

McNish, A.G., 1940. Physical representations of the geomagnetic field. *Trans. Am. geophys. Un. 21:* 287–91.

Runcorn, S.K., 1959. On the theory of the geomagnetic secular variation. *Annls. Géophys. 15:* 87–92.

Thellier, E., 1981. Sur la direction du champ magnétique terrestre, en France, durant les deux derniers millenaires. *Phys. Earth Planet. Ints. 24:* 89–132.

Thomas, G.B., 1960. *Calculus and Analytic Geometry* (3rd edn).

Thompson, R., 1983. Global Holocene magnetostratigraphy. *Hydrobiologia 103:* 45–51.

Thompson, R. and Barraclough, D.R., 1981. Cross validation, cubic splines and historical secular variation. Abs. in *IAGA Bulletin 45:* 120.

Thompson, R. and Turner, G.M., 1979. British geomagnetic master curve, 10,000–0 yr B.P. for dating European sediments. *Geophys. Res. Lett. 6:* 249–52.

Thompson, R., Turner, G.M., Stiller, M. and Kaufman, A., in prep. Near East palaeomagnetic secular variation recorded in sediments from the Sea of Galilee (Lake Kinneret).

Tolonen, K., Siiriäinen, A. and Thompson, R., 1975. Prehistoric field erosion sediment in Lake Lojärvi, S. Finland and its palaeomagnetic dating. *Ann. Bot. Fennici 12:* 161–4.

Tukey, J.W., 1977. *Exploratory Data Analysis.*

Veinberg, B.P. and Shibaer, V.P. (editor-in-chief: A.N. Pushkov). 1969. Catalogue. The results of magnetic determinations at equidistant points and epochs, 1500–1940, IZMIRAN, Moscow. Translation No. 0031 by Canadian Department of the Secretary of State, Translation Bureau, (1970).

Vestine, E.H., Laporte, L., Cooper, C., Lange, I. and Hendrix, W.C., 1947. Description of the Earth's main magnetic field and its secular change, 1905–1945 (Carnegie Inst. Wash. Publ., No. 578).

6 Stratigraphic changes in algal remains (diatoms and chrysophytes) in the recent sediments of Blelham Tarn, English Lake District

Elizabeth Y. Haworth

Introduction

BLELHAM TARN, a small, mildly enriched lake in the English Lake District, has been a site of scientific study for several decades. This was initially due to its proximity to the original laboratory of the Freshwater Biological Association, Wray Castle, but a major reason for continued interest in this site has been its position as one of the richest lakes in Pearsall's Lake District series (Jones 1972). The tarn has a surface area of 0.11km², a maximum depth of 14.5m, and a mean depth of 6.8m. The catchment area is 4.3km² (Ramsbottom 1976) and is composed mainly of improved or rough grassland, with some woodland (fig. 6.1); the height above sea level is between 45 and 245m (see also Bonny 1978). Several small streams flow through farmland and the two hamlets of Outgate and Wray before entering the lake and at least two of them have been artificially straightened or culverted.

Studies of the lake's environmental history began with chemical studies of the sediment by Mackereth (1966) and now include analyses of pollen, diatoms, animal microfossils, radio-isotopes, and palaeomagnetism (Harmsworth 1968, Evans 1970, Oldfield 1970, Pennington and Bonny 1970, Pennington *et al.* 1976, 1977, Haworth 1980, Thompson 1975). Such intensive study of a lake can be both advantageous and disadvantageous to the environmental historian. Long-term phytoplankton monitoring has enabled us to make a direct comparison of diatom-population changes and their stratigraphic sequence in recent sediments (Haworth 1980) but the disturbance of the mud surface by earlier sampling and the location of several Lund Tubes (Lund 1972) clearly limits the choice of a suitable sediment sampling site.

In making the comparison between recent diatom stratigraphy and phytoplankton records (Haworth 1980) it was only possible to consider those diatom taxa that were recorded from the live populations and the corresponding depth of sediment. This study revealed that the algal

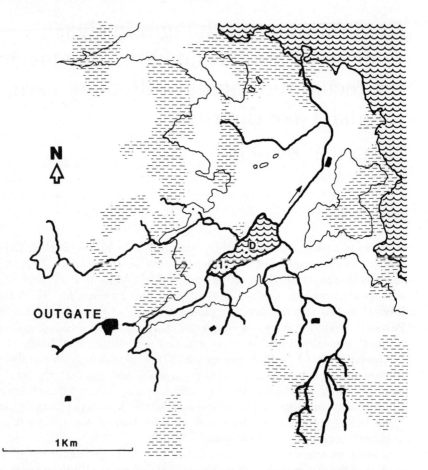

Fig. 6.1   Map of Blelham Tarn showing surface stream drainage including outflow to Windermere (→). 1 = site of WD 78-1, D = site of 74-D. 60m contour line and positions of farms and Outgate hamlet are also marked. Areas of woodlands are shaded.

changes observed by Lund (1979) between AD 1945 and 1977 and related by him to the enrichment of the lake were only a part of an older sequence. The aim of this paper is to present the diatom and chrysophyte evidence for these changes, to identify the origin of the sequence, and to correlate this with other evidence for changes upon the catchment that have affected the lake environment.

These recent changes are only the latest in a series, for Pennington *et al.* have shown (1977, fig. 1) that the Blelham Tarn sediments record several responses to events occurring upon the catchment. During the early post-glacial there was a decline in alkalinity of the lake due to the natural leaching of bases from the soils. Some later man-induced changes were temporary, such as that which resulted from partial tree clearance c. 200 BC, whilst others, such as the onset of ploughing c. AD 1500, had a more lasting impact on sediment accumulation due to the erosion of disturbed soils. Chemical evidence of an increased contribution of aquatic material as well as a change in the aquatic fauna, from one dominated by *Tanytarsus* and *Sergentia* spp. to one dominated by *Chironomous*, suggested enrichment of the lake at a time of deforestation c. AD 1000. This latter was apparently accompanied by changes in the diatom assemblages (Evans 1970) with a sharp decrease in the percentages of *Cyclotella comta* (Ehr.) Kützing and the re-appearance of *Stephanodiscus astraea* var. *minutula* (Kützing) Grunow in the same proportion as in the more alkaline, early post-glacial. As this assemblage remained dominated by *Cyclotella* spp., however, it cannot be regarded as one indicative of nutrient-rich or eutrophic conditions. It is only within the top 20cm of sediment that diatom indicators of eutrophication, such as *Fragilaria crotonensis*, are frequent and Stockner (1972) originally used an undated core to demonstrate the change from centric to araphid diatoms that he used to identify levels of mesotrophy and eutrophy.

Methods

One-metre cores of recent sediment were collected in 1974 and 1978 using Mackereth minicorers (Mackereth 1969). Pennington et al. (1976) analysed the pollen, geochemistry, and ^{137}Cs, ^{210}Pb, and ^{14}C isotopes in core 74-D, taken from the central part of the lake (D in fig. 6.1). In 1978 a minicore of 10cm diameter (78-1) was obtained from the western end (1 in fig. 6.1), this size of sampler collected sufficient material for analysis of 0.5cm sections of the core (Haworth 1980). The site of core 78-1 (fig. 6.1) was originally chosen for study because accumulation of post-1972 sediment containing *Stephanodiscus* valves was known to be greatest there (Haworth 1979) and it therefore offered a good site at which to investigate the fine details of the biostratigraphic record within contiguous 0.5cm slices of sediment.

A full account of the preparation of volumetric samples for diatom analysis can be found in Haworth (1979). Briefly, a 0.5cm^3 of wet sediment is cleaned with a chromic acid mixture to remove organic matter and the resulting diatom suspension diluted so that the amount dried and mounted on a glass slide is equivalent to 0.0001cm^3 of wet sediment. The total numbers of the diatom taxa or chrysophyte scales can then be related to a known volume of sediment (fig. 6.18).

Radio-isotope analysis of ^{210}Pb, made by A.E.R.E., Harwell, primarily provides the time scale for the pre-1945 sediment (table 6.1), since the more recent material can be correlated with the phytoplankton records (Haworth 1980).

Correlation with the earlier studies of Pennington *et al.* (1976, 1977) necessitated comparable percentage analyses of the diatom assemblages in both cores (figs. 6.16 and 6.17) and, later, analyses of certain chemical variables in core 78-1 were made by Mrs J.P. Lishman (fig. 6.20), also for comparison with those from core 74-D.

The study of the littoral diatoms in Blelham Tarn was undertaken by Miss M. Fawcett, who collected and identified benthic and epiphytic diatoms during the summer of 1978.

Taxonomy

Several diatom taxa in Blelham Tarn sediments have either proved difficult to identify, or provided useful information on the variability of the valve morphology of the taxon.

The taxonomy of *Stephanodiscus astraea* var. *minutula* (fig. 6.2) has been discussed previously (Haworth 1976) in respect of an obvious change in frustule size that occurred at population maximum. Measurements of the infrequent forms at 30cm sediment depth vary between 12 and 25μm in diameter (the mean diameter of 50 specimens being 16μm), whereas those at the population maximum at 4.5cm vary between 9.5 and 11μm (mean of 100 = 10.3μm); the most recent forms include ones with even smaller diameter, 7.5 to 11μm (mean of 100 = 9.6μm). The morphology of some of the smallest forms lack the distinctive undulation of the valve and they resemble the forms that occur in nearby Esthwaite Water and which were identified as *S. hantzschii* Grunow (Haworth 1981). This provides additional evidence for the close relationship of forms that appear to differ in nearby lakes. It was originally thought that the finely patterned forms (fig. 6.2b), where the puncta are unresolvable, belonged to a separate taxon, possibly *S. subtilis* van Goor or *S. invisitatus* Hohn and Hellerman, but studies of whole frustules show that both morphologies can occur in the same cell (see also, Stoermer *et al.* 1979). It is supposed that auxospores, formed prior to the autumn of 1972, produced forms that were better able to take advantage of apparently favourable conditions in Blelham Tarn but, although auxospores were identified in the sediments, there was no obvious increase in their numbers in this horizon.

Cyclotella praetermissa Lund (a colonial form of *C. comta*-type, Lund 1950 and fig. 6.3) has a less obvious size reduction in the upper sediments, where it is scarce. The valves have a wider range of diameter during the earlier period when they were more frequent (fig. 6.18).

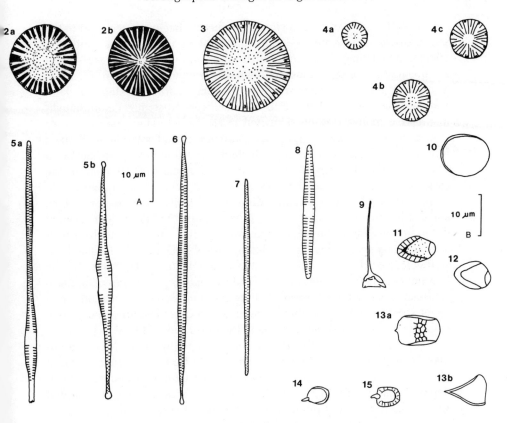

Figs. 6.2–6.15   Light microscope drawings to scales indicated: 5–7 as scale B.
2. *Stephanodiscus astraea* var. *minutula*; (a) normal valve. (b) finely marked valve. (Shaded striae are actually less well defined than can be demonstrated by ink drawings.)
3. *Cyclotella praetermissa*
4. *Cyclotella pseudostelligera*: (a) glomerata-type, (b) pseudostelligera-type, (c) woltereckii-type
5. *Fragilaria crotonensis*, (a) fine, rod-shaped form, (b) coarse, inflated-ends form.
6. *Synedra* cf. *S. radians*
7. *Synedra* cf. *S. tenera*
8. *Synedra* cf. *S. minuscula* (*Synedra* sp. D)
9. *Rhizosolenia eriensis*, in girdle view
10. *Mallomonas caudata* scale
11. *Mallomonas crassisquama* scale
12. *Mallomonas* cf. *alpina* scale
13. *Mallomonas reginae*; (a) body scale, (b) tail scale
14. *Synura spinosa* scale
15. *Synura uvella* scale

Evans (1970) found that *Cyclotella glomerata* Bachman occurred in the mid-post-glacial sediments. Its decline coincided with a clay band at c. 190cm, which may have resulted from a land-clearance phase, and it re-appeared in the uppermost sediments. Lund (personal communication) also identified *C. glomerata* in the Blelham Tarn plankton but later decided that it was probably *C. pseudostelligera* Hustedt, a taxon thought to prefer a moderately nutrient-rich environment. These taxa are very similar in morphology so that this confusion is understandable. A third form, *C. stelligera* var. *tenuis* Hustedt, recorded from sediments in Finnish lakes (Meriläinen 1969), appears identical to *C. glomerata* (the earliest specific epithet if these are considered to be synonymous). Study of numerous valves in Blelham sediments indicate a variable morphology and under the light microscope it is possible to distinguish several different types including those typical of both taxa (figs. 6.4a, 6.4b) as well as occasional specimens that could be defined as *C. stelligera* Hustedt, *C. stelligeroides* Hustedt, and *C. woltereckii* Hustedt (fig. 6.4c). In the original counts the forms were divided into either *glomerata*- or *pseudostelligera*-types which showed that the sediments above 17cm contained fewer specimens but that they were mainly of *pseudostelligera*-type (fig. 6.4b), whereas the greater numbers in the lower sediments were predominantly of *glomerata* (*stelligera* var. *tenuis*)-type (fig. 6.4a). More careful inspection reveals a shift in the diameter size from 3.5–7.0μm (mean of 100 = 4.8μm) at 19.5cm, to 4.0–9.0μm (mean = 6.8μm) at 8.0cm sediment depth. *C. glomerata* is said to have an even arrangement of marginal striae, a central ring of puncta, and no marginal spines while *C. pseudostelligera* has marginal striae of unequal lengths, a central ring of striae, and very obvious spines. The description of *C. glomerata* coincides mainly with valves of 4.0 to 5.0μm diameter whilst the latter type is mainly over 6.5μm in diameter. From the number of valve forms and the electron-microscope confirmation that spines occur on all specimens, it would seem that these are all part of one, variable taxon but that the populations can vary, perhaps in response to changes in environment; however, more critical comparison is needed.

The taxon *Fragilaria crotonensis* Kitton is known to include two distinct forms (Schmidt 1873–1959, plate 299) and both are found in Blelham sediments. The finer needle-like form with rod-shaped ends (fig. 6.5a) occurs first, with a maximum around 11cm sediment depth, and a more coarsely striate form that has inflated ends (fig. 6.5b) appears above 10cm and forms the major part of the second maximum (fig. 6.18). Under the electron microscope there appears to be no difference between these two forms, other than shape, but Canter and Jaworski (1978) observed that the rod-shaped forms are more susceptible to a species of chytrid and that the flared-form lacks mucilage.

Differentiation between the finer *Fragilaria crotonensis* form and some *Synedra* species was sometimes difficult except that, where sibling valves of the former remain together the shape is characteristic as the frustules are only joined around the central area. Three *Synedra* species have only tentatively been identified, despite their frequency in the diatom assemblage. The one most frequent in the lower sediments is most like *S. radians* Kützing (fig. 6.6 and as *S. acus* Kützing on figs. 6.16 and 6.17) but it is rather short in length. Its percentage declines at the same level as *C. praetermissa*, and where *F. crotonensis* first appears. Another long form may be *S. tenera* W. Smith (fig. 6.7), as the valve shape conforms to Hustedt's (1930–66) description and the central area is also absent, but there are fewer striae, c. 18 in 10μm. This form occurs in the sediments above 40cm (at frequencies of 1–2% only). A third, much shorter form, occurs in the upper 35cm at frequencies of 1–4% and was recorded from phytoplankton samples as *Synedra* D by J.W.G. Lund (unpublished data). It is tentatively assigned to *S. minuscula* Grunow (fig. 6.8) because specimens vary between 18 to 37μm long and 2μm wide, with round ends, and c. 18 striae in 10μm. It is very similar to a form ascribed to *S. rumpens* Kützing by Battarbee (1978) but does not have capitate ends. It also appears similar to *Fragilaria vaucheriae* (Kützing) Petersen but the central area is indefinite.

The counts of *Tabellaria flocculosa* var. *asterionelloides* (Grunow) Knudson also include a small number of short specimens of *T. flocculosa* (Roth) Kützing. These are more frequent in the lower sediments, where there are 30 to 50 specimens per slide, than in the upper ones, with less than 10 on a slide. The short form is more frequently found in littoral samples rather than the plankton and therefore represents the contribution of the former to deeper water sediments.

In spite of a belief that *Rhizosolenia eriensis* H.L. Smith is too lightly silicified to be preserved in sediments, a number of valves were found in the samples (fig. 6.9) and, on inspection with the electron microscope, these were found to be well preserved. The resulting profile (fig. 6.18) does not agree particularly well with the phytoplankton records but this may reflect the low level of productivity of this form.

To date 213 diatom taxa have been identified from recent sediments or from live littoral assemblages. Only a few are apparently confined to the deepest sediment samples but most only occur in low percentages and so can only be considered in presence-or-absence terms. The two *Achnanthes* species (figs. 6.16 and 6.17) are the only non-planktonic taxa to account for up to 10% of the assemblage.

Recently both scales and cysts of silicified chrysophytes have been studied with the electron microscope (Cronberg and Kristiansen 1980, Takahashi 1978), which allows palaeolimnologists to identify the forms found in lake sediments (Munch 1980, Smol 1980). Chrysophyte scales are less representative of actual specimens than diatom valves, as numbers per cell may vary, but it seems reasonable that a small aliquot of homogenized sample would not be biased by the presence or absence of single specimens and that an obvious increase in numbers of scales would reflect an increase in cell numbers. Lund's algal records show that several chrysophyte taxa occurred in Blelham Tarn but that the siliceous ones only formed a small part of the phytoplankton and, with the exception of *Mallomonas caudata* Iwanoff (fig. 6.10), were more frequent before 1950. Chrysophyte scales in sediments appear easier to find and identify than live cells in water samples, for several taxa do not appear in the plankton counts, e.g. *Mallomonas crassisquama* (Asmund) Fott (fig. 6.11), *M. alpina* Pascher and Ruttner (fig. 6.12), and *Synura spinosa* Korshikov (fig. 6.14). This suggests that close correlation between phytoplankton records and fossil assemblages is not possible with such small populations but the decline of *Mallomonas reginae* Teiling (fig. 6.13) is readily identified, as is the recent increase in *M. caudata* (fig. 6.18). *Dinobryon divergens* Imhof is the most abundant chrysophyte in Blelham (Lund 1979) but is not preserved, and the tiny scales of *Mallomonas akrokomos* Ruttner are also rarely found.

Results

BIOSTRATIGRAPHY

The percentage analyses of diatoms in the two cores (figs. 6.16 and 6.17) has revealed remarkable similarity in the profiles and thus correlates the more recent minicore, 78-1, with the 74-D core for which radio-isotope, pollen, and chemical analyses had previously been made (Pennington *et al.* 1976, 1977). Correlation at levels, A, B, C, and D was made by comparison of the *Asterionella* and *Cyclotella* profiles but absence of intermediate samples of the older material preclude a more precise match.

In his diatom profile of the complete post-glacial sequence in Blelham Tarn sediments, Evans (1970) showed that *Fragilaria* and *Melosira* were originally the most important taxa and were replaced by *Cyclotella* and *Achnanthes*, taxa of more acidic and less nutrient-rich water, in the mid-post-glacial. Due to these changes in dominant taxa, it is difficult to identify any period of apparent long-term stability with which to compare the more recent changes, especially as there were several periods when man's activities on the catchment area also caused changes in the sedimentation pattern in the lake (Pennington *et al.* 1977, fig. 1).

Within the upper metre, with which we are presently concerned,

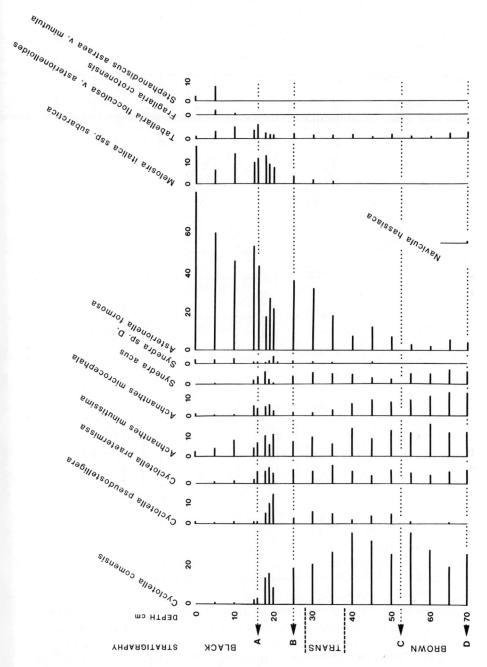

Fig. 6.16   Percentages of the major diatom taxa in core 78-1.

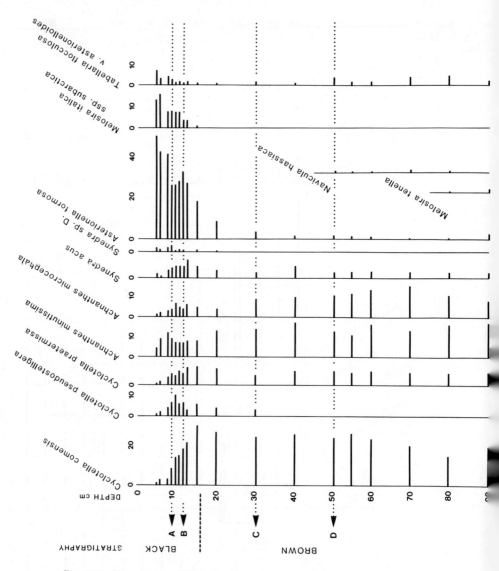

Fig. 6.17   Percentages of the major diatom taxa in core 74-D.

changes in sediment composition, accumulation rate, and the occurrence of a series of anomalous ^{14}C dates at c. 60cm in core 74-D were interpreted by Pennington et al. (1976 fig. 6, and 1977) as the result of ploughing post-AD 1000. In this lower section there is little change in the dominant diatom taxa, *Cyclotella comensis* Grunow, *Achnanthes minutissima* Kützing, and *A. microcephala* (Kützing) Grunow (fig. 6.17), although the proportions of planktonic forms increase slightly. *Melosira tenella* Nygaard is present at the base of this core (at 80 and 90cm), relating it to the top metre of the complete post-glacial sequence (Evans 1970). Minor taxa appear or disappear throughout all the section below D (fig. 6.17); *Achnanthes flexella* var. *alpestris* Brun (at 55, 70, and 90cm), *A. sublaevis* Hustedt (60 and 70cm), *Eunotia meisteri* Hustedt (60 and 70cm), *Fragilaria construens* var. *venter* (Ehrenberg) Grunow (40 to 80cm), *Navicula cari* var. *angusta* Grunow (55, 60, and 90cm), *N. bryophila* Petersen (70 and 80cm), *N. hassiaca* Krasske (50 to 90cm), *N. mediocris* Krasske (80 and 90cm), *N. subtilissima* Cleve (60 and 90cm), *Neidium affine* var. *amphirynchus* (Ehrenberg) Cleve (70 to 90cm), *Pinnularia subcapitata* (Gregory) (80 and 90cm), and *Stauroneis thermicola* (Petersen) Lund (55 and 70cm). The maximum percentages of *Navicula* spp., *Anomoeoneis* spp., and *Eunotia* spp. all occur in this lowest section. All are benthic or epiphytic taxa and their presence represents an increased contribution from the littoral, where most of them still occur (Fawcett 1978).

Horizon D is identified by the upper limit of *Navicula hassiaca* (figs. 6.16 and 6.17) and this correlates the bottom sample of 78-1 with 50cm in 74-D. A suspension of fine clay particles, noticed during preparation of the 50 to 70cm (section C–D) of core 78-1, is probably related to the increases in percentage of total minerals and in sodium and potassium content which Pennington *et al.* (1977) interpreted as the result of soil erosion on the catchment. There are no significant changes in the diatom assemblages within this section in either core, merely the continued increase in planktonic percentage and the dominance of *Cyclotella comensis*. One suggestion of change is the re-appearance of *C. pseudostelligera*, which apparently disappeared during the mid-post-glacial (Evans 1970). The C–D section is c. 20cm thick in both cores.

Horizon C is identified by the initial increase in *Asterionella* percentages and in the >1% occurrence of *C. pseudostelligera* (mainly of *glomerata*-type). Within the B–C section *Rhizosolenia eriensis* also appears (fig. 6.18), total diatom plankton increases to over 60%, and there is an increase in *Navicula* spp. percentages in both cores. In core 78-1 the numbers of large diatoms ($>30 \times 10\mu m$), such as *Pinnularia*, *Navicula* and *Gomphonema* spp., increase from approximately 100 to 150 valves per slide at 45cm and then drop to less than 80, indicating a change in the contribution of the littoral region to the diatom assemblage.

The initial rise in *Asterionella* percentages at horizon C is inversely related to a decline in most other taxa (figs. 6.16, 6.17). Rough estimates based on comparative transects of slides from core 78-1 (fig. 6.19) indicate that a steeper rise and a six-fold increase in *Asterionella* numbers do not occur until 35cm, which is within the transition from brown to black mud. *Melosira italica* ssp. *subarctica* O. Müller also increases in numbers

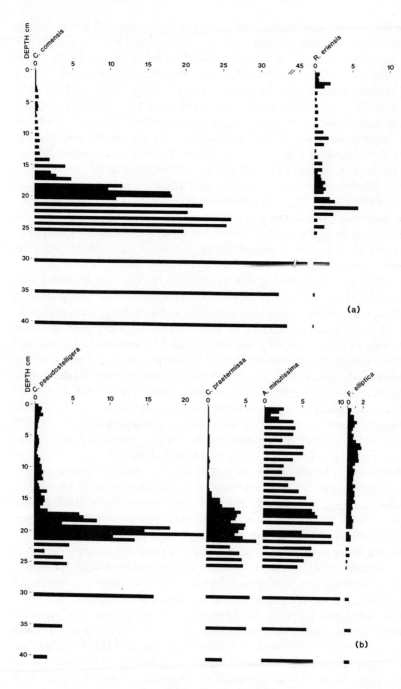

(a)

(b)

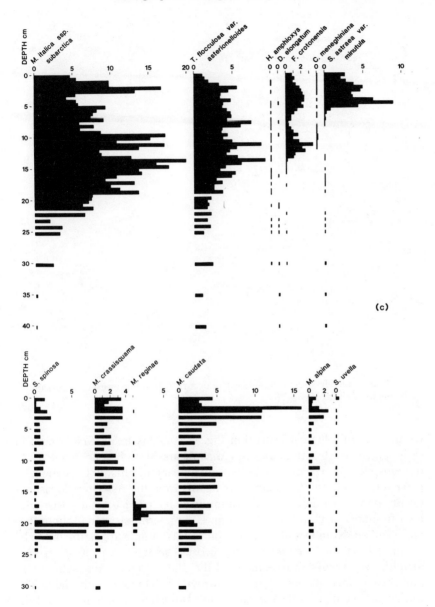

(c)

(d)

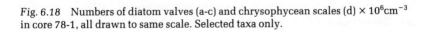

Fig. 6.18 Numbers of diatom valves (a-c) and chrysophycean scales (d) × 10⁶cm⁻³ in core 78-1, all drawn to same scale. Selected taxa only.

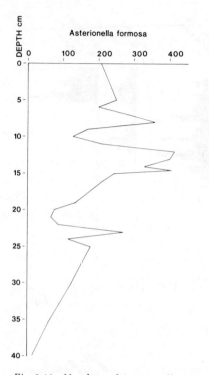

Fig. 6.19   Numbers of Asterionella valves occurring in single transects across the diameters of slides of core 78-1.

within this lithological transition (fig. 6.18). This taxon first occurs in early post-glacial sediments and disappears at the same level as *C. pseudostelligera* (Evans 1970) but re-appears in this B–C section. A percentage decrease in *C. comensis* results from the *Asterionella* rise, as the former reaches a numerical maximum at 30cm (fig. 6.18) together with high numbers of *C. pseudostelligera*, *C. praetermissa*, and *Achnanthes minutissima*. Numbers of *Asterionella* in this sample are estimated to be $1\frac{1}{2}$ times those of *C. comensis* (fig. 6.16). The sample at 30cm in 78-1 therefore represents an expansion of the diatom population, which was probably a consequence of some nutrient enrichment of the lake. The lithological change to darker sediment also occurs at this level and the change in organic component is indicated by the increase in carbon percentage (fig. 6.20). The depth of sediment between diatom horizons B and C differs at the two sites (figs. 6.16 and 6.17), with 27cm accumulating at 78-1 and only 18cm at 74-D. As Pennington et al. (1977) interpreted the changes in sediment composition within this section of 74-D as representing restabilization of mineral soils after the alterations of Fishpond Beck, it is suggested that the extra material in 78-1 may come from one of the inflows at the western end. Comparison of the diatom profiles (figs. 6.16 and 6.17) rules out any suggestion of redeposition of lake sediment.

A temporary decrease in *Asterionella* percentage correlates section A–B in the two cores (figs. 6.16 and 6.17). The profiles suggest that sediment accumulation at site 78-1 was more than double that at 74-D. A numerical decline of *Cyclotella comensis* begins around 25cm and appears to be inversely related to *Asterionella formosa*, as numbers of the former remain high during the temporary decrease of *Asterionella* and decline sharply at 18cm, where the latter increases again (figs. 6.18 and 6.19). Between 23 and 17cm there is a succession of taxa with numerical maxima, including minor as well as dominant constituents of the relic assemblage, and it is here that the numbers of chrysophyte scales occurring in the sediments increase significantly (fig. 6.18). Maximum numbers of *Synura spinosa* scales and *Rhizosolenia eriensis* valves occur at 20 and 21cm, and spores of the latter are also found in slightly greater numbers in the same samples. These are followed by increased numbers of *Mallomonas crassisquama*, *Cyclotella pseudostelligera*, and *M. reginae*. There are also further increases in *Melosira* (at 22cm) and *Melosira* and *Tabellaria* (at 18cm). *C. pseudostelligera* numbers decline sharply at 17cm and *C. praetermissa* at 16cm, the latter being the level identified as 1950 from algal records (Haworth 1980). Additional chrysophyte evidence refines the dating of the lower end of the algal time scale, as the decrease of *Mallomonas reginae* scales above 17cm identifies this layer as 1949 material (Lund 1979). This suggests that the 18.5cm increases in both *Melosira* and *Tabellaria* might date from 1946.

The more recent stratigraphic changes of diatom-valve numbers and their relationship with the plankton records have already been discussed (Haworth 1980). Post-1950 fluctuations in *Asterionella*, *Melosira*, and *Tabellaria*, and the sequential appearance of *Fragilaria crotonensis*, *Cyclotella meneghiniana*, and *Stephanodiscus astraea* var. *minutula*, dated to between 1960 and 1973 (fig. 6.18), are accompanied by a considerable increase in the numbers of *Mallomonas caudata* scales in the upper 3cm of sediment. Algal records (Lund 1979) indicate that populations of the latter increased around 1975, just before a second increase in *Stephanodiscus* numbers. Scales resembling those of *Mallomonas alpina* also occur in the more recent sediments (fig. 6.18) but it is difficult to distinguish these from *M. crassisquama* (figs. 6.11 and 6.12). *Synura uvella* Ehr. emend Korshikov is more frequent in surficial deposits.

Diatom plankton has been extensively studied (Lund 1979) but less is known of the non-planktonic taxa. Round (1960) found that littoral forms flourished best between February and June in the English Lake District

and that Navicula, Nitzschia, and Pinnularia are the best represented
epipelic genera. Diatom assemblages among aquatic vegetation in
Blelham Tarn (Fawcett 1978) are dominated by Achnanthes microcephala, A. minutissima, A. pusilla (Grunow) de Toni, Eunotia lunaris (Ehr.)
Grunow, Fragilaria vaucheriae (Kütz.) Petersen, Gomphonema parvulum
Kützing, Navicula cryptocephala Kützing, Nitzschia palea (Kütz) W.
Smith, Synedra cf. S. minuscula, S. radians, and Tabellaria flocculosa.
Achnanthes lanceolata (Breb.) Grunow, A. microcephala, Cymbella
ventricosa (Agardh) Kützing, and Meridion circulare (Grev.) Agardh are
the most frequent of stream vegetation and A. lanceolata, A. microcephala, Amphora ovalis var. pediculus (Kütz.) Van Heurck, Cymbella sinuata
Gregory, C. ventricosa, and Nitzschia dissipata (Kütz.) Grunow are most
frequent on sand.

GEOCHEMISTRY

Results of the chemical analysis (fig. 6.20) do not show the obvious
division of the profile that is suggested by the diatom analysis. They are
therefore best considered independently. Analyses of contiguous samples
over the 10–20cm section showed considerable variation, some of which
has been reduced by the averaging of adjacent samples. Such small-scale
variation can be compared to the year-by-year fluctuations in the phytoplankton and is probably due to annual changes in type and amount of
sedimentation produced by seasonal floods or prolonged dry periods.

Over the upper 50cm of 78-1 the potassium level gradually decreases
upwards and, as this element is derived from the catchment soils, this
suggests an increased sedimentation rate due to autochthonous accu-

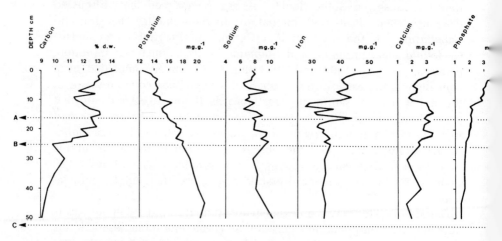

Fig. 6.20   Analysis of certain geochemical variables in core 78-1.

mulation. The increase in percentage carbon supports this interpretation. A decrease also occurs in the dominant forms of *Achnanthes* (fig. 6.16), both of which are very common on stones and aquatic plants of the streams and lake shores, and form part of the littoral and allochthonous components. Sodium also decreases throughout the profile but not so greatly or consistently. The carbon percentage rises at the level (38cm) of the transition to black mud, with a temporary decline between 14 and 9cm and the same trend occurs in the percentage increase of *Asterionella* (fig. 6.16). The temporary decrease in carbon at 25cm is accompanied by an increase in clay particles observed on the diatom slides and in a decrease in the numbers of many taxa (fig. 6.18), which suggests dilution by inwashed minerals. Both calcium and sodium rise slightly at this level. As the organic fraction (denoted by carbon) increases, there is a similar overall trend in iron, phosphate, and calcium, especially above 15cm (post-1950 according to the algal time scale), with iron exhibiting the same temporary decrease as carbon, and the diatom *Tabellaria* (fig. 6.18) increasing in numbers at the same level. The *Fragilaria crotonensis* increases and the sudden increase in *Stephanodiscus* all occur within the same section as the higher levels of phosphate.

RADIO-ISOTOPE ANALYSES

Calculation of the annual geochemical input is entirely dependent on the annual rate of sediment accumulation as indicated by the algal or ^{210}Pb and ^{137}Cs time scales. The ^{137}Cs analysis of core 78-1 shows a close relationship between the 1963 peak of this isotope and the 1964 increase in *F. crotonensis* (Haworth 1980). A similar relationship has been assumed from the information available for core 74-D (Pennington *et al.* 1976) but the percentages of *Fragilaria* are too low to show this change. Both cores have been analysed for ^{210}Pb and the resulting time scales (table 6.1 and Pennington *et al.* 1976) show some divergence in the dates of apparently similar diatom horizons. In the upper sediments, where algal records have been used to identify post-1949 horizons in core 78-1, both ^{210}Pb analyses are in good agreement with the algal time scale and, according to the diatom correlation, with each other. Below this, the dates of horizons B, C, D, and the stratigraphic change become increasingly older for 78-1 than for 74-D, with a difference of 80 years in the dates of horizon C which would appear to be beyond any standard statistical error. Where the sampling interval was widely spaced, there could be some reconsideration of the exact correlation of the diatom horizons (figs. 6.16 and 6.17) but the maximum latitude would still not bring the discrepancy in these dates within acceptable limits of variation. Since the total residual unsupported ^{210}Pb content of all cores analysed from Blelham

Table 6.1.   Time scales based on phytoplankton records and the ^{210}Pb calculations made by P.G. Appleby (personal communication).

Horizon	Core 78-1			Core 74-D	
	Depth (cm)	Algal date	^{210}Pb date	Depth (cm)	^{210}Pb date
Melosira	2.0–3.0	1976	1975		
Stephanodiscus	4.5–5.5	1973	1972		
Tabellaria	7.5–8.5	1970	1967		
Tabellaria	11.0–11.5	1965	1962		
Tabellaria	13.5–14.0	1961	1956		
Tabellaria	15.5–16.5	1954	1950		
Cyclotella/A	16.0–17.0	1950	1949	9–10	1952
Mallomonas	17.0–17.5	1949	1946		
B	25.0–25.5		1918	12–13	1944
Stratigraphic change	28.0–38.0		1850–1900	17–18	1926
C	c. 52		c. 1800	30–31	1875
D				50–51	~1750

Tarn was found to be comparable and to be consistent with the rate of supply from the atmosphere (P.G. Appleby, personal communication), and there is thus no reason to suspect any irregularity in the ^{210}Pb accumulation, the c.r.s. (constant rate of supply) method of calculating the time scale was used (Appleby and Oldfield 1978). The discrepancy between the ^{210}Pb dates of the earlier horizons in the two cores (table 6.1) is not resolved by using other methods, such as the c.i.c. (constant initial concentration) calculation and the recalculated time scale for core 74-D showed insignificant variation from the original c.i.c. calculation of Pennington et al. (1976). This is consistent with there being no major change in the sedimentation rate of that core during the period c. 1860–1960. Pennington et al. (1976) showed that the sedimentation rate in core 74-D suggests a temporary increase in sediment accumulation after deposition of a paler material c. 1870, followed by a decline c. 1900 to the original mid-nineteenth century value. Between c. 1940 and 1965 there appears to have been little variation. The ^{210}Pb time scale for core 78-1 suggests a significant increase in sediment accumulation between 1900 and 1920, with a decline until 1948, followed by a rapid increase to a peak value c. 1964.

Discussion

Comparison of the microfossil analyses of both cores clearly suggests that there has been greater sediment accumulation at the site of core 78-1, without any suggestion of redeposition, since the same pattern of diatom percentages occur over a greater depth of sediment (figs. 6.16 and 6.17).

Age calculations based upon ^{210}Pb analyses confirm this for the past 30 to 40 years. Prior to this time, however, the ^{210}Pb appears to indicate similar overall rates of deposition in both cores, resulting in widely differing dates for the older microfossil horizons. This suggests that there are either considerable errors in the older ^{210}Pb dates or that the apparent microfossil horizons represent non-synchronous events. It is unfortunate that this difference in the cores only occurs below the level where algal records provided independent dates. However, since the fidelity of the diatom stratigraphic record has already been demonstrated in this lake (Haworth 1979, 1980), it is difficult to envisage the same pattern of changes occurring at differing times.

It has been shown (P.G. Appleby, personal communication) that ^{210}Pb profiles from the western basin (78-1 and B in Pennington *et al.* 1976) both tend to have older dates for identifiable horizons than those from central or eastern locations (D and A in Pennington *et al.* 1976). It is therefore possible that inwash of terrestrial material, not containing ^{210}Pb, has diluted the unsupported ^{210}Pb to near equilibrium level with background [226]Ra and has thus produced an older date due to unrecorded ^{210}Pb in deeper sediment. The increased sediment accumulation in 78-1 may be due to the proximity of the two major inflows at the western end of the lake but this in itself should not cause difficulty, especially as the c.r.s. calculations were designed to accommodate changes in sedimentation.

This difference in time scale in cores from different parts of the lake makes it difficult to calculate diatom or chemical accumulation rates for 78-1 or to relate positively sedimentary horizons to events known to have occurred on the catchment.

Census records for 1841 and 1861 indicate that about 200 persons lived in the catchment area of Blelham Tarn and that there were eight or nine farms. The two small villages are composed of domestic dwellings and there are now only four farms. Changes that have taken place within the last 180 years have been identified from old maps, records, and from personal recollections. The earliest of these events was the lowering of the lake level c. 1830 (Oldfield 1970) by straightening the outflow channel and thus reclaiming areas of peat moss for pasture, especially at the east end. The inflow streams were also channelled around this time, for the 1847–8 Ordnance Survey map includes the partially completed covered culverts, or 'drains', of Fordwood beck and its tributaries. Presumably work was then in progress, as the drainage pattern of the 1888 edition is similar to that at present (T.T. Macan, personal communication). Fish Pond beck was also realigned prior to the 1888 survey so the amount of suspended mineral material in at least two of the inflow streams probably increased. The date of the construction of the Fish Pond is unknown but was certainly pre-1830. No records of nineteenth-century land use have yet been found but in the early twentieth century most farms ploughed some fields and this acreage was increased during the Second World War. Local residents recall that several fields near the lake were regularly ploughed until 1950.

In 1900 the hamlet of Outgate had not grown appreciably since 1700 but several new houses were soon added and these installed septic tank drainage whilst older properties had earth closets (T.T. Macan, personal communication). A piped water supply was brought to the hamlet in 1950. This increased the domestic and farm usage and overloaded the existing drainage system. The tributaries of Fordwood beck became polluted and so a small sewage treatment plant had to be installed in 1962 (Macan 1962, Lund 1979).

Pennington *et al.* (1977) showed that a higher input of mineral soils, $8mm \, yr^{-1}$ at c. 40cm in 74-D, could be correlated with the alteration of the inflows, especially Fish Pond beck which is near to the core site. *Cyclotella pseudostelligera* re-appears and the *Asterionella* percentages begin to increase at this same level c. 1870 (fig. 6.17), perhaps due to increased soil nutrients. The same D–C section of core 78-1 also includes greater numbers of large diatoms indicative of shallow water or stream benthos and, by correlation with 74-D, would have a sediment accumulation rate of c. $5mm \, yr^{-1}$ due to the increased input of mineral, rather than organic, soil. According to the ^{210}Pb analysis, however, this D–C section dates to pre-1800 with an accumulation rate of only c. $2mm \, yr^{-1}$. This was prior to the records of land improvement or alterations to stream drainage but is not old enough for medieval ploughing at c. AD 1500, so that there is no known reason for any change in lake environment that would cause these changes in diatom assemblages.

The steeper rise in *Asterionella* and the rise in *Tabellaria* and *Melosira* percentages occurred c. 1930 in 74-D and this, together with the stratigraphic change to black sediment, suggests lake enrichment due to a combination of land improvement and an increase in domestic drainage in streams due to the new septic tank systems. The reduced sedimentation rate between 1900 and 1940 represents a decrease in soil disturbance (Pennington *et al.* 1977). In contrast, accumulation in 78-1 remained high, according to the correlation of microfossil horizons, or low, according to the ^{210}Pb time scale; the latter suggests that the transition from brown to black mud took place between 1860 and 1910 and was related to the period of stream realignment, namely to erosion of mineral soil rather than organic nutrient enrichment. The organic-related carbon percentage did increase at 35cm, however. This correlation of a lithological change with a big increase in *Asterionella* percentage has now been found in three cores from Blelham Tarn (figs. 6.16, 6.17 and Haworth 1980) – although Evans (1970) claimed that this was not so in the cores that he studied – and also in the recent sediments of Windermere, c. 1850 (Pennington 1973). The similarity in the changes in the two lakes at differing times in their recent history suggests that these are more likely to have been the result of domestic rather than agricultural drainage, as the increase in population occurred earlier around the shores of Windermere.

Between the stratigraphic colour change at 17cm and the horizon A at 9cm (1920 50), the 74-D accumulation rate is 2–3mm yr^{-1} and the related diatom horizons of 34 and 17cm in 78-1 represent an initial accumulation of 7mm yr^{-1}, decreasing to 3mm yr^{-1} around 1950, which could be related to the cessation of ploughing of fields on the catchment. The ^{210}Pb time scale, however, suggests that the temporary increase in accumulation rate occurred between 1900 and 1920 in the 78-1 core. The increased accumulation rate within this section, indicated by either time scale, is substantiated by the decrease in carbon and in the numbers of several diatom taxa (fig. 6.18), together with the observed increase in clay particles, and suggests that there was some addition of mineral material to the sediments. Comparison of percentage carbon profiles (fig. 6.20 and Pennington *et al.* 1977) suggests that this was mainly restricted to the western end of the lake as there is no similar decrease in the carbon profile of 74-D. This mineral inclusion may also be responsible for the extended transition zone of change from brown to black mud in core 78-1.

Rapid changes in the chrysophyte assemblages above 22cm in 78-1 are dated as c. 1930 or 1945, by direct ^{210}Pb or by correlation to 74-D by diatom stratigraphy respectively. They reflect some environmental change towards enrichment of the lake. According to the ^{210}Pb time scale, a second, more rapid increase in accumulation rate occurred within the upper 17cm, when the rate changed from 3 to 10mm yr^{-1}, without the decrease during the 1950s suggested by the algal time scale. The diatom populations were also changing and increasing at this time, along with other algal taxa (Lund 1979). The decrease in *Cyclotella* spp. and increase in *Asterionella*, *Tabellaria*, and *Melosira* occurred during a period when Macan (1962) recorded changes in aquatic insect populations in Fordwood beck and its tributaries and related them to organic pollution. This was due to the installation of a new piped water supply in the hamlet of Outgate which caused overloading of the existing septic tanks. Nutrients may have been added to the lake in soluble form, not making any direct contribution to the sediments so that the accumulation rate was not increased.

In 1962–3 the installation of a small sewage treatment plant reduced the pollution in the stream but increased the input of mineral matter in the lake (Lund 1979). There were also increases in the use of fertilizers and detergents at this time. The effect on the lake is reflected in the appearance of *Fragilaria crotonensis* (fig. 6.18) as an important component of the phytoplankton and of *Cyclotella meneghiniana* Kützing, and in the continued increase in sediment-accumulation rate, which includes higher proportions of organic matter and phosphorus, while the potassium content remains the same indicating that there was no increase in soil input (Pennington *et al.* 1977, Pennington and Lishman 1984).

Conclusions

There is a major problem in deciding which time scale should be adopted for core 78-1, based on the foregoing results. However, a number of different variables have been analysed in both cores and all can therefore be considered in the correlations. These include changes in the diatom assemblages, the visible stratigraphic colour change from brown to black, the geochemical variables, and radio-isotope analysis. In many studies only one or two of these variables may be used for critical correlation of different sites or cores and normally radio-isotope time scales are available on only one core. Battarbee (1978) had two profiles from Lough Neagh dated by ^{210}Pb and which were found to be in good agreement with respect to the diatom assemblages but one only dated from 1900 onwards. It is only because of the extensive and intensive use of Blelham Tarn as a limnological and palaeolimnological site that such a wealth of detail is available (Pennington 1977).

The evidence in this study may be summarized as follows:

1. Geochemical variables include both an allochthonous and autochthonous component (material from outside and within the lake) and are, in the present context, the least conclusive evidence for correlation as they are best compared in terms of annual accumulation (Pennington *et al.* 1977) and therefore dependent on the accumulation rates which are in question. However, the carbon proportion rises above 10% at the stratigraphic colour change in both cores although the obvious temporary decrease at 25cm in 78-1 is hardly observable at the appropriate level in 74-D.

2. The upper part of the recent sediments in Blelham Tarn is black whereas the lower part is brown but the exact boundary of this colour change is often difficult to identify visually. Furthermore precise lack of knowledge as to how the change has been, or is being, made also allows for speculation. If the oxygen level at the mud surface is a critical factor, for example, then the colour change might occur earliest in sediments in deeper parts of the lake, due to deoxygenation, or at different times in the various basins within the

lake. At present, therefore, the synchronism of this horizon is only certain if correlated with another variable.

3. Time scales based on ^{210}Pb indicate similar deposition rates for both cores. This isotope is formed in the atmosphere and falls onto land and water surfaces but the precise process of its accumulation in lake sediments is still not entirely understood (Oldfield and Appleby, ch. 3 above). The time scale is produced by integration between measured samples, but the effect of close sample-to-sample variation is also unknown.

4. The planktonic diatoms provide one set of variables that are truly autochthonous and as these form the dominant component of the relic diatom assemblage in Blelham sediments, they provide an excellent record of any major change in that assemblage. Plankton in lakes of oceanic climate tends to be of similar composition throughout the lake due to mixing by wind and water currents. It is therefore most likely that any obvious change in the plankton would accrue to sediments throughout the lake, at the same time. The synchronism of these diatom assemblages has been amply demonstrated by the stratigraphic records of the distribution of *Stephanodiscus* valves found in different parts of Blelham Tarn (Haworth 1979).

Even allowing for some latitude in percentages, there is good correlation of several horizons of diatom changes in the two cores. The positions of the increase in percentage of *Asterionella*, in relation to both an increase in carbon and the horizon of colour change in both cores, lends weight to the suggestion that these diatom horizons are the same.

Despite the environmental changes and disturbances that have occurred throughout the post-glacial, it would appear possible to identify the beginning of the present series of changes in diatoms and chrysophytes as that period when *Cyclotella pseudostelligera* re-appeared in the assemblage analysis and *Asterionella* percentages first increased. This change probably occurred during the mid-nineteenth century, due to improvement in drainage and farm land on the catchment. Stockner's (1972) suggestion that drainage and liming was the original cause of the diatom changes is probably correct. However, a comparison shows that the assemblages only reached his postulated mesotrophic level (Araphidineae/Centric ratio of 1.0) in more recent times, at about 25cm, above the colour change in core 78-1 which is probably c. 1935. This is the level where the increase of diatom numbers, with *Asterionella* becoming a dominant 35% of the assemblage, reflects the enrichment from domestic drainage. His eutrophic A/C ratio of 2.0 has been correlated with the 1950 horizons in both cores, when the improved water supply was installed. Although the A/C ratio is indeed related to changes in this lake the numerous exceptions noted elsewhere now prevent this being regarded as a general indicator of levels of enrichment. This investigation illustrates that although a wealth of stratigraphic detail may be revealed by sampling at close intervals, correlations between different cores is not always straightforward. Nevertheless, the cross-checks provided by close correlations of several sets of variables may, as in this study, reveal the degree to which changes have occurred within different areas of a lake in responses to differences in environmental factors and also reveal the complexities surrounding the understanding of the processes by which material is brought into the lake.

Acknowledgments

The Freshwater Biological Association is grant aided by the Natural Environmental Research Council.

I am indebted to all those who have helped in fieldwork, preparation, and discussion resulting in this paper, especially Peter V. Allen for assistance and support throughout this study, Jean P. Lishman for chemical analyses, John D. Eakins and Roger S. Cambray of A.E.R.E. Harwell for ^{210}Pb and ^{137}Cs analyses, Peter G. Appleby of Liverpool University for ^{210}Pb recalculations and interpretation, Margery Fawcett for the analyses of littoral diatom assemblages, Gertrud Cronberg of Lund University, Sweden, for advice on chrysophytes, Andrew Moss and Trevor Furnass for some of the diagrams, Anne Bonny for her review of the manuscript and, finally, to Professor Winifred Tutin, F.R.S., who made the older material and analyses available to me, by whom I was trained in palaeolimnology, and with whom I have had many useful discussions.

References

Appleby, P.G. and Oldfield, F., 1978. The calculation of lead-210 dates assuming a constant rate of supply of unsupported ^{210}Pb to the sediment. *Catena 5*: 1–8.

Appleby, P.G. and Oldfield, F., 1983. The assessment of ^{210}Pb data for use in limnochronology. *Hydrobiologia 103*: 29–35.

Battarbee, R.W., 1978. Observations on the recent history of Lough Neagh and its drainage basin. *Phil. Trans. R. Soc. B. 281*: 303–435.

Bonny, A.P., 1978. The effect of pollen recruitment processes on pollen distribution over the sediment surface of a small lake in Cumbria. *J. Ecol. 66*: 385–416.

Canter, H.M. and Jaworski, G.H.M., 1978. The isolation, maintenance and host range studies of a chytrid *Rhizophydium planktonicum* Canter emend., parasitic on *Asterionella formosa* Hassall. *Ann. Bot. 42*: 967–79.

Cronberg, G. and Kristiansen, J., 1980. Synuraceae and other chrysophyceae from Småland, Sweden. *Bot. Notiser 113*: 595–618.

Evans, G.H., 1970. Pollen and diatom analyses of Late-Quaternary deposits in the Blelham Basin, North Lancashire. *New Phytol. 69*: 821–74.

Fawcett, M., 1978. A comparative study of the littoral diatoms of Blelham Tarn in the English Lake District (unpublished report, Freshwater Biological Association Library).

Harmsworth, R.V., 1968. The developmental history of Blelham Tarn (England) as shown by animal microfossils, with special reference to the Cladocera. *Ecol. Monogr. 38*: 3, 223–41.

Haworth, E.Y., 1976. The changes in the composition of the diatom assemblages found in the surface sediments of Blelham Tarn in the English Lake District during 1973. *Ann. Bot. 40*: 1195–205.

Haworth, E.Y., 1979. The distribution of a species of *Stephanodiscus* in the recent sediments of Blelham Tarn, English Lake District. *Nova Hedwigia Beiheft 64*: 395–410.

Haworth, E.Y., 1980. Comparison of continuous phytoplankton records with the diatom stratigraphy in the recent sediments of Blelham Tarn. *Limnol. Oceanogr. 25*: 1093–103.

Haworth, E.Y., 1981. A note concerning Grunow's *Stephanodiscus hantzschii*. In Florilegium Florinis Dedicatum, *Striae 14*: 191–221.

Hustedt, F., 1930–66. Die Kieselalgen Deutschlands, Österreichs und der Schweiz. In *Krytogamen-Flora*, Bd 7, ed. L. Rabenhorst (Leipzig).

Jones, J.G., 1972. Studies on freshwater micro-organisms: phosphatose activity in lakes of differing degrees of eutrophication. *J. Ecol. 60*: 777–91.

Lund, J.W.G., 1951. Contributions to our knowledge of British algae. XII A new planktonic *Cyclotella* (*C. praetermissa* n. sp.). *Hydrobiologia 3*: 93–100.

Lund, J.W.G., 1972. Preliminary observations on the use of large experimental tubes in lakes. *Verh. int. Verein. theor. angew. Limnol. 18*: 71–7.

Lund, J.W.G., 1979. Changes in the phytoplankton of an English lake, 1945–1977. *Hydrobiol. J. 14(1)*: 6–21.

Macan, T.T., 1962. Biotic factors in running water. *Schweiz. Z. Hydrol. 24*: 386–407.

Mackereth, F.J.H., 1966. Some chemical observations on post-glacial lake sediments. *Phil. Trans. R. Soc. B. 250*: 165–213.

Mackereth, F.J.H., 1969. A short core sampler for subacqueous deposits. *Limnol. Oceanogr. 14*: 1.145–51.

Meriläinen, J., 1969. The diatoms of meromictic Lake Valkiajärvi, in the Finnish Lake District. *Ann. Bot. Fennici. 6*: 77–104.

Munch, C.S., 1980. Fossil diatoms and scales of Chrysophyceae in the recent history of Hall Lake, Washington. *Freshwater Biol. 10*: 61–6.

Oldfield, F., 1970. The ecological history of Blelham Bog, National Nature Reserve. In *Studies in the Vegetational History of the British Isles*, ed. D. and R.G. West, 141–57.

Pennington, W., 1973. The recent sediments of Windermere. *Freshwater Biol:* *3*: 363–82.

Pennington, W. and Bonny, A.P., 1970. Absolute pollen diagram from the British late-glacial. *Nature, Lond. 226:* 871–3.

Pennington, W., Cambray, R.S., Eakins, J.D. and Harkness, D.D., 1976. Radionuclide dating of the recent sediments of Blelham Tarn. *Freshwater Biol. 6:* 317–31.

Pennington, W., Cranwell, P.A., Haworth, E.Y., Bonny, A.P. and Lishman, J.P. 1977. Interpreting the environmental records in the sediments of Blelham Tarn. *Rep. Freshwat. biol. Assoc. 45:* 37–47.

Pennington, W. and Lishman, J.P., 1984. The post-glacial sediments of Blelham Tarn: geochemistry and palaeoecology. *Arch. Hydrobiol. Suppl. 69:* 1–54.

Ramsbottom, A.E., 1976. Depth charts of the Cumbrian lakes. *Scient. Publ. Freshwat. biol. Assoc. 33:* 1–39.

Round, F.E., 1960. Studies on bottom-living algae in some lakes of the English Lake District. IV. The seasonal cycles of the Bacillariophyceae. *J. Ecol. 45:* 529–47.

Schmidt, A., 1873–1959. *Atlas der Diatomaceenkunde* (Leipzig).

Smol, J.P., 1980. Fossil synuracean (Chrysophyceae) scales in lake sediments, a new group of palaeoindicators. *Can. J. Bot. 58:* 458–65.

Stockner, J.G., 1972. Palaeolimnology as a means of assessing eutrophication. *Verh. Int. Verein. theor. angew. Limnol. 18:* 1018–30.

Stoermer, E.F., Kingston, J.C. and Sicko-Goad, L., 1979. The morphology and taxonomic relationships of *Stephanodiscus binderanus* var. oestrupii, A. Cl. *Proc. 5th Symposium on Recent and Fossil Diatoms*, 65–78 (Vaduz).

Takahashi, E., 1978. *Electron microscopical studies of the Synuraceae (Chrysophyceae) in Japan* (Tokyo).

Thompson, R., 1975. Long period European geomagnetic secular variation confirmed. *Geophys. J.R. astr. Soc. 43:* 847–59.

Population census returns for England and Wales. 1841–1861 (Public Record Office).

7 The preservation of algal remains in recent lake sediments

David Livingstone

Introduction

A VARIETY of preserved biogenic remains can be recovered from lake sediments, although only pollen grains and diatom frustules are commonly found and recorded. The most complete stratigraphical record is for flowering plants, since pollen grains are highly resistant to decomposition and physical damage. Palynology, however, is concerned with long-term changes in the lake catchment and not directly with those in the water body itself. The aquatic environment reflects short-term changes in addition to longer time scales and hence can provide valuable palaeo-limnological evidence of annual or seasonal fluctuations. Unfortunately, the autochthonous flora of many lakes is poorly represented in the sedimentary record and commonly restricted to the siliceous cell components of the diatoms, the Chrysophyceae; and certain highly resistant taxa. Amongst the latter, species of *Pediastrum* and *Staurastrum* are frequently recovered in lake sediments (e.g. Nipkow 1927; Birks, H.J.B., 1976) and may be found in pollen preparations (e.g. Messikomer 1938; Alhonen and Ristlluoma 1973; Kadota 1976). In certain highly organic deposits, however, a wide variety of non-siliceous algal remains has been recovered, notably from a series of lakes in the U.S.S.R. (Korde 1960, 1966). This type of sediment, known as 'sapropel', is characterized by an organic component greater than 50% and contains many identifiable remains of microscopic organisms (Korde 1960). The algae recovered from these sediments include representatives of the Chlorophyceae, Cyanophyceae, and Dinophyceae, although some sapropels consist almost entirely of a single alga (e.g. *Phacotus* or *Botryococcus*). Similar types of organic oozes have also been described from the U.S.A. and Africa by Bradley (1966) and are also dominated by autochthonous remains, particularly blue–green algae.

The sediments of a shallow man-made lake, Upton Broad, and two lakes in the Shropshire–Cheshire Plain, Ellesmere Mere and Rostherne Mere, are similar to sapropel deposits in that they contain many algal remains preserved in autochthonous oozes. The preservation of the algal

Table 7.1.　Physical characteristics of the lakes studied.

Lake	Size (km²)	Maximum depth (m)	Grid ref.	Date cores taken	
Upton Broad	0.5	1.0–1.7	TG 389134	Aug. 1978	
Ellesmere Mere	4.61	18.8	SJ 406350	May 1978	
Rostherne Mere	4.87	27.5	SJ 745843	deepwater:	March 1977
					Aug. 1977
				shallow:	June 1978
					Sept. 1978

Further descriptions of sediments are given by the following authors:

Upton Broad: Lambert and Jennings (1951); Moss *et al.* (1979).
Ellesmere Mere: Reynolds (1979),
Rostherne Mere: Reynolds (1979); Tattersall and Coward (1914); Brinkhurst and Walsh (1967); Gaskell and Eglinton (1976); Livingstone and Cambray (1978); Thompson and Eglinton (1978); Livingstone (1979).

flora in the permanent sediments of these contrasting lake types is considered and compared with other sites of algal preservation. Cores were taken with a one-metre Mackereth corer (Mackereth 1969) and cut into one-centimetre sections. Fresh sediment was diluted with glycerol/water (ratio 1:1) and aliquots taken for identification and enumeration. Details of the lakes studied are given in table 7.1.

Upton Broad, Norfolk

Upton Broad is a small shallow lake in the Bure Valley, Norfolk, which has no direct connection with the River Bure. It is fed by a slow-flowing dyke that drains farmland. The sediments are uncompacted, organic, calcareous muds underlain by peat (Lambert and Jennings 1951; Moss *et al.* 1979). The sediment cores taken in August 1978 showed three distinct layers. The topmost 20cm was a highly flocculant, pungent, bright green sediment which overlay a light brown marl deposit. The bottom of the cores (below c. 70–80cm) consisted of dark brown granular peat. The sediments above the peat layer were dominated by the remains of colonial, non gas-vacuolate, blue–green algae, *Aphanothece* spp., which retained their green pigmentation in the upper sections. Below 20cm the cells showed more degradation and the colonies became diffuse cells in a yellow–orange mucilage (fig. 7.1). Other identifiable

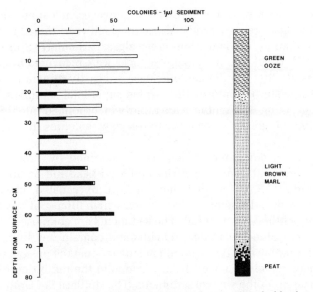

Fig. 7.1   Abundance of *Aphanothece* colonies per 1μl of fresh sediment in a core from Upton Broad. Colonies retaining their pigmentation are represented by the unshaded portion of each bar.

algal remains included *Pediastrum, Scenedesmus, Staurastrum,* and *Cosmarium*; the latter was common in the peaty deposits. Cells of *Scenedesmus* also remained undegraded and were frequently recovered from the topmost 20cm.

The flooding of the peat excavations to form the Norfolk Broads occurred around the fourteenth and fifteenth centuries AD and hence a nominal date of AD 1400 ±50 years can be assigned to the peat/lake sediment transition (Moss *et al.* 1979). Thus the remains of the *Aphanothece* throughout the core indicate that the algae have been present, living on the lake bottom, for over 500 years and their identifiable remains show long-term preservation. In recent years Upton Broad has, in common with other Norfolk Broads, shown an increase in fertility (Moss 1977). Moss *et al.* (1979) present evidence that the Broad was probably a moderately calcareous lake until around 1935, since when it has become increasingly rich, perhaps due to farm drainage. Diatom concentrations and the rate of deposition in the core taken by Moss *et al.* closely followed changes in the sedimentation rate. There has been an increase in the rate of sediment accrual from around 1935 to the 1970s. The green *Aphanothece* layer in the core probably represents only 10–15 years of the 500 or so years in which the deposits have been accumulating. The lack of degradation of the *Aphanothece* colonies in the upper 20cm is taken to indicate a very slow rate of mineralization in the sediments. However, it cannot be inferred that all the algae are preserved because in recent years there have been abundant populations of filamentous green algae, particularly *Spiro-gyra* spp. (Moss *et al.* 1979), and no remains of these were recovered from the core. It is possible that the filamentous algae are easily mineralized whilst in either the water column or the aerobic superficial sediments, whereas the bottom-living *Aphanothece* are rapidly incorporated into the sediments and the progressive overgrowth of new colonies retards decomposition.

Upton Broad can be compared to four sites examined by Bradley (1966) – Lakes Victoria and George in Africa and Mud Lake and Saddle-bags Lake in Florida, U.S.A. – in that they all have oxygenated, clear water and are fringed by a dense mat of vegetation which filters out most of the inflowing allochthonous material. The sediments are composed of gelatinous oozes from blue–green algal remains and contain few living bacteria and fungi. Bradley and Beard (1969) postulated that the preservation of *Aphanothece* remains in Mud Lake was due to the inhibition of bacterial mineralization in the surface sediments. The similarity of Upton Broad to these lakes suggests that algal preservation is the product of similar factors. The low rate of mineralization is restricted to the sediments since the non-siliceous phytoplankton of all these lakes is not preserved.

Ellesmere Mere and Rostherne Mere

The Shropshire–Cheshire Meres are a group of about 50 small, fertile lakes typically occupying hollows in thick glacial deposits (Reynolds 1979). The basins are predominantly fed by nutrient-rich ground water, having only small or no inflows. Ellesmere and Rostherne Meres are among the largest and deepest of the meres and have a similar phyto-plankton composition (Reynolds 1979). The deep-water sediments are organic with a high autochthonous component. The profundal sediments of Rostherne Mere remain anoxic throughout the year, despite the mixing and aeration of the overlying waters during winter (Brinkhurst and Walsh 1967).

Cores taken in the deepest area from both meres contained not only siliceous diatom remains but many non-siliceous algae, dominated by the remains of *Microcystis, Ceratium, Anabaena, Aphanizomenon, Pedias-trum,* and *Microcystis* (figs. 7.2 and 7.3). The remains of *Microcystis* were visually similar to those of the *Aphanothece* recovered from Upton Broad in that diffuse and degraded cells embedded in yellow/orange mucilage, were easily recognizable. The colonies in the superficial sediments were

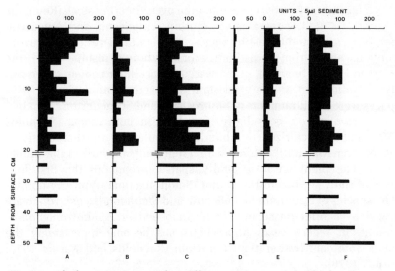

Fig. 7.2   Algal remains in a core from Ellesmere Mere per 5μl of fresh sediment: (A) *Ceratium* cysts, (B) *Microcystis* colonies, (C) *Anabaena* akinetes, (D) *Aphanizomenon* akinetes, (E) *Staurastrum* cells, (F) *Pediastrum* colonies.

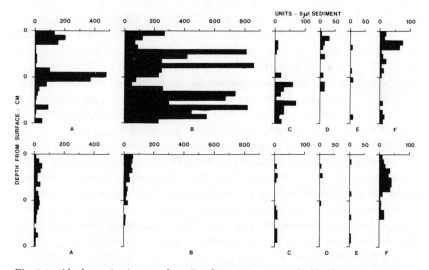

Fig. 7.3   Algal remains in cores from Rostherne Mere per 5μl of fresh sediment. The diagrams show the difference in abundance between a core taken in the profundal sediments (top) to one taken in the shallow water sediments (bottom). (A) *Ceratium* cysts, (B) *Microcystis* colonies, (C) *Anabaena* akinetes, (D) *Aphanizomenon* akinetes, (E) *Staurastrum* cells, (F) *Pediastrum* colonies.

bright green and may have represented the overwintering stock (Reynolds and Rogers 1976). The remains of *Ceratium*, *Anabaena*, and *Aphanizomenon* were restricted to resting stages. The *Ceratium* cysts were devoid of contents apart from a distinctive orange/brown residue while the blue–green algal akinetes often possessed intracellular contents. The cell walls of *Pediastrum* are very resistant and were common in the sediments, as were small numbers of *Staurastrum* and *Scenedesmus* remains.

The phytoplankton periodicity is similar in both meres (Reynolds 1978), being characterized by abundant populations of *Microcyctis* or *Ceratium* from May until autumn which represents the bulk of the annual biomass. The plankton is generally sparse throughout the first four months of the year, before the vernal diatom maxima (typically dominated by species of *Asterionella*, *Melosira*, and *Stephanodiscus*), which are followed by a brief phase in which *Anabaena* or *Aphanizomenon* is dominant. In recent years *Microcyctis* has become increasingly the summer dominant, rather than *Ceratium* (Reynolds and Rogers 1976; Reynolds 1978).

The algal stratigraphy in Rostherne Mere accurately reflects the documented phytoplankton records (Livingstone and Cambray 1978). In particular, the fluctuations between *Microcyctis* and *Ceratium* are reproduced well in the core as are several diatom maxima enabling dates to be assigned to particular horizons and layers in the core. This 'algal chronology' provided the first independent evidence for the validity of caesium-137 dating (Livingstone and Cambray 1978). The stratigraphic remains of *Anabaena* and *Aphanizomenon* do not reflect the abundance of these algae in the phytoplankton (Reynolds 1978). This may be due to the fact that neither produce akinetes annually, unlike *Ceratium* where akinete production may be essential for its continued existence (Lund 1965). The cysts also apparently require a dormancy period and thus remain in the sediments over winter (Livingstone 1979; Chapman et al. 1981). Even if *Anabaena* and *Aphanizomenon* populations sporulate, the akinetes are capable of either rapid germination within the same season (Rother and Fay 1977) or overwintering in the sediments (Rose 1934; Wildman et al. 1977), thus altering the relationship between akinetes produced and those found in the sediments. Those that remain viable in the sediments may ensure the long-term survival of the alga. The akinetes of both *Aphanizomenon* and *Anabaena* from Rostherne Mere sediments were found to be viable from horizons 18 and 64 years old respectively (Livingstone and Jaworski 1980).

Although representing the bulk of the annual biomass in both meres, the algal remains found in the sediment only represent a small number of the recorded taxa (table 7.2). Sediment traps in Rostherne Mere also failed to recover remains of such algae as *Chlorella*, *Cryptomonas*, and *Rhodo-*

Table 7.2.   Algae (excluding diatoms) recovered from the sediments of Ellesmere Mere, compared to the phytoplankton composition (Reynolds 1973).

1. *Cyanophyceae*	ANABAENA	_____	Akinetes preserved
	APHANIZOMENON	_____	Akinetes preserved
	Aphanocapsa		
	Coelosphaerium		
	MICROCYSTIS	_____	Colonies preserved
2. *Chrysophyceae*	Chrysococcus		
3. *Chlorophyceae*	Ankistrodesmus		
	Botryococcus		
	Chlorella		
	Closterium		
	Didymocystis		
	Elakatothris		
	Eudorina		
	PEDIASTRUM	_____	Colonies preserved
	Phacotus		
	Raphidonema		
	SCENEDESMUS	_____	Colonies preserved (at surface only)
	STAURASTRUM	_____	Cells preserved
	Tetraedron		
4. *Euglenophycaea*	Trachlemonas		
5. *Dinophyceae*	CERATIUM	_____	Cyst wall preserved
	Peridinium		
6. *Crytophyceae*	Cryptomonas		
	Rhodomonas		

monas (Livingstone and Reynolds 1981). Some remains of the large *Oscillatoria* crop of 1978 were found in the traps but the filaments were typically degraded and the trap recoveries did not reflect the planktonic populations (Livingstone and Reynolds 1981). It is unlikely that this alga will be preserved in the deep-water sediments.

In addition to the deep-water cores, the shallow-water sediments under c. 8m of water were also sampled. The algal remains found in these sediments that become aerobic during the year, were far fewer than those recovered from the profundal zone and did not reflect the recorded phytoplankton. They showed no evidence of long-term preservation (fig. 7.3).

Discussion

The study of algal preservation in these contrasting lake types has shown that only certain taxa are resistant to decomposition. The dominant remains recovered were colonial blue–green algae, the resting stages from filamentous blue–green algae and a dinoflagellate, and a small number of green algae, notably *Pediastrum* and *Staurastrum*. Korde (1960, 1966) records a similar list of preserved algae from Russian sapropel deposits. The similarity of Upton Broad to the lake studied by Bradley (1966) has already been mentioned. In addition to *Microcyctis* and *Aphanothece*, Korde (1966) also recovered the blue–green *Gloeocapsa*, *Gloeothece*, and *Aphanocapsa* in several sapropels. However, the remains of *Ceratium* were exclusive to the superficial sediments in sapropel whereas cysts in Rostherne Mere were identifiable in material over 100 years old (Livingstone 1979). Korde (1966) also recorded *Tetraedron*, *Phacotus*, and *Botryococcus* but these algae have not been frequently recorded in water samples from the lakes studied.

Algal oozes composed predominantly of *Botryococcus* are known in Russia, but fossil deposits are more widespread and known as 'boghead coal' (Britain), 'kerosene shale' or 'coorongite' (Australia), and 'balkhashite' (U.S.S.R.) (Blackburn and Timperely 1939, Zalessky 1914, 1926).

In known sites of algal preservation the taxa which are not recovered beyond the superficial sediments, are typically smaller unicellular algae (e.g. *Chlorella* and the delicate cryptomonads in the meres) and filamentous algae (e.g. *Spirogyra* in Upton Broad and *Oscillatoria* in Rostherne Mere). These algae appear to be greatly mineralized, some may rapidly lyse in the water column or be photo-oxidized on the surface when present as a 'bloom', while others may be rapidly consumed by herbivores. In some sediments microscopic algal remains may be absent but organic compounds can be preserved and analysed (see Philp *et al.* 1976).

Chlorophyll-degradation products have been analysed and correlated with lake fertility (e.g. Gorham 1960; Belcher and Fogg 1964), but their specific origin cannot be determined. In contrast, other compounds are highly specific, such as myxoxanthophyll to blue–green algae (Züllig 1961) and oscillaxanthin to the genus *Oscillatoria* (Brown and Colman 1963). At Esthwaite Water in the English Lake District, the permanent sediments do not contain algal remains (Livingstone 1979) but Griffiths (1978) has shown the concentration of oscillaxanthin in sediment cores closely reflects the abundance of *Oscillatoria* in the phytoplankton.

The two meres studied and Upton Broad represent two contrasting lake types. Both the meres are relatively deep and fertile with an anaerobic hypolimnion during the summer, whereas Upton Broad is also fertile, but is shallow and oxygenated. Pennington (1978) has postulated that high productivity and seasonal hypolimnetic oxygen deficiency lead to the survival of autochthonous organic matter in the permanent sediment. This is confirmed for Ellesmere Mere and, in particular, Rostherne Mere (where the profundal sediments remain anoxic throughout the year), both of which contain much autochthonous organic matter and where recognizable algal remains survive. Conversely, the shallower, oxygenated sediments of Rostherne Mere contain far fewer preserved remains and so it is inferred that the oxygen level is a critical factor in the preservation of algal remains. However, Upton Broad appears to represent a second type of environment in which algae may be preserved in the sediments, namely a shallow, clear-water lake, dominated by benthic colonial algae that form a gelatinous ooze. It is possible that the bottom-living algae may restrict the decomposition of the algae beneath, which are subsequently incorporated into the permanent sediments. This view is consistent with that of Bradley (1966) and Bradley and Beard (1969), who found very slow rates of decomposition in similar lake sediments and postulated some form of bacterial inhibition. Water bodies which fulfil the above conditions, in that they are fertile with an anoxic hypolimnion or are shallow with a benthic flora, must also contain algae which are fairly resistant to decomposition. Hence, a highly eutrophic pond, such as a sewage-oxidation pond, which may be considered as an ideal site for algal preservation, will typically contain many small Chlorophyceae that, seemingly, are rapidly mineralized and are rarely found in the sediments.

Conclusions

The sediments of Upton Broad, Rostherne Mere, and Ellesmere Mere show many similarities to other sites from the U.S.S.R., Africa, and America described in the literature. The sediments could be described as 'sapropel' deposits on the basis that they are dominated by autochthonous particles containing recognizable remains. However, Korde (1960) makes the differentiation between lake muds and sapropels at an organic content of 15%. Lake muds have less than 15% organic matter while sapropels have over 50% and deposits with 15–50% are described as 'depleted' sapropels. The organic matter of the mere sediments and the Upton Broad sediments is c. 40% and so they are 'depleted' sapropels. The comparison of algal stratigraphy in cores taken from Rostherne Mere and phytoplankton records shows that some preserved taxa can be used to give quantitative estimates of abundance in the water column. However, it cannot be concluded that an alga has not been present in the plankton by its absence in the sediments.

In the cases studied considerable information has been obtained by the examination of the sediments for non-siliceous algal remains. For example, in Upton Broad the lake has been devoid of diatoms for over 500 years (Moss *et al.* 1979), but during that time period there has been a benthic flora of *Aphanothece* present. Similarly, in Rostherne Mere the change of summer dominance between *Microcystis* and *Ceratium* has been shown to have commenced c. 1958 (Livingstone 1979), which confirms the estimate of Reynolds and Rogers (1976). The lack of *Microcystis* beyond the 1958 horizons demonstrates that the alga is only a recent dominant rather than an intermittent one – a possibility considering the fragmentary phytoplankton data available before 1962.

The mechanism of algal preservation is unknown but from both this study and others it may be inferred that the lack of oxygen in the sediments is an important factor, whether working directly or indirectly. Although not all the algal flora will be recovered from sediments that do preserve non-siliceous remains, the information provided must increase the potential palaeoecological interpretation. There are obviously many lake sediments that are dominated solely by allochthonous material but the coring of further sites with deposits comprising autochthonous oozes will probably give a more complete palaeolimnological algal record.

Acknowledgments

I would like to thank Professor W. Tutin, F.R.S., and Dr J.W.G. Lund, F.R.S., for their supervision of the research. I am also indebted to Dr E.Y. Haworth, Dr C.S. Reynolds, Mr P.V. Allen, and many other members of the Freshwater Biological Association's staff for assistance and discussion.

References

Alhonen, P. and Ristlluoma, S., 1973. On the occurrence of subfossil *Pediastrum* algae in a Flandrian core at Kirkkonummi, S. Finland. *Bull. Geol. Soc. Finland* 45: 73–7.

Belcher, J.H. and Fogg, G.E., 1964. Chlorophyll derivatives and carotenoids in the sediments of two English lakes. In *Recent Researches in the fields of Hydrosphere, Atmosphere and Nuclear Geochemistry*, ed. Y. Miyake and T. Koyama, 39–48 (Tokyo).

Birks, H.J.B., 1976. Late-Wisconsinian vegetational history at Wolf Creek, central Minnesota. *Ecol. Monogr.* 46: 395–429.

Blackburn, K.B. and Timperely, B.N., 1939. *Botryococcus* and the algal coals (parts I and II). *Trans. R. Soc. Edinb.* 58: (3), no. 29, parts 1–2.

Bradley, W.H., 1966. Tropical lakes, copropel, and oil shale. *Bull. geol. Soc. Am.* 77: 1333–8.

Bradley, W.H. and Beard, M.E., 1969. Mud Lake, Florida; its algae and alkaline brown water. *Limnol. Oceanogr.* 14: 887–97.

Brinkhurst, R.O. and Walsh, B., 1967. Rostherne Mere, England: a further instance of guanotrophy. *J. Fish. Res. Bd. Can. 24:* 1299–1309.

Brown, S.R. and Colman, B., 1963. Oscillaxanthin in lake sediments. *Limnol. Oceanogr. 8:* 352–3.

Chapman, D.V., Livingstone, D. and Dodge, J.D., 1981. An electron microscope study of the encystment and early development of the dinoflagellate *Ceratium hirundinella. Br. phycol. J. 16:* 183–94.

Gaskell, S.J. and Eglinton, G., 1976. Sterols of a contemporary lacustrine sediment. *Geochim. cosmochin. Acta. 40:* 1221–8.

Gorham, E., 1960. Chlorophyll derivatives in surface muds from the English Lakes. *Limnol. Oceanogr. 5:* 29–33.

Griffiths, M.G., 1978. Specific blue–green algal carotenoids in sediments of Esthwaite Water. *Limnol. Oceanogr. 23:* 777–84.

Kadota, S., 1976. A quantitative study of the microfossils in a 200m long core sample from Lake Biwa. In *Paleolimnology of Lake Biwa and the Japanese Pleistocene,* ed. S. Horie, 4: 297–307.

Korde, N.V., 1960. Biostratification and classification of Russian sapropels. (Akademiya Nauk S.S.R.: Moscow.) (English translation by J.E.S. Bradley, published by the British Library Lending Division, Boston Spa.)

Korde, N.V., 1966. Algal remains in lake sediments – a contribution to the development history of lakes and the surrounding regions. *Ergebn. Limnol. 3:* 1–38. (English translation by Fisheries Research Board of Canada, no. 1371.)

Lambert, J.M. and Jennings, J.N., 1951. Alluvial stratigraphy and vegetational succession in the region of the Bure valley broads. II Detailed vegetational–stratigraphical relationships. *J. Ecol. 39:* 120–48.

Livingstone, D., 1979. Algal remains in recent lake sediments (Ph.D. thesis, University of Leicester).

Livingstone, D. and Cambray, R.S., 1978. Confirmation of Cs-137 dating by algal stratigraphy in Rostherne Mere *Nature, Lond. 276:* 259–61.

Livingstone, D. and Jaworski, G.H.M., 1980. The viability of akinetes of blue–green algae recovered from the sediments of Rostherne Mere. *Br. phycol. J. 15:* 357–64.

Livingstone, D. and Reynolds, G.S., 1981. Algal sedimentation in relation to phytoplankton periodicity in Rostherne Mere. *Br. phycol. J. 16:* 195–206.

Lund, J.W.G., 1965. The ecology of the freshwater phytoplankton. *Biol. Rev. 40:* 251–93.

Mackereth, F.J.H., 1969. A short core sampler for subaqueous deposits. *Limnol. Oceanogr. 14:* 145–51.

Messikomer, V.E., 1938. Beitrage zur Kenntnis der fossilen und subfossilen Desmidiaceen. *Hedwigia 78:* 107–201.

Moss, B., 1977. Conservation problems in the Norfolk Broads and rivers of East Anglia, England – phytoplankton, boats and the causes of turbidity. *Biol. Conserv. 12:* 95–114.

Moss, B., Forrest, D.E. and Phillips, G., 1979. Eutrophication and palaeolimnology of two small mediaeval man-made lakes. *Arch. Hydrobiol. 85:* 409–25.

Nipkow, F., 1927. Über das Verhalten der Skelette planktischer Kieselalgan im geschichteten Tiefenschlamm des Zürich-und Baldeggersees. *Schweiz. Z. Hydrol. 4:* 71–120.

Pennington, W. (Mrs T.G. Tutin), 1978. Responses of some British lakes to past changes in land use on their catchments. *Verh. int. Verein. theor. angew. Limnol. 20:* 1,636–41.

Philp, R.P., Maxwell, J.R. and Eglington, G., 1976. Environmental organic geochemistry of aquatic sediments. *Sci. Prog. 63:* 521–45.

Reynolds, C.S., 1973. Phytoplankton periodicity of some north Shropshire meres. *Br. phycol. J. 8:* 301–20.

Reynolds, C.S., 1978. Notes on the phytoplankton periodicity of Rostherne Mere, Cheshire, 1967–1977. *Br. phycol. J. 13:* 329–35.

Reynolds, C.S., 1979. The limnology of the eutrophic meres of the Shropshire–Cheshire Plain: a review. *Fld Stud. 5:* 93–173.

Reynolds, C.S. and Rogers, D.A., 1976. Seasonal variations in the vertical distribution and buoyancy of *Microcystis aeruginosa* Kütz, emend. Elenkin in Rostherne Mere, England. *Hydrobiologia 48:* 17–23.

Rose, E.T., 1934. Notes on the life history of *Aphanizomenon flos-aquae. Stud. nat. Hist. Iowa Univ. 16:* 129–41.

Rother, J.A. and Fay, P., 1977. Sporulation and the development of planktonic blue–green algae in two Salopian meres. *Proc. R. Soc. B. 196:* 317–32.

Tattersall, W.N. and Coward, T.A., 1914. Faunal survey of Rostherne Mere: I Introduction and methods. *Mem. Proc. Manchr. lit. phil. Soc. 58:* 1–21.

Thompson, S. and Eglington, G., 1978. The fractionation of a recent sediment for organic geochemical analysis. *Geochim. cosmochin. Acta 42:* 199–207.

Wildman, R., Loescher, J.H. and Winger, C.L., 1975. Development and germination of akinetes of *Aphanizomenon flos-aquae. J. Phycol. 11:* 96–104.

Zalessky, M.D., 1914. On the nature of *Pila,* the yellow bodies of Boghead, and on Sapropel of the Ala-Kool of Lake Balkhash. *Bull. Comité Geol. Petersbourg 33*(248): 495–507.

Zalessky, M.D., 1926. Sur les nouvelles algues découvertes dans le sapropélogène du Lac Beloe (Hauteurs de Valdai) et sur une algue sapropélogène, *Botryococcus braunii* Kützing. *Rev. Gén. Bot. 38:* 31–42.

Züllig, H., 1961. Die Bestimmung von Myxoxanthophyll in Bohrprofilen zum Nachweis vergangener Blaualgenentfaltungen. *Verh. int. Verein. theor. angew. Limnol. 14:* 263–70.

8 Stress, strain, and stability of lacustrine ecosystems

Edward S. Deevey, Jr

Introduction

ECOLOGY'S notorious lack of theoretical substance was addressed, in plenary sessions, by the First International Congress of Ecology in 1974. Some 'unifying concepts in ecology' were brought together under that title by van Dobben and Lowe-McConnell (1975). Regarding the whole enterprise, and the 'trophic-level concept' in particular, as 'symptomatic of the sickness of modern ecology', Frank Rigler dared to notice that *concepts*, unifying or divisive, are not *theories*. 'Concepts, being general notions . . . are not predictive and are not falsifiable' (Rigler 1975: 15). To this unfalsifiable charge the editors rejoined that concepts need not be falsifiable to be *useful*.

More charitable by temperament than Rigler, I think of theoretical ecology as 'primitive' (Deevey 1969b, 1976) rather than 'sick'. Like Winifred Pennington, to whom this essay is dedicated, I am a historian of ecosystems, and have studied lake sediments without much reference to theoretical formulations. Historians are not expected to propound theories; in human history, indeed, 'unscientific' is the mildest term normally applied to a generalization of any sort. Lacustrine history is not 'mere' history, however. Introduction of the term *ontogeny* – which I may have used more freely than has Pennington, despite our common debt to W.H. Pearsall – transforms the discourse. If lakes as different as Windermere and Linsley Pond share similarities in development (Deevey 1955), that development must be at least as lawful as the succession of plant communities on land. And, if formulation of lawfulness permits prediction of the future state of a Lake District, a theory of a lake's ontogeny is at least as scientific as Keynesian economics (Keynes 1936), falsification of which is still being debated.

This paper presents nothing so ambitious as the General Theory of a lake's ontogeny. Sketching instead what Margalef (1968) called 'a context in which it is possible to speak of a theory', I first delimit the (physical) context to the lake and its basin, catchment, or *paralimnion* (confusingly called the 'watershed' by American ecologists). Also with Margalef (1968, 1969, 1975), I restrict the time-span of interest to a few millennia, the 'short-run' or 'ecological' time during which community succession occurs but macroevolutionary change does not. The objective of ecosystem theory I take to be explanatory description and prediction of a system's *trajectory* in time. As trajectories include but extend well beyond successional or other ontogenetic stages, the concept (*pace* Rigler) that most needs explication is that of *stability*. The concept – or anyway, the term – is thickly overlain by semantic difficulties (Orians 1975). As one result, much ink has been spilled, and much computer time wasted, on the naive engineers' notion that stability is intrinsic to ecosystems, as it is to various mechanical devices.

When stability is defined in the ordinary way, as resistance to perturbation, *resistance*, including the modes called *resilience* by observers, can be thought of as a property of the ecosystem (Holling 1973). Stability, however, resides neither in the 'recipient' ecosystem nor in the 'donor' environment, but comprises a class of interactions between them. Moreover, for large, complex ecosystems that are durable enough to be interesting, presence or absence of stability can be demonstrated (the concept is operational) only when environments change. Donor and recipient systems are not forbidden to exchange roles during interaction, a point that is particularly evident when the donor is lake water. It follows that no useful concept of stability can be formed while 'environment', like the surround of a *mappamundi*, is held to be an abode of dragons, yclept perturbing forces. Whereas stability of certain biotic constituents of ecosystems (populations, guilds, trophic levels, etc.) can be approached deductively (May 1981) and perhaps even measured, 'environment' is itself a complex system of problematic stability, and a deductive analysis of the wider interaction is bound to fail.

A useful alternative approach is to examine some historically known trajectories that permit some perturbations, not only to be demonstrated, but to be estimated quantitatively. By this methodology (Deevey 1969a), history is 'coaxed' to perform experiments that *stress* ecosystems with informative results, of the kind known to engineers as states of *strain*. Measuring strain turns out to be a formidable problem, as it usually is in engineering, and a quantitative theory of ecosystem stability is a challenge for the future – though not, I think, an idle dream. Anyway, for ecosystems that preserve records of their own trajectories, as lakes do, the matter of falsification takes care of itself. Any generalization that begins to resemble a theory is testable by further study of the historical record.

Unlike some philosophers of science (Peters 1976), I am untroubled by 'tautology' or 'circularity' in historical science, where most prediction happens to be 'post-diction'. As is also true in astrophysics, for example, it is not a trivial achievement to predict, successfully, 'what will be found to have occurred' (Deevey 1967).

Semantics and epistemology aside, some thorny procedural obstacles bar the way to generality in historical ecology. In this paper I grapple with some of them – some that may seem to theoreticians to have been dragged in, in flagrant disregard of parsimony – with my perverse insistence that lake and paralimnion be considered together. When limnologists focus on the lake and its biota as a dynamic system, the paralimnion is treated as a black box. In eutrophication theory (Vollenweider 1969), as long as the paralimnion's most interesting output ('nutrients') is seen as the lake's most essential input, it is unnecessary to disaggregate the box, or to consider non-nutrient outputs. For small lakes on normal hydrologic gradients, where outflow of water balances inflow and residence times are short, outflow of nutrients also balances inflow – with one qualification. Retention of certain inputs by the sedimentary sink is sometimes measured, for example in sediment traps, but is more often estimated by subtraction of outflow (Canfield and Bachmann 1981). The dynamic structure of the ecosystem can then be seen, in terms borrowed from macroeconomic theory (Richey *et al.* 1978), as a set of measurable fluxes (net flows) of materials between 'compartments', among which the sink is the least interesting.

Material flow, also known as *mineral cycling*, is *nutrient loading* in more general form. In pointing more directly to the need to disaggregate various black boxes, if models are to gain generality, it also suggests analytical procedures for doing so. Comparable advantages were seen, a generation ago, for *energy flow* (Lindeman 1942); but the 'energy content' of a 'compartment' varies with ecologists' definitions of 'energy', 'content', and 'compartment', and has proved impossible to measure without ambiguity.

I suggest, then, that a general theory of a lake's trajectory will begin, not with energy, but with material flow through ecosystems. If it comes to resemble a 'sectoral' macroeconomic model like that of Gossling (1972), I shall not be surprised (though I do not expect to understand it). Appropriate sectors or compartments are unlikely to be taxonomic categories, populations, or trophic levels, but plainly must include (1) the lake, (2) surrounding terrestrial ecosystems, and (3) the human society of the paralimnion. Because the material-moving power of the latter is enormously enhanced by technology, human residents need not be riparian to influence a lake's development in fundamental ways. Pennington (1943, 1947) was among the first to recognize that this has been true nearly everywhere since the Neolithic; many palaeolimnologists – notably Frey (1955), Davis (1976), and Deevey *et al.* (1979) – have followed her lead in measuring human disturbance of paralimnia.

Modelling the dynamic structure of ecosystems and component sectors is not the ultimate objective of theoretical ecology, however; it is only a necessary first step. Difficult as it is turning out to be, the joyful moment of its achievement will be marred by the historian's reminder that 'this, too, will pass away', for the most realistic mathematical models of dynamic systems, in economics and geophysics as in ecology, are deliberately, almost arrogantly, *ahistorical*. The steady-state asumptions, so convenient in permitting the solution of many simultaneous differential equations, are built into the models. Assuming dynamic systems to be immortal, they tend to wave aside historical questions – about ontogeny, about long-term stability, about the form and future of trajectories when states are no longer steady.

These are questions that can be answered, clearly in particular cases if not as yet with perfect generality or rigour, by direct examination of the sedimentary sink. As a sink for materials, it may be quantitatively negligible, though it is not so in the lakes where it has been studied. As a source of information it is unrivalled, and the information recorded refers, not to a countryside infested with abstract dragons, but to real, observable environments where limnologists live and work.

'Local' and 'global' stability

Trajectories of real ecosystems move in many dimensions; a three-dimensional picture, like that of fig. 8.4, collapses biological reality into two dimensions in order to emphasize time as one of three. Some two-dimensional simplifications are helpful in clarifying the concept of stability.

Several stable and unstable forms of two-population equilibrium are graphed, after Holling (1973), in fig. 8.1. In the phase or state plane defined by positive values of populations x and y, a stable node toward which most or all values of x, y converge (figs. 8.1B and 8.1F) is mechanically the simplest, and ecologically among the least probable, of stable solutions. When some values converge, as in fig. 8.1B, the stability depicted is *local*; when all values converge, as in fig. 8.1F, stability is *global*. Ecologically, we are more interested in limit cycles – dashed lines in figs. 8.1C, 8.1E and 8.2A – corresponding to oscillatory equilibria. A stable limit cycle is a closed curve, defining a *domain of attraction* (fig. 8.1C) toward which extreme values of x, y converge, and within which values either spiral inward to a node, or continue to cycle in one or several oscillatory pathways. A single, continuously oscillatory trajectory (fig. 8.1E) is commoner in physics than in population biology, where abstract dragons called *stochastic processes* tend either to damp the oscillations or to drive them beyond the domain of attraction (fig. 8.1C). Multiple equilibria with oscillations of different amplitude and period (fig. 8.1D), for which stability is *neutral*, are conceivable but biologically very unlikely.

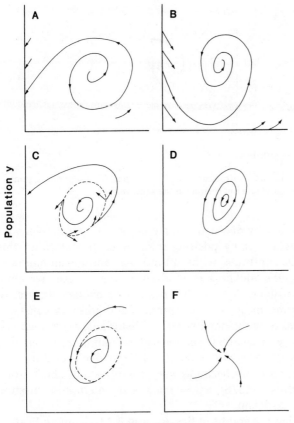

Population x

Fig. 8.1 Examples of possible behaviour of systems in a phase plane, from Holling 1973, fig. 2: (A) unstable equilibrium, (B) stable equilibrium, (C) domain of attraction, (D) neutrally stable cycles, (E) stable limit cycle, (F) stable node.

The equilibria of fig. 8.1 are copied from Holling's review; their fairly straightforward solutions can be grasped by an historian with a smattering of calculus. Injection of some ecological realism, as in fig. 8.2, leads at once to what systems modellers call 'hand-waving'. The cases shown are alternative domains of attraction, either in different regions of the state plane (fig. 8.2A) or in larger and smaller segments of the same region (fig. 8.2B). Intuitively, by extension to many dimensions, the first alternative is extremely common in ecology, and may correspond to the 'almost-intransitivity' of Lorenz (1968), by which climatologists (e.g. Mitchell 1972) attempt to model glacial and interglacial climates of the Pleistocene. No illustration so exotic is needed for the case of fig. 8.2B, for 'winter' and 'summer' populations serve well in two dimensions.

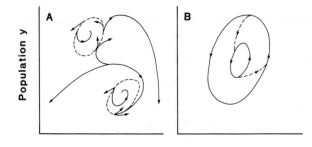

Population x

Fig. 8.2   Alternative domains of attraction in a phase plane: (A) in separate regions of the plane, (B) in larger and smaller regions of the same plane.

By adding dimensions, by distinguishing populations qualitatively, for example by their niches, or by positing transitional probabilities, this graphical game can be continued without limit. To continue *ad nauseam* is to experience Rigler's 'sickness of modern ecology'. Cases far more elaborate than those of fig. 8.2 can be formulated mathematically, of course, but the pyramid of *ad hoc* assumptions and arbitrary constants has only coincidental resemblance to reality. Historians, experimenters, and other empiricists who search for ecological validation are doomed to increasing frustration. As May (1978: 20) put it: 'once one is dealing with many interacting species, life becomes very complicated'. May's irony might be lost on Berlinski (1976), whose book is an entertaining mixture of cogent and unfair criticism.

One notices now that the graphs of figs. 8.1 and 8.2 lack any indication of scale, and 'parameterization' raises the spectre of the mathematician Liapunov (1892). According to Lewontin (1969), his is the 'usual approach' to stability, at least in classical mechanics. As described by Webster *et al.* (1975), 'a [Liapunov] system is stable if, following displacement from equilibrium, its subsequent behaviour is restricted to a bounded region of state space' [a domain of attraction]. Moreover, 'a stronger stability concept involves return to equilibrium following initial displacement'. 'Just so,' as Berlinski might remark; but the magnitude of the 'bounded region of state space' is specified, in Liapunov systems, by making displacements arbitrarily small. When perturbations are realistically large, and especially when the system of equations describing them is non-linear, Liapunov solutions lose generality, and the distinction between local and global stability cannot be maintained. In other words (Lewontin, 1969: 17), 'global stability must be defined in terms of the universe of interest'. The prospect of such a definition for n-dimensional ecosystem space, given that ecosystems are imbedded in other n-dimensional systems, is obviously negligible. At most, ecologists can hope to define some form of local stability, sufficiently general in many dimensions to confer observable (therefore falsifiable) *trajectory stability* in time. Liapunov, global, and 'asymptotic' stability, 'however relevant for classical stability analysis, may well be theoretical curiosities in ecology' (Holling 1973: 3).

Among theoretical ecologists who refuse to accept this view, several contributors to the work of Halfon (1979), including in their several chapters Siljak, Goh, Harte, Jeffries, and Webster, suggest various mathematical candidates for Liapunov functions. Empirical validation is not seriously discussed by these authors; their common pre-occupation with multi-species equilibria suggests that many modellers undervalue the distinction, drawn long ago by Tansley, between biotic communities and ecosystems.

Coming closer to a limnologist's concern with mineral cycling, Webster *et al.* (1975) applied classical stability theory to several 'real' ecosystems – well-studied examples in which perturbations of nutrient input have been measured and can be simulated. In their critique of this effort Harwell *et al.* (1977) found that the simulations cannot be duplicated on a different computer, and suggest that one source of difficulty is the imperfect analogy between material flow and behaviour of a forced mass-spring system. A stronger criticism would emphasize the falsity of any mechanical analogy, mechanical stability being the result of design for a specified environment; moreover, the analogy forced Webster *et al.* to assume linearity for the clearly non-linear equations of material flow. Concealed in the polemics of this controversy is a valuable conjecture, as noted by Watson and Loucks (1979: 360). 'Small compartments with rapid cycling of nutrients (i.e. short turnover times) contribute to system *resilience* . . . Large compartments with long turnover times . . . seem to increase the system's *resistance* to perturbation' [my italics]. Not only because this is intuitively attractive, but because one ecosystem of Webster *et al.* (1975) is 'the lake', their model deserves careful consideration. Shown in graphical form as fig. 8.3, it is discussed with other lacustrine models in the next section.

Lacustrine models

In temperate and high latitudes, and in seasonally arid tropical regions, most ecosystems change far more from winter to summer than they do from year to year. Averaged over several years, therefore, however complex the dynamic structure, a large number of its constituent reactions have rates of change demonstrably equal to zero. This innocent

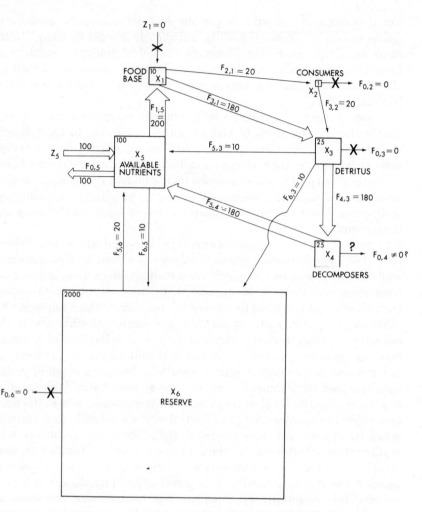

Fig. 8.3   Simulation model of a lake ecosystem, after Webster *et al.* 1975, fig. 3. Compartment sizes and flow rates are shown to the approximate scales of numerical data in *ibid.*, table 1.

observation – that the metabolism of the total system is in steady state – both validates and constrains the analytical procedures whereby differential equations can be written for innumerable short-term non-zero rates of change. If these, too, are to average out to zero, many coefficients must change both magnitude and sign from day to night and/or with the march of the seasons. Such a system is self-evidently non-linear, and homeostasis is maintained by cybernetic mechanisms (Margalef 1968, Patten and Odum 1981). Constructed on these principles, models of large ecosystems simulate reality about as well as do general circulation models of the atmosphere (Gates 1976) – which is to say 'with low resolution', 'with excessive aggregation', or simply 'not very well'. Fortunately, in considering questions of historical development and long-term stability, we are permitted to look past details of dynamic structure, in order to concentrate on the steadiness of steady states.

NUTRIENT LOADING

A lake's degree of eutrophy is among the most familiar of steady states, and one that is famously perturbed by marginal increments of certain nutrients. In the standard approach to eutrophication (Vollenweider 1968, 1969), where accelerated nutrient loading is the stress, the magnitude of the strain is a function of the *size* and *depth* of the lake, the *relative areas* of lake and of natural and disturbed upland, and, most critically, of *flushing rate*, the reciprocal of which is *residence time* (of water). Summed inputs minus summed exports (to the sediments, to the outlet, and, for some inputs, to the atmosphere) define the stationary concentration, and at steady state the loading rate is the input rate per unit area of lake. Thus, for the case of phosphorus, in the notation of Dillon and Rigler (1975),

$$\frac{d[P]}{dt} = \frac{J}{V} - \sigma[P] - \rho[P] \tag{1}$$

where [P] is concentration of phosphorus, J is the annual supply of P, V is lake volume, σ is the sedimentation-rate coefficient, and ρ is the flushing rate (per year). On integration, at steady state,

$$[P] = \frac{L}{\bar{z}(\sigma + \rho)} \tag{2}$$

where L is the loading rate (annual rate of supply per unit lake area), and $\bar{z}$ is the mean depth.

With this simple model, after measurement or estimation of L, σ, and ρ, concentrations of phosphorus and standing crops of phytoplankton are predicted with fair to excellent success in many lakes. When measurements are reasonably accurate, a major source of imprecision is insolubility, namely varying biological availability under field conditions, of oxidized forms of phosphorus. The available fraction, estimated as $71 \pm 5\%$ of total phosphorus (P_{tot}) income to the Finger Lakes (Schaffner and Oglesby 1978) is surely higher in sewage outfalls, and probably much lower for soil-borne apatites delivered directly to lakes as colluvium (Deevey *et al.* 1980).

An important practical difficulty, not yet confronted in northern lake districts, is that of measuring flushing rates of lakes without surface outlets. In the karst country of El Peten, Guatemala (Deevey *et al.* 1980), as elsewhere in the Yucatan Peninsula and in peninsular Florida, most lakes are closed, but solutes do not accumulate by evaporation as they do in arid regions. Water leaks slowly downward to the regional aquifer, giving flushing rates 10–20 times slower than expected for drainage lakes (Hughes 1974, Deevey, in prep.); but phosphorus and other conserved solutes may not leave the lake basin at the leakage rates estimated by hydrologic methods. This question is under intensive study in the Florida lake district.

Despite difficulties of application in particular cases, the Vollenweider-Dillon–Rigler model is a soundly-based paradigm for material input to lakes, and is capable of extension in several directions: (1) to the natural and disturbed terrestrial ecosystems of the paralimnion; (2) to non-nutrient materials; (3) to the dynamic structure sustained by material flow

DISTURBED UPLANDS

Given estimates of sedimentary phosphorus income to the Peten lakes (Deevey *et al.* 1979), with archaeological estimates of population densities (Rice 1978, Rice and Rice 1980) and strong indications that most soil phosphorus is anthropogenic and is delivered, slowly but quantitatively, by colluviation (Deevey and Rice 1980), it is possible to estimate per capita excretory output over two millennia of Mayan occupation. In the resulting model (Deevey *et al.* 1979: 305),

$$\dot{D} = RO_M\bar{X}_r \tag{3}$$

where $\dot{D}$ is the delivery rate of phosphorus ($=\dot{L}$, in $kg \cdot m^{-2} \cdot yr^{-1}$), $\bar{X}_r$ is mean effective (riparian) population per km^2 of lake area, O_M is Mayan output ($\sim 0.5 kg\, P$ per capita-year), and R (found to be unity for the Mayan case) is the fraction of soil-borne phosphorus eventually delivered to the lake. So far, with complete data-sets limited to two lakes, Yaxha and Sacnab, it is not possible to substitute O_s (societal output) for O_M in equation (3):

$$O_s = O_a + O_b + \ldots + O_j \tag{4}$$

where O_a is the physiological output, and $O_b \ldots O_j$ are increments from food waste, fertilizers, detergents, and so on.

Where subsistence agriculture is injected into a forested region, as in the essentially Neolithic conditions of the Maya Preclassic, one of the first results of land clearance should be a downhill flush of nutrients.

Following Iversen's classic work on *Landnam* (Iversen 1941), the point is implicit in pollen diagrams throughout the North Temperate Zone (Cushing and Wright 1967, Birks and West 1973, Tsukada 1972, etc.). Experimentally, with explicit attention to release and export of nutrients, the consequences of forest clearance are amply documented at Hubbard Brook (Likens and Davis 1975, Likens *et al.* 1977). In terms of equation (3), D is accelerated by deforestation, either by a temporary decrease in soil retention increasing R, or (equation (4)) by addition of a temporary component O_c (output due to clearance) of societal output O_s.

SILTATION

But delivery of non-nutrient materials is also accelerated, less temporarily, by disturbance. As a rough generalization, supported by soil-erosion measurements reviewed by Deevey and Rice (1980), siltation of lakes ($\dot{D}_{SiO_2}$, by analytical convention) is increased by one order of magnitude by *Landnam*, by two or three orders of magnitude by intensive agriculture, and by four or five orders of magnitude by urbanizing activities. Whereas equation (3) describes a constant per capita influence on delivery of phosphorus, as might be expected for a Neolithic economy, the high exponents found for delivery of inorganic matter express the amplified power of technology to influence material flow. Thus one might write

$$\dot{D}_{SiO_2} = R O_s X^n \qquad (5)$$

where the exponent n of population density X is characteristic of a particular socioeconomic mode or intensity of land use.

TROPHIC STATE

Because standing crops of phytoplankton are correlated, from lake to lake, with phosphorus concentrations, the nutrient-loading model predicts the annual mean of either measure with low residual variance, expressed as $1 - R^2$. (Resolution on shorter time scales, for example predicting the relative size and timing of a plankton bloom, is the familiar but distinct problem of forecasting rain on the climatologists' picnic.) When phytoplankton or its surrogate, chlorophyll concentration, is joined statistically with total phosphorus, nutrient loading predicts a combination of both, usually with lowered residual variance $1 - R^2$. What is then estimated is a multivariate *index*, bivariate in this example, to the degree of eutrophy or *trophic state*. Other, more loosely correlated variates can be added, subject to the usual criteria of significance, by principal-components analysis.

A seven-parameter index, derived from measurements on 55 Florida lakes by Shannon and Brezonik (1972a) is constructed as follows:

$$TSI = f(PP, P_{tot}, TON, Chl, Cond, T_S, Ca:Na) \tag{6}$$

TSI, the trophic-state index, being unconstrained numerically, is a sensitive if not very accurate measure of strain, a static property. Contributing variates are, respectively, primary productivity; total phosphorus, total organic nitrogen, and chlorophyll concentration; specific conductance; Secchi disc transparency; and the inverse Pearsall ratio of divalent to monovalent cations. A simpler but serviceable index (Carlson 1977) combines phosphorus, chlorophyll, and transparency.

Given evidence (Shannon and Brezonik 1972b) that TSI is highly correlated ($r^2 = c. 0.65$) with inputs of phosphorus and nitrogen,

$$TSI = f(\dot{L}) \tag{7}$$

or
$$S_T = m(TSI) = k(L) \tag{8}$$

where S_T is trophic state, or strain; $\dot{L}$ is nutrient loading (stress) as before; and f, m, and k are constants.

Among limnological ambiguities overriden by these multivariate statistics, there is a probable dichotomous relationship between transparency and other correlatives of productivity. In clear lake waters, transparency is diminished in proportion to plankton and autochthonous seston; but allochthonous organic matter, particulate or dissolved, inhibits photosynthesis and indicates, not eutrophy, but the lowered productivity associated with dystrophy. In clear and moderately coloured lakes, transparency accounts for a small proportion of the variance in TSI, and the point is of little consequence. Once organic seston becomes organic sediment, however, the meaning of a Secchi disc reading is not the issue. More troublesome is a fundamental ambiguity of provenience: the old problem of the difference between gÿttja and dy (Lundqvist 1927). Gÿttja is autochthonous and a valid measure of former productivity, whereas dy (lake peat) is allochthonous, produced in the paralimnion, and probably inimical to lacustrine productivity. As it clouds the interpretation of a lake's ontogeny as inferred from the organic content of sediments, this problem is noticed again in a later section.

Despite its inelegance, which is curable – for example, by finding a better measure of annual productivity than the sum of faulty diurnal estimates of carbon fixation by phytoplankton – a multivariate TSI is a valuable heuristic device. It is useful for historical purposes in that correlatives of trophic state are rather well-known in sediments. When rates of sedimentation are known, as by isotopic and/or palynologic dating, a static index like that of equation (6) can be cast in dynamic form, as a rate of change:

$$dS_T/dt = m \cdot d(TSI)/dt = k' \cdot dL/dt \tag{9}$$

where k' is a new constant, evaluated statistically as are its components f, m, and k in equations (7) and (8). With this model, applied to a carefully selected list of Florida lakes, we are attempting to measure changing trophic state over the last few decades of increasing (but still moderate) cultural disturbance of their basins. The project is still in its data-gathering stages, the only publication so far being that of Flannery *et al.* (1982) on the most recent sediments recovered in short cores.

Dynamic structure

A lake's trophic state is the resultant of an array of interactions comprising the ecosystem's metabolism. How far can a material-flow approach take us into the metabolic details? Pre-occupied with taxonomic, demographic, or trophic-level categories, which are subdivisions of biotic communities but not necessasrily functional components of an ecosystem, most theoretical ecologists treat materials as undifferentiated 're-sources' to be 'allocated'. This is a microeconomic view, productive in its context, but tending to evade some crucial issues called macroeconomic in other contexts. For example, if 'lakes, separated from terrestrial systems by real boundaries, tend to draw their organization from rocks and dead wood' (Deevey 1969b: 314), one needs to ask 'which rocks?', and to specify, if possible, the proportions and fates of inputs from lithosphere, atmosphere, and terrestrial biosphere. To judge from two major works previously cited (May 1981, Halfon 1979), such questions, raised a century ago by such pioneers of limnology as Forbes and Birge, and answered in part by the classical work of Forel, Pearsall, Thienemann, and Hutchinson, have escaped the notice of a whole generation of systems ecologists.

A few exceptions – mathematized models of a lake's dynamic structure as organized around inflowing materials – seem worthy of brief notice. Like all such models, they are ahistorical. Other, perhaps better examples of the modeller's art can be found in the literature, but the three chosen for mention illustrate rather different conceptual schemes, all of which may need further exploration if a full-dress macroeconomic model is ever to contain lake, upland ecosystems, and riparian society as sectors. As it happens the three models use arithmetic, calculus, and matrix algebra and can be called 'primitive', 'classic', and 'baroque', thus conforming to another historical paradigm. Of the three, that of Webster *et al.* (1975) is most seriously flawed by limnological ignorance. The WINGRA III model (Watson and Loucks 1979) is sounder, but is unduly concerned (for our purposes) with short-range prediction. The structural model of Richey *et al*. (1978) takes a giant step in the right direction, by treating carbon flow as a problem in input–output econometrics (Leontief 1966).

The 'lake ecosystem' model of Webster *et al*. is one of eight supposedly symmetrical models of nutrient cycling, and its worst misconceptions are traceable to the forced imposition of symmetry on forest, grassland, and aquatic systems. In graphical form (fig. 8.3), with compartment sizes and flow rates scaled as in the authors' table 1, it suggests at once that 'reserve' compartment X_6 is in fact standing timber, hardly an appropriate analogue of a body of lake sediments. Nutrients in that compartment are not regenerated in a century (at 20 of 2000 units per year), or indeed in tens of millennia; and if the much smaller nutrient pool at the mud–water interface regenerates phosphorus at one-fifth the external input rate ($F_{5,6} = 20$; $z_5 = X_5 = 100$, by primary assumption), the limnological community (Golterman 1977) will be greatly surprised. By assumption, this lake is airless ($z_1 = 0$) and loses no nutrients to groundwater ($f_{0,6} - 0$), flow-rates are freely manipulated to make the model 'run', apparently toward the grotesque 10-compartment parody of Cedar Bog Lake invented by Williams (1971); and there are other near-absurdities. Even if the model is purely didactic, its main conceptual weakness may be to suppose that, because large compartments that turn over slowly (like the Federal Reserve) form the visible regulatory flywheel of many dynamic systems, lakes must contain such flywheels.

But perhaps they do, and perhaps the 'reserve' compartment is neither timber nor sediments, but a melange of all relatively non-labile components of other compartments: detritus and fresh sediments refractory to microbial attack, and the bodies of living organisms. If so, we have been misled by some numbers in fig. 8.3; nutrients in such a reserve are certainly not 2000 times the nutrients in consumers (X_2), but the defect may be arithmetic, not conceptual. Disaggregation of a compartment so ill-defined would be a sterile exercise, but models are paper simulacra and are fated to be discarded.

The concept that each metabolic compartment contains a fast, labile fraction and a slow, refractory flywheel is explicit in WINGRA III, as expounded by Watson and Loucks (1979). Fair simulation of nutrient flow through Lake Wingra, Wisconsin, is achieved by disaggregation into 18 compartments (not including pondweeds): four pairs, labile and 'structural', for algae; three similarly divided pairs for dissolved, suspended, and settled organic detritus; and four consumer compartments. Two important assumptions add realism to the model: conservation of mass, and constant C:N:P ratios through time. These constraints guarantee the non-linearity and complexity, if not the fidelity to all field observations, of the system of equations describing the behaviour of 49 state variables.

As turnover times are computed and simulated on a daily basis, the model triumphantly reproduces summer conditions when given spring conditions as input. Like meteorological models, which do well with tomorrow's weather but make rude noises when asked to predict climate, lake models that aim at short-term resolution can tell us nothing about long-term stability.

Lake Wingra is one of four North American lakes intensively studied during the International Biological Programme. The four are treated together in the notably concise, penetrating account of carbon flow by Richey *et al.* (1978). Details of cycling through consumer compartments (zooplankton, insects, and fish) need not detain us, though animal ecologists will find them of enormous interest. What concerns us is the conceptual schema – the elegant approach, *via* the matrix of transfer coefficients, to the synthesis of particulate organic carbon from incoming bicarbonate and its subsequent cycling and recycling through all compartments comprising the dynamic structure. 'Particulate organic carbon' is of course a complex, but is readily disaggregated into living phytoplankton (rapidly recycled on a time scale of years) and suspended and settled refractory particles of specifiable provenience; for allochthonous components of input, dissolved and particulate, were monitored in the inlets of these lakes.

Of familiar and novel inferences that can be drawn from a structural model, Richey *et al.* (1978) devote attention to one of general significance. Expressing the proportion of cycled flow of carbon as a fraction of total carbon throughflow, the authors find this fraction to vary from 0.661 in Mirror Lake, New Hampshire, to 0.031 in Marion Lake, British Columbia. The fraction is an index to the summed magnitudes of inflow and outflow between compartments, namely to the biotic complexity supported by an inorganic nutrient. Strongly eutrophic Lake Wingra has the second-highest index, 0.572, but the ranking of indices in these four lakes is clearly related, not to eutrophy as such, but to flushing rate. Water resides in Mirror Lake for 1.0 years, and in Lake Wingra for half a year; but Findlay Lake, Washington, with renewal time seven weeks, and especially Marion Lake, renewed every five days, are literally wide places in a stream. Such lotic environments cannot be expected to sustain a structurally complex lenitic ecosystem.

This point brings us back to equation (2), where annual concentration of phosphorus is a defined function of flushing rate; so, by equations (7) and (8), is trophic state. If the proportion of cycled carbon is also a definable function, and if C:N:P ratios are either held constant by the system or vary (as they do) in measurable ways, general statements can soon be made about the relative importance in lacustrine inputs of three of the six essential constituents of the biosphere (Deevey 1970). Relative importance of C, N, and P in Lake Wingra is admirably discussed by Watson and Loucks (1979).

Ontogeny of lakes

From pre-Darwinian days the phylogenetic speculations of comparative anatomy have been tested against the facts of embryonic development. Like plant ecology, comparative limnology has long made analogous deductions, historical/developmental *process* being inferred from contemporaneous lake types mentally arranged as *stages*. So English lakes now called *oligotrophic* and *eutrophic* were thought of as *primitive* and *evolved* by Pearsall (1921). Looking beyond structural complexities, in ecosystems that are usually modelled as instantaneous slices through time, I turn to consider some developmental trajectories. I attempt no comprehensive account of palaeolimnology, the subject of several international conferences (Frey 1969, Polsk. Arch. Hydrobiol. v. 25, no. 1/2, 1978) and reviews by Frey (1974) and Wetzel (1975). Little explicit reference can be made to major contributions from the English Lake District, spanning much of the research career of Winifred Pennington (1943, 1981). The selective account that follows is intended as explication of the simplistic graphical model in fig. 8.4, and of the two processes, *eutrophication* and *paludification*, believed to generate such trajectories.

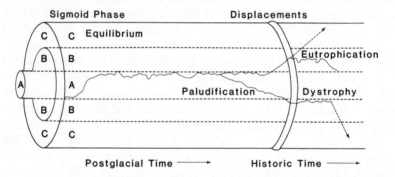

Fig. 8.4   Stability of trophic equilibria in space-time, as inferred from ontogenies of eutrophic lakes like Linsley Pond, dystrophic lakes like Cedar Bog Lake, and culturally eutrophied lakes like Lake Washington. A similar ontogeny of a temperate deciduous forest like that at Hubbard Brook is similarly graphed by Bormann and Likens 1979, figs. 6–9.

HOLOCENE EUTROPHICATION

As formulated by G.E. Hutchinson and his students (Deevey 1938, 1942; Hutchinson and Wollack 1940), the ontogenetic process observable in lake sediments is distinguished only verbally, as *typological*, from other kinds of ecological *succession*. In small lakes on newly deglaciated terrain, and relatively free of allochthonous organic inputs, lake communities 'react on the habitat', if only by dying and filling it, preparing it for other communities that replace them. Clementsian language ('ecologists' cant' to Charles Elton and many others) had largely been discarded by the 1940s, but progression from oligotrophy to eutrophy, in nondystrophic lakes called 'harmonic' by Thienemann (1931), could well have been called a *prisere*. The appropriateness of the term *climax* was also in doubt, but Linsley Pond, after a sigmoid increase in net deposition of organic matter, was said to have reached *typological equilibrium*, implying that the lake had become as eutrophic as its geologic circumstances permitted. This stage outlasted several palynologically documented forest climaxes in southern New England, and endured until the recent cultural disturbance reported by Brugam (1978a, 1978b) and Norvell (1977).

Sedimentary histories of many north European lakes, though demonstrated by the non-chemical methods of Lundqvist (1927), tended to confirm the ontogenetic deductions from comparative limnology (Lundbeck 1934, Groschopf 1936). The history of Windermere (Pennington 1943, 1947) provided a more unambiguous test of Hutchinson's thesis. When these studies were reviewed at a conference in Pallanza (Deevey 1955), many lakes were omitted from consideration for lack of pertinent data. Some well-known ontogenies were also omitted, because the history of Cedar Bog Lake (Lindeman 1941), like countless peat sections described by palynologists, raised the separable problem of the genesis of dystrophy. With certain qualifications, dictated by methodological difficulties and by geologic differences between lake districts, eutrophication is accepted today as an ecological truism.

The qualifications can be briefly listed as follows:

1. Sigmoid growth, in an ecosystem as in an organism, requires both positive and negative acceleration; as to the latter, 'biotic growth tendencies are bounded by resource availability' (Webster et al. 1975: 2). Early Holocene self-acceleration of Linsley Pond's productivity, at a time when nutrient loading was either constant or diminishing (Livingstone 1957), was supposedly generated by intensified recycling of phosphorus as increasingly reduced mud accumulated under an increasingly anoxic hypolimnion (Vallentyne and Swabey 1955). But this feedback mechanism, not re-evaluated by modern methods, may not generate logarithmic growth in the absence of stratification. In shallower lakes the nutrient feedback from pondweeds (Carpenter 1981) may promote logarithmic

increase in sedimentation, but this mechanism has not been shown to operate in an early Holocene ontogeny.

2. Replacement of an 'oligotrophic' (*Tanytarsus*) by a 'eutrophic' (*Chironomus*) benthic association – documentation of which makes Gams (1927) a progenitor of eutrophication theory – is to be expected in any moderately deep lake whose hypolimnion is gradually obliterated by sediments. Thus chironomid succession ordinarily records *morphometric* eutrophication only; *edaphic* eutrophication, conceived as a real increase in productivity, must be demonstrated by other methods.

3. Organic matter accumulates in lakes in consequence of a positive balance between production and diagenesis. Both the amount and the proportion of organic carbon so sequestered vary between lakes and between lake districts, for climatic and hydrologic reasons that have no necessary bearing on productivity. Historical changes in net deposition of biochemical moieties (such as pigments) are therefore easily overinterpreted. In particular lakes, however, sigmoid increase of organic matter is routinely corroborated by changes in numbers of identifiable fossils – organic and siliceous 'particles', many of which are known not to undergo further diagenesis after burial.

4. A high proportion of the world's lakes lie in glaciated terrain, and most of these resemble Linsley Pond in being floored by till. Hydrologic tightness of their basins is enhanced by an overlying stratum of silty clay, the fine-grained fraction of regolith resorted by the early lake. Many glacial lakes developed on sandy outwash lack the silty stratum; as in Cedar Bog Lake (and in the sandhill lakes of Florida, underlain by Pleistocene beach sand: Deevey and Brenner 1978), the section of organic lake deposits overlies coarse sand with sharp contact. The implication is that oxidative leaching has condensed the organic record to its most resistant, ligneous, and impermeable residue. At and below the contact, any record of the lake's logarithmic growth is likely to have been removed by diagenesis, assisted by molar action of the sand itself. Here, I consider the organic deposit to be chiefly autochthonous; but allochthonous peaty matter often overlies sand with sharp contact, proving that outwash plains, loess landscapes, and dune fields were paludified by expansion of dystrophic lakes.

CULTURAL EUTROPHICATION

In historical and theoretical perspective, the 'accelerated maturation' that began in many north European lakes in the Neolithic is a positive increment in rate of change of trophic state, $dS_T/dt > 0$. From equation (9) it follows, as held by limnologists generally, that among many ways in which riparian society can deflect a lake's trajectory, accelerated nutrient loading is most readily identified. On this view, the Holocene trophic equilibrium, when the rate of maturation was zero, was nudged by disturbance – specifically, by sewage or fertilizers – toward a more mature state, but cannot reach a new equilibrium while $d\dot{L}/dt$ continues positive. A decrease of S_T to a less mature state is conceivable – acid rain, heavy metals, petroleum, and peat can all have this effect – but such a process cannot be called eutrophication. The semantic restriction inheres in the meaning of *eutrophic* as 'fully nourished'.

In Clementsian language, a culturally eutrophied lake is probably *postclimax*, but this self-contradictory term compounds confusion over the nature and especially the duration of *climax*, a subject of continuing controversy among plant ecologists (McIntosh 1980, 1981). Cultural eutrophication resembles *secondary succession*, usually approached by modellers as a problem in floristics, but differs in that its substratum is by no means a *tabula rasa*, what Shelford et al. (1935) once called 'denuded water'. Curtailing a tedious discussion, I have to notice the astonishing idea, expressed by Margalef (1968) and Odum (1969), that eutrophication is not maturation, but rejuvenation. Like a flawed jewel, this misconception mars two of the most brilliant works of theoretical ecology.

Margalef's contention that 'oligotrophic lakes are certainly more mature systems than eutrophic lakes' – because their specific diversity is higher – is an obsolete deduction from a productive but faulty premise, also reflecting the ancient confusion between morphometric and edaphic eutrophy. (The large deep lakes of Europe, for example the Alpenrandseen, are 'dilute eutrophic lakes' (Deevey 1969b), and their recently accelerated maturation has begun to reduce their specific diversity). Odum also employed the argument from specific diversity, adding to it the then-fashionable idea that 'energy flow' must be maximal in early stages of succession. The confusion, here, is the easy equation between energy, information, and negentropy, and the unexamined relation between any of these abstractions and nutrient flow.

PALUDIFICATION

Sediments of Cedar Bog Lake illustrate another feature that is common on sandy glacial drift: truncated at the bottom by excessive diagenesis, its section is expanded near the top by incomplete diagenesis of allochthonous, peaty matter. Such paludification, extending to entire landscapes, began in earnest in many northern countries with the 'Sub-Atlantic climatic deterioration' of c. 600 BC. That such landscapes may be less eutrophic than the lakes and swamps they replace was recognized by Lindeman and built into his model of ecological succession. Peatlands are not restricted to high latitudes; recent expansion of Florida's cypress swamps (Watts 1969, 1971, 1975) is thought to be related, not to climatic change, but to the Holocene sea level rise that also created the Great Dismal Swamp in Virginia (Kirk 1979). Although paludification is an ecological truism to European ecologists, it is widely ignored south of the taiga belt in the United States (Heinselman 1975). Recent ecological reviews (Moore and Bellamy 1974; Good *et al.* 1978) are curiously ahistorical discussions of nutrient dynamics.

Increased accumulation of peat has two kinds of consequences for lacustrine production: (1) sequestering of nutrients in peat, an increasingly oligotrophic habitat as it thickens; (2) downhill movement of large, sometimes disastrous amounts of particulate organic matter, as indigestible by lakes as cellulose is by most animals. For ontogenetic studies, the paradoxical result is that a decrease of trophic state is signalled by an increase in net deposition of organic carbon.

Separation of allochthonous from autochthonous organic matter is therefore a methodological problem of great urgency. Allochthonous income to existing lakes can of course be monitored, as in the four I.B.P. lakes of Richey *et al.* (1978), but last year's measurements cannot be extrapolated to the early Holocene. In sediments, the non-quantitative microscopic methods of Lundqvist (1927) may distinguish dy from gÿttja, at least in extreme cases; but thousands of small, non-dystrophic lakes like Linsley Pond receive significant amounts of particulate matter from tree-leaves. No satisfactory resolution of the problem of provenience has been devised, but chemical studies of plant pigments (Sanger and Gorham 1970), alkanes (Cranwell 1973), and monocarboxylic acids (Cranwell 1974) appear to have much promise (Pennington *et al.* 1977).

CONCLUDING DISCUSSION: THE MODEL

A graphical model of trophic equilibrium and its displacements (fig. 8.4) melds the stability concept of Margalef (1969: fig. 2) with schematic lacustrine trajectories. The latter move rightward in time as does the similar but wholly inferential ontogeny of the Hubbard Brook ecosystem (Bormann and Likens 1979: figs. 6–9). The vertical coordinate, carefully left unlabelled in such diagrams, plainly represents biomass in the concept of Bormann and Likens, and may stand here for trophic state. Within a central cylinder of space–time, A, are shown a sigmoid early ontogeny, a long-lasting but irregular trophic equilibrium, and recent displacements toward, into, and through a tubular space B, from which resilient responses by the system can in principle return it to space A. By convention, upward displacements are eutrophication, while decreases in trophic state, e.g. toward dystrophy, are shown as downward displacements. Beyond B lies another tubular region, C, in which instability is unidirectional and collapse of the system is certain. Space A and some unmapped inner portion of B (the atoll of Liapunov) comprise a domain of attraction. Space C is an abode of dragons, some lineaments of which have been sketched in earlier discussion.

That cultural eutrophication is a real departure toward a new equilibrium, but is reversible on reduction of the driving stress, is shown by the recent history of Lake Washington (Edmondson 1974, 1977). If we deduce from this that stability space A is bounded by the magnitude of geochemical inputs, it seems likely that the boundaries between resilience space B and instability space C are also set by modes and rates of material flow. To say this, however, is not to quantify stability, even though the rate of return to an earlier equilibrium measures something ordinarily thought of as resilience. Eutrophied Lake Washington had a more objectionable phytoplankton in the 1960s, as Lake Erie had and still has a less desirable fishery, but neither system can be called less stable after perturbation than it was before. The ultimate limits of lacustrine stability are evidently not tested by eutrophication; presumably more can be learned about them by looking at downward displacements. Unlike the fires that terminate systems like Hubbard Brook's, paludification may be only one of several ecological processes that lead through orderly senescence to collapse.

Although such a model is too vague to be worth casting in mathematical form, it contains a clear linkage (equation (9)) between trophic equilibria and material inputs. As these inputs are the outputs of terrestrial systems (Hasler 1975), including disturbed and undisturbed segments of the Hubbard Brook forest, upland and lacustrine trajectories could rather easily be coupled in a sectoral model (Loucks 1975). Among the ten co-authors of the carbon-flow model of Richey *et al.* (1978), one can be sure that at least six limnologists have considered its validation in the ontogeny of Mirror Lake (Likens and Davis 1975). Unfortunately, no single body of sediment will separate mixed inputs from clear-cut and undisturbed forest stands upstream; and although particulate organic matter was measured in Mirror Lake's inlet, its proportionate contribution to the sedimentary sink will be difficult to estimate.

Summary

The objective of general ecosystem theory is explanatory description and prediction of a system's trajectory in time. A deductive theory of stability may be possible for biotic communities, but is held to be unattainable for ecosystems, because their long-term stability – defined as resistance to perturbation – is a class of historical interactions with 'environment', itself a complex system of problematic stability. Most theoretical ecology, including the frankly ahistorical kind called systems ecology, is preoccupied with the roles of taxonomic or trophic categories in a supposedly ongoing metabolic structure, and takes a microeconomic view of throughflowing inorganic materials as resources to be allocated. For an adequately macroeconomic approach I urge the necessity (and imminent feasibility) of input–output models representing nutrient and non-nutrient material flow through compartments of three or more coupled sectors: the lake, natural and disturbed upland ecosystems, and riparian human society.

Near the downstream end of such a system, accumulating lake sediments are both a sink for materials and a source of information about changing rates of material flow, as modulated by the perturbed dynamic structure of all interacting sectors. So interpreted, a stratigraphic history provides a test, for some sectors the only available test, of theories of a large ecosystem's trajectory.

A lake's trophic state is treated as *strain*, for which the rate of supply of nutrients is *stress*. In the context of the Mayan experiment with urbanization I show in outline how a simple nutrient-loading model can be extended to nutrient and certain non-nutrient outputs of human society, as well as into the lake's dynamic structure at steady state. In other, more northern contexts, including that of an experimentally modified upland ecosystem (Hubbard Brook, where measured export of materials is the input to Mirror Lake), one can glimpse other features of a sectoral model that couples terrestrial and aquatic trajectories. Reviewing the long-familiar Holocene ontogenies of such lakes as Linsley Pond and Cedar Bog Lake, I suggest that trophic state is stressed by two partly or sequentially opposed processes, *eutrophication* and *paludification*, which can be very difficult to separate in the sedimentary record. According to this model, the limits of a lake's stability are not tested by eutrophication; but paludification, perhaps one of several processes that reduce a lake's trophic state, can abolish the system's resilience through orderly but irreversible senescence.

Acknowledgment

Supported by the National Science Foundation, DAR-79-24812.

References

Berlinski, D., 1976. *On Systems Analysis* (Cambridge, Mass.).

Birks, H.J.B. and West, R.G. (eds), 1973. *Quaternary Plant Ecology*, British Ecol. Soc., Symp., *14* (Oxford and New York).

Bormann, F.H. and Likens, G.E., 1979. *Pattern and Process in a Forested Ecosystem* (New York).

Brugam, R.B., 1978a. Human disturbance and the historical development of Linsley Pond. *Ecology 59*: 19–36.

Brugam, R.B., 1978b. Pollen indicators of land-use changes in southern Connecticut. *Quaternary Res. 9*: 349–62.

Canfield, D.E. and Bachmann, R.W., 1981. Prediction of total phosphorus concentrations, chlorophyll *a*, and Secchi depths in natural and artificial lakes. *Can. J. Fish. Aquat. Sci. 38*: 414–23.

Carlson, R.E., 1977. A trophic state index for lakes: *Limnol. Oceanogr. 22*: 361–9.

Carpenter, S.R., 1981. Submersed vegetation: an internal factor in lake ecosystem succession. *Am. Nat. 118*: 372–83.

Cranwell, P.A., 1973. Chain-length distribution of *n*-alkanes from lake sediments in relation to post-glacial environmental change. *Freshwater Biol. 3*: 259–65.

Cranwell, P.A., 1974. Monocarboxylic acids in lake sediments: indicators, derived from terrestrial and aquatic biota, of paleoenvironmental trophic levels. *Chem. Geol. 14*: 1–14.

Cushing, E.J. and Wright, H.E., 1967. *Quaternary Paleoecology* (Proc. INQUA Cong., 7th).

Davis, M.B., 1976. Erosion rates and land use history in southern Michigan. *Environ. Conserv. 3*: 139–48.

Deevey, E.S., 1938. Typological succession in Connecticut lakes (Ph.D. thesis, Yale University).

Deevey, E.S., 1942. Studies on Connecticut lake sediments. III. The biostratonomy of Linsley Pond. *Am. J. Sci. 240*: 233–64, 313–38.

Deevey, E.S., 1955. The obliteration of the hypolimnion. *Mem. Ist. Ital. Idrobiol. Suppl. 8*: 9–38.

Deevey, E.S., 1967. The reply: letter from Birnam Wood. *Yale Rev. 56*: 631–40.

Deevey, E.S., 1969a. Coaxing history to conduct experiments. *BioScience 19*: 40–3.

Deevey, E.S., 1969b. Review of Margalef, R., *Perspectives in ecological theory. Limnol. Oceanogr. 9*: 1–11.

Deevey, E.S., 1970. Mineral cycles. *Sci. Amer. 223* (Sept. 1970), 148–58.

Deevey, E.S., 1976. [Introduction] *Air, Water, and Land: the Global Ecosystem* (San Francisco).

Deevey, E.S. and Brenner, M., 1978. Sedimentary history of Spanish Pond; Appendix E. In Harris, L.D., *El Pantano de los Españoles* (U.S. Nat. Park Serv., Contract no. CX 5000041666).

Deevey, E.S. and Rice, D.S., 1980. Coluviación y retención de nutrientes en el distrito lacustre del Petén central, Guatemala. *Biotica 5.3*: 129–44.

Deevey, E.S., Rice, D.S., Rice, P.M., Vaughan, H.H., Brenner, M. and Flannery,

M.S., 1979. Mayan urbanism: impact on a tropical karst environment. *Science* 206: 298–306.

Deevey, E.S., Brenner, Mark, Flannery, M.S. and Yezdani, G.H., 1980. Lakes Yaxha and Sacnab, Peten, Guatemala: limnology and hydrology. *Arch. Hydrobiol. Suppl.* 57: 419–60.

Dillon, P.J. and Rigler, F.H., 1975. A simple method for predicting the capacity of a lake for development based on lake trophic status. *J. Fish. Res. Bd. Can.* 32: 1519–31.

Edmondson, W.T., 1974. The sedimentary record of the eutrophication of Lake Washington (Proc. U.S. Nat. Acad. Sci., 71: 5093–5).

Edmondson, W.T., 1977. The recovery of Lake Washington from eutrophication. In *Recovery and Restoration of Damaged Ecosystems*, ed. J. Cairns, K.L. Dickson and E.E. Herricks, 102–9 (Virginia).

Flannery, M.S., Snodgrass, R.D. and Whitmore, T.J., 1982. Deepwater sediments and trophic conditions in Florida lakes. In *Interactions between Sediments and Fresh Waters, 1981. Hydrobiologia* 91/92: 597–602.

Frey, D.G., 1955. Längsee: a history of meromixis. *Memorie Ist. Ital. Idrobiol.*, Suppl. 8: 141–64.

Frey, D.G. (ed.), 1969. *Symposium on paleolimnology. Mitt. int. Verein. theor. angew. Limnol.* 17.

Frey, D.G., 1974. Paleolimnology. *Mitt. int. Verein. theor. angew. Limnol.* 20: 95–123.

Gams, H., 1927. Die Geschichte der Lunzer Seen, Moore, und Wälder. *Int. Rev. ges. Hydrobiol. Hydrogr.* 18: 304–87.

Gates, W.L., 1976. Modelling the Ice-Age climate. *Science* 191: 1138–44.

Golterman, H.L. (ed.), 1977. *Interactions between Sediments and Fresh Waters* (The Hague and Wageningen).

Good, R.E., Whigham, D.F., Simpson, R.L. and Jackson, C.G. (eds), 1978. *Freshwater Wetlands. Ecological Processes and Management* (New York).

Gossling, W.F., 1972. *Productivity Trends in a Sectoral Macro-economic Model.*

Groschopf, P., 1936. Die postglaziale Entwicklung des Grossen Plöner See in Ostholstein auf Grund pollenanalytischer Sedimentuntersuchungen. *Arch. Hydrobiol.* 30: 1–84.

Halfon, Efraim (ed.), 1979. *Theoretical Systems Ecology: Advances and Case Studies* (New York).

Harwell, M.A., Cropper, W.P. and Ragsdale, H.L., 1977. Nutrient recycling and stability: a re-evaluation. *Ecology* 58: 660–6.

Hasler, A.D. (ed.), 1975. *Coupling of Land and Water Systems* (New York).

Heinselman, M.L., 1975. Boreal peatlands in relation to environment. In *Coupling of Land and Water Systems* ed. A.D. Hasler, 93–103 (New York).

Holling, C.S., 1973. Resilience and stability of ecological systems. *Ann. Rev. Ecol. Syst.* 4: 1–23.

Hughes, G.H., 1974. Water balance of Lake Kerr – a deductive study of a landlocked lake in north-central Florida. *Fla. Bur. Geol., Rep. Invest.* no. 73.

Hutchinson, G.E. and Wollack, A., 1940. Studies on Connecticut lake sediments. II. Chemical analysis of a core from Linsley Pond, North Branford. *Am. J. Sci.* 238: 493–517.

Iversen, J., 1941. Landnam i Danmarks Stenalder. *Danm. Geol. Unders. ser.* 2.

Keynes, J.M., 1936. *The General Theory of Employment, Interest, and Money.*

Kirk, P.W. (ed.), 1979. *The Great Dismal Swamp* (Charlottesville, Va.).

Leontief, W.W., 1966. *Input–output Economics* (New York and Oxford).

Lewontin, R.C., 1969. The meaning of stability. In *Diversity and Stability in Ecological Systems*, 13–24. Brookhaven Symp. Biol., no. 22; BNL-50175 (C-56); Springfield, Va., NTIS, TID-4500).

Liapunov, M.A., 1982. *Problème Générale de la Stabilité du Mouvement*, repr. as *Annals of Mathematical study*, no. 17 (Princeton).

Likens, G.E. and Davis, M.B., 1975. Post-glacial history of Mirror Lake and its watershed in New Hampshire U.S.A.: an initial report. *Verh. int. Verein. theor. angew. Limnol. 19:* 982–93.

Likens, G.E., Bormann, F.H., Pierce, R.S., Eaton, J.S. and Johnson, N.M., 1977. *Biogeochemistry of a Forested Ecosystem* (New York).

Lindeman, R.L., 1941. The developmental history of Cedar Bog Lake. *Am. Midl. Nat. 26:* 101–12.

Lindeman, R.L., 1942. The trophic-dynamic aspect of ecology. *Ecology 23:* 399–418.

Livingstone, D.A., 1957. On the sigmoid growth phase in the history of Linsley Pond. *Am. J. Sci. 255:* 364–73.

Lorenz, E.N., 1968. Climatic determinism. In *Causes of Climatic Change* (Meteorol. Monogr. 8, no. 30), 1–3.

Loucks, O.L., 1975, Models linking land-water interactions around Lake Wingra, Wisconsin. In *Coupling of Land and Water Systems*, ed. A.D. Hasler, 53–63 (New York).

Lundbeck, J., 1934. Uber den 'primär oligotrophen' Seetypus und den Wollingster See als dessen mitteleuropäischen Vertreter. *Arch. Hydrobiol. 27:* 221–50.

Lundqvist, G., 1927. *Bodenablagerungen und Entwicklungstypen der Seen*. In *Die Binnengewässer 2*, ed. A. Thienemann (Stuttgart).

Margalef, R., 1968. *Perspectives in Ecological Theory* (Chicago).

Margalef, R., 1969. Diversity and stability: a practical proposal and a model of interdependence. In *Diversity and Stability in Ecological Systems*, 25–37 (Brookhaven Symp. Biol., no. 22; BNL-50175 (C-56); Springfield, Va., NTIS, TID-4500).

Margalef, R., 1975. Diversity, stability, and maturity in natural ecosystems. In *Unifying Concepts in Ecology*, ed. W.H. van Dobben and R.H. Lowe-McConnell, 151–60 (The Hague).

May, R.M., 1978. Factors controlling the stability and breakdown of ecosystems. In *The Breakdown and Restoration of Ecosystems* ed. M.W. Holdgate and M.J. Woodman, 11–25, (NATO Conf. Ser. 3: New York).

May, R.M., 1981. *Theoretical Ecology: Principles and Applications* (2nd edn, Sunderland, Mass.).

McIntosh, R.P., 1980. The relationship between succession and the recovery process in ecosystems. In *The Recovery Process in Damaged Ecosystems*, ed. J. Cairns, 11–62 (Ann Arbor).

McIntosh, R.P., 1981. Succession and ecological theory. In *Forest Succession: Concepts and Application*, ed. D.C. West, H.H. Shugart and D.B. Botkin, 10–23 (New York).

Mitchell, J.M., 1972. The natural breakdown of the present interglacial and its possible intervention by human activities. *Quaternary Res. 2*: 436–45.

Moore, P.D. and Bellamy, D.J., 1974. *Peatlands.*

Norvell, W.A., 1977. Rapid changes in the composition of Linsley and Cedar Ponds (North Branford, Connecticut). *Arch. Hydrobiol. 80*: 286–96.

Odum, E.P., 1969. The strategy of ecosystem development. *Science 164*: 262–70.

Orians, G.H., 1975. Diversity, stability, and maturity in natural ecosystems. In *Unifying Concepts in Ecology*, ed. W.H. van Dobben and R.H. Lowe-McConnell, 139–50 (The Hague).

Patten, B.C. and Odum, E.P., 1981. The cybernetic nature of ecosystems. *Am. Nat. 118*: 886–95.

Pearsall, W.H., 1921. The development of vegetation in the English Lakes, considered in relation to the general evolution of glacial lakes and rock basins. *Proc. R. Soc. B. 92*: 259–84.

Pennington, W., 1943. Lake sediments: the bottom deposits of the north basin of Windermere, with special reference to the diatom succession. *New Phytol. 42: 1 27.*

Pennington, W., 1947. Studies of the post-glacial history of British vegetation. VII. Lake sediments: pollen diagrams from the bottom deposits of the north basin of Windermere. *Phil Trans. R. Soc. B. 233*: 137–75.

Pennington, W., 1981. Records of a lake's life in time: the sediments. *Hydrobiologia 79*: 197–219.

Pennington, W., Cranwell, P.A., Haworth, E.Y., Bonny, A.P. and Lishman, J.P., 1977. Interpreting the environmental record in the sediments of Blelham Tarn. *Rep. Freshwat. biol. Assoc. 45*: 37–47.

Peters, R.H., 1976. Tautology in evolution and ecology. *Am. Nat. 110*: 1–12.

Rice, D.S., 1978. Population growth and subsistence alternatives in a tropical lacustrine environment. In *Pre-Hispanic Maya agriculture*, ed. P.D. Harrison and B.L. Turner, 35–61 (Albuquerque).

Rice, D.S. and Rice, P.M., 1980. The northeast Peten revisited. *Am. Antiq. 45*: 432–54.

Richey, J.E., Wissmar, R.C., Devol, A.H., Liken, G.E., Eaton, J.S., Wetzel, R.G., Odum, W.E., Johnson, N.M., Loucks, O.L., Prentki, R.T. and Rich, P.H., 1978. Carbon flow in four lake ecosystems: a structural approach. *Science 202*: 1183–6.

Rigler, F.H., 1975. The concept of energy flow and nutrient flow between trophic levels. In *Unifying Concepts in Ecology*, ed. W.H. van Dobbin and R.H. Lowe-McConnell, 15–26 (The Hague).

Sanger, J.E. and Gorham, E., 1970. The diversity of pigments in lake sediments and its ecological significance. *Limnol. Oceanogr. 15*: 59–69.

Schaffner, W.R. and Oglesby, R.T., 1978. Phosphorus loadings to lakes and some of their responses. Part I, A new calculation of phosphorus loading and its application to 13 New York lakes. *Limnol. Oceanogr. 23*: 120–34.

Shannon, E.E. and Brezonik, P.L., 1972a. Limnological characteristics of north and central Florida lakes; *Limnol. Oceanogr. 17*: 97–110.

Shannon, E.E. and Brezonik, P.L., 1972b. Relationships between lake trophic state and nitrogen and phosphorus loading rates. *Environ. Sci. Technol. 6*: 719–25.

Shelford, V.E., Weese, A.O., Rice, L., Rasmussen, D.I., McLean, A. and
Markos, H.C., 1935. Some marine biotic communities of the Pacific coast of
North America. *Ecol. Monogr.* 5: 249–354.

Thienemann, A., 1931. Tropische Seen und Seetypenlehre. *Arch. Hydrobiol.
Suppl.* 9: 205–31.

Tsukada, Matsuo, 1972. The history of Lake Nojiri, Japan. In *Growth by
Intussusception: Ecological Essays in Honor of G. Evelyn Hutchinson*, ed.
E.S. Deevey, 337–65; *Connecticut Acad. Arts Sci. Trans.* 44.

Vallentyne, J.R. and Swabey, Y.S., 1955. A reinvestigation of the history of
Lower Linsley Pond, Connecticut. *Am. J. Sci.* 253: 313–40.

van Dobben, W.H., and Lowe-McConnell, R.H. (eds), 1975. *Unifying Concepts
in Ecology* (The Hague and Wageningen).

Vollenweider, R.A., 1968. *Recherches sur l'aménagement de l'eau;* (OECD:
Paris. DAS/CSI/68.27).

Vollenweider, R.A., 1969. Möglichkeiten und Grenzen elementarer Modelle
der Stoffbilanz von Seen. *Arch. Hydrobiol.* 66: 1–36.

Watson, V. and Loucks, O.L., 1979. An analysis of turnover times in a lake
ecosystem and some implications for system properties. In *Theoretical
Systems Ecology: Advances and Case Studies*, ed. E. Halfon, 355–83 (New
York).

Watts, W.A., 1969. A pollen diagram from Mud Lake, Marion County,
north-central Florida. *Bull. geol. Soc. Am.* 80: 631–42, 1 pl.

Watts, W.A., 1971. Postglacial and interglacial vegetation history of southern
Georgia and central Florida. *Ecology* 52: 676–89.

Watts, W.A., 1975. A late Quaternary record of vegetation from Lake Annie,
south-central Florida. *Geology* 3: 344–6.

Webster, J.R., Waide, J.B. and Patten, B.C., 1975. Nutrient recycling and the
stability of ecosystems. In *Mineral Cycling in South-eastern Ecosystems*,
ed. F.G. Howell, J.B. Gentry and M.H. Smith, 1–27 (U.S. ERDA, UC-11;
Springfield, Va., NTIS, CONF-740513).

Wetzel, R.G., 1975. *Limnology* (Philadelphia).

Williams, R.B., 1971. Computer simulation of energy flow in Cedar Bog Lake,
Minnesota based on the classical studies of Lindeman. In *Systems Analysis
and Simulation in Ecology*, v. 1, ed. B.C. Patten, 543–82 (New York).

9 Pollen recruitment to the sediments of an enclosed lake in Shropshire, England

Anne P. Bonny and Peter V. Allen

Introduction

INVESTIGATION of the way in which pollen assemblages are formed in deposits of peat and lake sediment can provide information of potential use when the fossil pollen record is interpreted in terms of vegetation history. The experimental studies initiated and guided by Professor Winifred Tutin for the Freshwater Biological Association, for example, have served to emphasize the importance – first recognized by Pennington (1964) – of the streamborne supply of pollen to lakes in the English Lake District. The input of airborne pollen directly to streamfed lakes is also significant, however, although rates of deposition from the air appear to be generally low in relation to rates of pollen deposition at the sediment surface. This may be due partly to the fact that, previously, the aerial pollen influx to lakes has been measured at more than 30m offshore, which may be beyond the trajectory of much of the local pollen emitted from marginal vegetation. Since rates of airborne pollen deposition in vegetation on land are now known to be substantially greater than in the offshore region of lakes, it seems highly probable that there is a steep gradient in the intensity of pollen input away from the shore of a lake.

The rate of aerial pollen supply to the littoral water surface may be important in determining both the absolute abundance of pollen in a lake sediment and its percentage composition. The investigation detailed here was concerned primarily to compare absolute rates of pollen input from the air at different distances from a lake shore, and to assess the extent to which local pollen supplied to the inshore water surface is circulated subsequently within the lake. To this end, traps for airborne pollen were put out along a transect from the shore to the centre of a lake, where a submerged trap was also installed. Pollen catches were monitored every few weeks for 15 months. The lake chosen is fed by groundwater rather than by surface drainage so there was no streamborne pollen component to consider. The extent to which pollen airborne from various sources becomes integrated in this enclosed lake was investigated by comparing pollen spectra from different points over the mud surface.

Crose Mere and its environs

THE LAKE

Crose Mere (national grid reference SJ 430305) lies 19km north of Shrewsbury in a group of meres which occupy hollows and kettle-holes in the Devensian drift of the region (see Reynolds 1979). Most of these lakes are maintained by groundwater and receive little or no surface drainage (Gorham 1957). Crose Mere appears to be fed primarily by subsurface flow through the gravels and clayey sands of the catchment which overlie impermeable till. Only small amounts of surface water are supplied, in particularly wet weather, *via* three ditches on the southern shore, or by direct precipitation on the lake, which is 15.2ha in area. A small channel at the eastern end conducts outflowing water some 500m to Sweat Mere (fig. 9.1), which drains *via* ditches to the River Roden.

Evidence (in Hardy 1939) suggests that Crose Mere, Sweat Mere, and another mere now represented by Whattall Moss (fig. 9.1) once formed one large lake. Drainage operations in 1864 are reputed to have lowered Crose Mere by two or three metres to its present level of 87m OD. Its previous extent is marked by a belt of peaty land around the perimeter, except along the north-eastern shore which abuts the gravel ridge now separating Crose Mere from the Whattall basin to the north.

The bathymetric survey of Reynolds (1973a) shows Crose Mere to be a simple basin with a maximum depth of 9.2m, to which the northern shore shelves quite steeply underwater. The eastern end of the mere is most subject to disturbance by wave action and the littoral sediments here include areas of sand, gravel, and stones. The less-exposed shores accumulate coarse-detritus mud in the littoral region. Fine sediment in deeper water is a highly-organic, black, thixotropic ooze of low mineral content.

VEGETATIONAL SURROUNDINGS

A detailed account of the vegetation around Crose Mere and Sweat Mere is given by Sinker (1962).

Emergent plants are almost absent from the exposed shores, but elsewhere there is a fringe of reedswamp, up to 10m wide, which is developed best at the sheltered, western end of the lake. This grades into a species-rich fen community on land, and then into *Alnus glutinosa* carr. Along parts of the shore this complete hydrosere sequence is telescoped or attenuated towards the lake, or passes on land into short fen meadow grazed by cattle. A fenced nature reserve along the southern shore (fig. 9.1) has been protected from grazing, however, and this area supports a

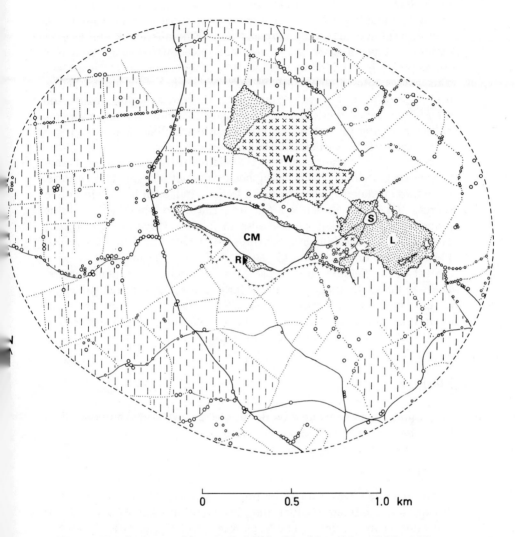

Fig. 9.1 Distribution map of vegetation within 1 km of Crose Mere. CM Crose Mere;
L Lloyds Wood; R nature reserve; S Sweat Mere; W Whattall Moss.
Stipple = deciduous woodland; cross-hatched = conifer plantation; vertical pecked
lines = arable land; blank = grassland; circles = mature wayside and hedgerow
trees; dotted lines = field boundaries; unbroken lines = roads, except for drainage
channels, where arrows mark the direction of flow. T-shaped symbols around Crose
Mere mark the escarpment which delimits the probable former extent of the lake.

luxuriant growth of tall herbs in the alder-carr zone. The reserve lies in the path of the prevailing wind from the south-west, so some of the pollen emitted by plants growing here is likely to be carried out over the mere.

Within 100m of the mere *Quercus*, *Ulmus*, *Fraxinus*, and *Pinus* occur as mature, isolated trees in the surrounding meadows. A major source of *Alnus* pollen is provided by the extensive alder carr around Sweat Mere. This grades into birch woodland with occasional oaks on the better-drained acid peat of Lloyd's Wood (fig. 9.1). Young conifers growing on the *Molinia* fen at the eastern end of Crose Mere are not yet producing pollen, but older plantations on Whattall Moss include mature stands of *Pinus sylvestris* as well as tall birch woodland.

An attempt was made to assess the relative incidence of the different tree genera growing by roads and in hedgerows within 1km of Crose Mere, since these isolated specimens, open to the wind, form important pollen sources. Of those shown on fig. 9.1, 235 were identified, and the proportions of different genera were as follows. *Quercus* 61%, *Ulmus* 11%, *Acer* 8%, *Aesculus* 6%, *Fraxinus* 6%, *Fagus* 2%, *Betula* 1%, *Pinus* 1%; *Carpinus*, *Cedrus*, *Populus*, *Taxus*, *Tilia*, and *Tsuga* all <1%.

Most of the fields around Crose Mere are under permanent pasture. Those to the south are grown chiefly for hay or silage and are usually cut in June, before the meadow grasses are in full flower, so the grassland around the mere is relatively unproductive of pollen. Arable land within 1km (fig. 9.1) is used mainly for cereal crops.

Methods

Procedures used are detailed by Bonny (1972, 1976, 1978) and described briefly below with notes on additional methods.

TRAPPING POLLEN

Airborne pollen was caught in traps constructed to Tauber's (1974) design, but modified so that the lids could be removed easily for collection of the contents. The traps were primed with 10% glycerol to prevent drying-out. Formaldehyde was added to stop the growth of micro-organisms. Each floating trap was mounted on a platform made from a block of expanded polystyrene clad in thin sheets of aluminium and covered with PVC tape. The traps were anchored on a south to north transect at 5m, 15m, 25m, 50m, and c. 150m from the southern shore. One further trap was mounted at a height of 1m on a post in the alder carr about 5m from the lake margin. Crose Mere supports a large population of water birds which contaminated some pollen catches with faeces. This problem was reduced, with the minimum of aerodynamic interference, by attaching some upward-pointing strands of copper wire to the Tauber traps to discourage roosting. The traps were installed in May 1978 and were sampled at intervals of a few weeks until August 1979.

The air trap farthest from the shore was anchored in the deepest part of the mere. A submerged trap was also installed here at 5.5m depth to catch pollen and other suspended particulate matter. The submerged trap, designed by Pennington (1974), consisted of a pair of straight-sided cylinders suspended from a vertical rope which was supported by a floating buoy and anchored by a concrete block. Normally this whole assembly is hauled up when the traps are emptied, but the surface sediments of Crose Mere are so soft that a block tends to sink into them. Lifting such an anchor would have resulted in local disturbance of the mud surface and contamination of the water column at the trapping station with particles of bottom sediment including settled pollen. A modified trap assembly was therefore designed to obviate this problem (fig. 9.2). With this system the traps are suspended from a second rope in parallel with the anchor-rope. The two are prevented from tangling by stainless steel shackles which are passed through the trap-rope and around the anchor-rope. This allows the trap-rope to be raised or lowered relative to the anchor-rope. At the surface the trap-rope is attached directly to a buoy which also supports a spring-loaded, jointed 'Jumar' clip of the type used in mountaineering. The anchor-rope passes through the clip and is held firmly in its jaws. Care must be taken when fitting the clip to the anchor-rope, however, as the tension created by the buoy, applied incorrectly, will open the jaws and release the rope. With the clip fitted correctly, the jaws of the clip can be opened by applying downward pressure to the arm of the clip, against the movement of the spring, and the anchor-rope will then run freely. The slack portion of the anchor-rope above the clip is buoyed up by corks. It has been found best during sampling to run the clip about 1m up the anchor-rope so that the buoy, with both ropes attached, can be secured inside the boat. Then the trap-rope is hauled in and made fast when the traps rise just clear of the water at the side of the boat, at which point they can be emptied conveniently. The reverse procedure re-instals the equipment.

COLLECTION OF SURFACE SEDIMENT AND MINICORES

Surface sediment was obtained for analysis from 32 points in the mere. Samples from six stations along the trap transect were collected by aqualung divers, and the rest were obtained with a small Ekman grab. The sediment, scooped into wide-mouthed jars, consisted of approximately the topmost 5cm of deposit plus some overlying water.

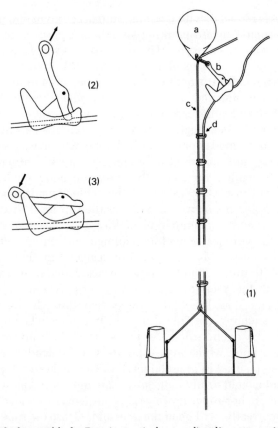

Fig. 9.2   Modified assembly for Pennington (submerged) sediment trap. (1) a = buoy; b = Jumar clip; c = trap-rope; d = anchor-rope, which leads to a concrete block (not shown), (2) clip in closed position, holding anchor-rope firmly, and (3) clip open, allowing anchor-rope to pass through the jaws.

A short core of sediment, which included the mud/water interface, was taken with a minicorer (Mackereth 1969) at two stations at the eastern end of the mere.

PREPARATION OF SAMPLES FOR POLLEN ANALYSIS

Trap samples were prepared for pollen counting by the method which involves addition of a known number of exotic 'marker' grains (Bonny 1972, 1976).

Samples of surface sediment were kept in a cold room at 5°C for two to three weeks after collection, by which time the material was assumed to have settled to a degree of compaction resembling the natural state. Free water was then removed with a pipette and the samples were stirred. Subsamples of 1cm³ were taken and prepared for pollen counting as in Bonny (1972). Pre-treatment with 8% HCl was required to remove particulate CaCO₃, and the samples were also simmered in 40% HF for as long as was necessary to remove most of the siliceous matter. Measured subsamples from the minicores were prepared in the same way. Some grab samples from the eastern end of the mere were found to contain an intractable sand fraction. These were decanted into fresh centrifuge tubes after HF treatment, leaving behind the largest mineral particles. This sand residue was inspected at ×20 magnification and was found to contain little or no visible organic matter, so it was concluded that no significant loss of pollen was likely to have been incurred by splitting the sample in this way.

POLLEN COUNTING AND IDENTIFICATION

Identification of pollen grains followed Pennington *et al.* (1972). A total of 500 determinable native pollen grains (TDP) was counted for each submerged trap sample and sample of surface sediment. Minicore counts were of 100 upland arboreal pollen (excluding *Alnus* and *Corylus*). The number of native grains in the sample was estimated from the proportion of exotic pollen encountered, as in Bonny (1972). Pollen counts of air trap samples varied according to the numbers of pollen present. The proportion of the total air-trap sample counted was 2.3% on average, but the actual proportions ranged from 0.3% (a large sample due to massive input of *Alnus* pollen to a trap in the nature reserve), to 3.8% (a small sample collected during a period of low input to the trap in mid-lake).

Results

POLLEN INPUT TO THE AIR TRAPS

Seasonal pollen input to the floating traps. Pollen catches in all the floating traps reflected the phenology of flowering, as is illustrated by results from the traps in mid-lake (fig. 9.3). The input of most tree pollen was over by early summer, and remained thereafter at a low level until leaf fall in November. Catches of non-tree pollen were greatest during the summer months. High winds in autumn and winter appear to have brought about some reflotation of grains first released earlier in the year so that the input of certain taxa (e.g. *Betula*, Gramineae) persisted from one flowering season to the next.

Annual pollen input to the floating traps. The total number of pollen grains caught annually cm⁻² of trap mouth decreased sharply along the offshore transect up to a distance of 50m from the lake margin (table 9.1).

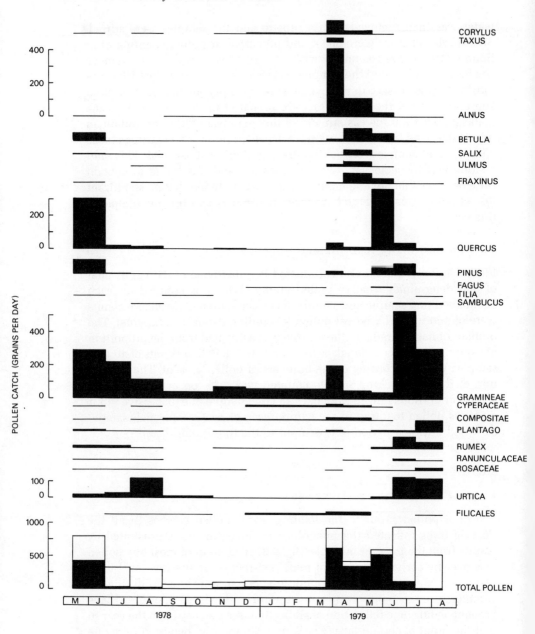

Fig. 9.3   Pollen catches of individual taxa in traps floating in mid-lake, May 1978 to August 1979. Catches shown are the mean number of grains per day in the traps at 50m and c. 150m offshore. The histogram for total pollen is divided to show catches of tree and shrub pollen (black) and of herbaceous pollen (blank).

Table 9.1. Pollen caught cm^{-2} in floating Tauber traps, 9 May 1978–16 May 1979.

Trap no.	1	2	3	4	5
Distance from shore	5m	15m	25m	50m	c. 150m
Alnus	7145	2565	1397[1]	666	830
Betula	487	256	281	293	306
Fraxinus	170	212	88	138	100
Pinus	328	223	251	255	250
Populus	17	47	12	3	6
Quercus	714	495	415	743	790
Taxus	46	22	35	12	40
Tilia	28	12	5	17	9
Ulmus	57	69	77	67	58
Other tree pollen	0	15	16	12	11
Corylus	171	94	164	174	128
Cupressaceae	12	0	3	6	0
Salix	130	171	37	56	86
Sambucus	25	21	9	0	5
Other shrub pollen	0	10	0	0	3
Artemisia	0	21	3	0	9
Calluna	12	9	10	11	5
Chenopodiaceae	0	5	27	12	8
Compositae undiff.	347	66	116[1]	116	101
Cruciferae	162	10	0	8	14
Cyperaceae	44	33	24	87	73
Filipendula	186	53	42	22	10
Gramineae	3628	1988	2454	2133	1941
Labiatae	16	0	3	0	0
Leguminosae	53	3	6	12	11
Plantago	56	19	36	58	40
Ranunculaceae	87	18	18	16	5
Rosaceae undiff.	41	23	16	9	9
Rumex	119	103	61	80	79
Umbelliferae	22	6	9	5	13
Urtica	1155	694	540	351	409
Other herb pollen	27	5	31	11	11
Pollen not identified	27	11	12	9	17
Total determinable pollen	15312	7279	6198	5382	5377
Total Filicales	102	43	58	139	94
Sphagnum spores	2	2	8	6	13

1. Includes an interpolated estimate, based on nearest trap results, for a sample which contained an anther of this taxon.

It seems that most of the local pollen was deposited within this range since results from the traps at 50m and 150m offshore were very similar. The gradient in pollen deposition was steepest for those taxa with a strong source in the nature reserve (e.g. *Alnus, Salix, Sambucus, Filipendula, Urtica*), input being highest to the traps at 5m or 15m from the margin of the lake. Catches of pollen taxa with extra-local sources at some tens or hundreds of metres from the lake (e.g. *Pinus, Quercus, Taxus, Tilia, Ulmus, Corylus, Calluna*) were distributed more evenly along the transect, although the input of some of these taxa was highest to the trap close inshore. Enhanced pollen deposition in this region may have been due to the reduction of wind speeds by the belt of fen, and to the reflotation of pollen impacted on plants exposed to rain splash at the lake's edge.

Pollen-input gradients across the littoral water surface. In order to investigate more thoroughly the evident sharp decrease in the intensity of pollen deposition across the littoral water surface, an additional trap (S) was installed, on a post 5m inside the nature reserve, to monitor the strength of local pollen deposition at source. This trap was put out at the start of the reflotation season (November–March) in 1978 and was left in position until the end of the main flowering season (March–August) in 1979.

Results showed that total pollen deposition decreased with distance from the nature reserve during both the reflotation season and the main flowering period. The rate of decrease was similar during both periods, input to the trap at 5m offshore being equivalent to less than 30% of the input to trap S (table 9.2). The absolute numbers of pollen deposited along the transect were far higher, however, during the flowering season than during the reflotation period.

Results for the whole period November 1978 to August 1979 showed that catches of those individual pollen taxa which were present locally in

Table 9.2. Absolute and relative catches of total determinable pollen along the trap transect, 1978–9.

	Reflotation period (1 Nov. 1978–21 Mar. 1979)		Flowering period (21 Mar. 1979–13 Aug. 1979)	
	catch cm^{-2}	% of catch S	catch cm^{-2}	% of catch S
Trap S (source)	8246	(100)	31494	(100)
Trap 1 (5m)	1295	15.7	8322	26.4
Trap 2 (15m)	1370	16.6	6898	21.9
Trap 3 (25m)	1689	20.5	5391	17.1
Trap 4 (50m)[1]	800	9.7	5545	17.6

1. Distances of traps offshore.

the reserve decreased most sharply along the transect. Of the tall pollen emitters in this group, results for *Alnus*, a strong source, were most marked: a plot of catches along the transect formed a hyperbolic curve (fig. 9.4). The very high catch of *Alnus* pollen in trap S is attributed to

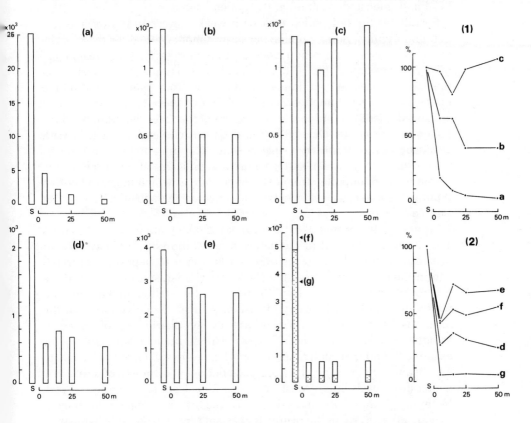

Fig. 9.4 Pollen input cm^{-2} to the Tauber traps in the nature reserve (S) and to those floating at 5m, 15m, 25m, and 50m offshore, for the period 1 November 1978 to 13 August 1979. The pattern of pollen input from tall sources is illustrated by (a) *Alnus*, (b) *Betula* plus *Fraxinus*, *Salix* and *Sambucus*, and (c) other non-local tree and shrub pollen; pollen input from sources of medium height is illustrated by (d) *Urtica*, (e) Gramineae, (f) other anemophilous NAP, and (g) insect-pollinated NAP. Line diagrams (1) and (2) show catches of pollen taxa a–g in the floating traps as percentages of the respective catches in trap S at 'source' in the reserve.

heavy deposition from overhanging branches of alder which protruded into the glade where the trap was sited. The results of Andersen (1970, 1974) from moss polsters and from Tauber traps similarly indicate that much of the pollen from woodland trees falls – as single grains or as clumps – nearly vertically to the ground, either as a dry deposit or in rainsplash droplets, and that horizontal pollen drift is relatively unimportant where wind speeds are low. The strip of nature reserve at Crose Mere is more open to the wind than is Andersen's woodland, but the sharply-falling curve for *Alnus* pollen dispersal suggests, nevertheless, local and intense deposition at the pollen source.

Results for the tall pollen taxa with weak sources in the reserve (i.e. *Betula, Fraxinus, Salix, Sambucus*) show a more gradual decrease in the intensity of pollen deposition along the trap transect (fig. 9.4). Presumably this is because none of these taxa was adjacent to trap S and deposition at this point depended on lateral pollen drift from sources within the reserve, but at some metres distance. Input of these taxa to trap S was enhanced, therefore, but to a far lower extent than was the input of *Alnus* pollen, for example. Generally, the dispersal curves for tall taxa with weak sources in the reserve suggest some dissemination of local pollen across the littoral water surface plus a significant addition, along the whole transect, of pollen from extra-local sources which increased pollen input to the traps farthest from the shore relative to deposition in trap S at 'source'.

Input of pollen from tall extra-local or regional plant taxa, namely trees and shrubs not present in the nature reserve, did not show a consistent decrease in intensity along the trap transect (fig. 9.4). Deposition in trap S and in the trap farthest from the shore was similar, suggesting that these catches resulted primarily from the fairly uniform fallout of 'background' pollen derived from a variety of distant sources. Some variation in the results is attributed to the natural but uneven distribution of pollen in turbulent air passing over the traps. Grosse-Brauckmann (1978), for example, found that such random variations were sufficient to give rise to significantly different results from traps put out at distances of only a few metres from each other. Most of the non-local pollen arriving at Crose Mere must, therefore, have been reasonably well mixed in transit, although the 'noise' in the results for certain taxa, e.g. *Populus* (table 9.1), suggests some inhomogeneities in the pollen content of air passing over the traps.

Of the results for wind-pollinated taxa of medium height (1–2m), those for *Urtica*, a strong local pollen source throughout the reserve, showed the highest pollen input to trap S as compared with input to the floating traps (fig. 9.4). Results for Gramineae, a weaker local source, and for other anemophilous NAP followed much the same pattern, but catches in the floating traps were, comparatively, higher than for *Urtica*, suggesting a greater addition of extra-local pollen from sources outside the reserve. In contrast, results for insect-pollinated NAP of medium height indicated very high input to trap S in relation to the floating traps. The form of this dispersal curve is undoubtedly related to the catch of insects, especially flies contaminated with pollen, which was much higher in trap S among food plants on shore than in the floating traps.

The relative pollen dispersal curves, in which catches in the floating traps are expressed as proportions of the catch at 'source' in trap S, show marked differences between taxa (fig. 9.4). The curve for *Alnus* pollen, for example, falls almost immediately to less than 20% of the value at source, indicating that a very high proportion of pollen from some plants – even tall emitters which are open to the wind – may be deposited close to the point of origin. Such intense, localized deposition is likely to result in part from the pollen being released from the anthers as clumps rather than as single grains which will be carried further by air. The dehiscence mechanism and the degree of stickiness of the pollen will therefore affect dispersal, as will the height of pollen emission, the extent of exposure to wind and absolute wind strength. Hence the gentle curve for *Urtica* (in contrast with the sharp curve for sticky, insect-pollinated NAP) indicates relatively efficient dispersal – which might perhaps be expected in view of the explosive mode of anther dehiscence in this taxon. Efficient dispersal also means that a higher proportion of pollen from one particular source will reach other areas as part of the non-local pollen component: as local influence decreases, extra-local and regional influences increase, as is demonstrated for, for example, by the relative uniformity along the transect of pollen input from non-local tree taxa in relation to input from strong local pollen sources in the reserve.

Possible effects of trap efficiency. In this interpretation of results from the air traps it has been assumed that all the traps along the transect were operating with similar efficiency. This assumption may not be wholly justified since the efficiency of the Tauber trap falls with increasing wind speed (Tauber 1974: fig. 3), and the wind velocity in the nature reserve was probably lower, at any given time, than the free wind velocity over the lake. It may be, therefore, that the high catches in the reserve trap reflected relatively efficient pollen collection at low wind speeds, whereas smaller catches from the floating traps were due partly to reduced collection efficiency at the higher wind speeds over the lake. The most marked reduction in free wind velocity in the reserve would be expected to occur when the plants are in full leaf, so the difference between wind speed in the reserve and over the lake should be greater during the flowering season than during the winter when deciduous vegetation offers little resistance to the wind. This seasonal variation was not apparent in the results, however, since the rate of decrease in total pollen deposition with distance from the reserve was similar during the flowering and reflotation periods. It seems, therefore, that any inequalities of trap efficiency which arose from differences in wind velocity did not materially affect the pattern of pollen deposition along the transect.

A second factor which may not have affected all the traps equally is the 'rainsplash error' which Krzywinski (1977) claims may augment the catches in a Tauber trap. He found that some of the pollen which settles on the collar of the trap in dry weather may be splashed into the mouth by droplet impact when rain falls. Traps sited under trees may also be subject to the addition of pollen in this manner if drips from vegetation fall on the collar: Tauber (1977) regards the latter as a greater potential hazard. No conclusive evidence of rainsplash error was found, however, when the Tauber traps at Crose Mere were paired for a period of five months with simple cylindrical vessels without collars, although the possibility that the catches had been augmented to a small extent could not be ruled out (Bonny and Allen 1983). However, the relative uniformity of catches along the transect of non-local tree pollen, for example, suggests that the effects of rainsplash, if any, did not introduce serious errors into the results.

COMPARISON OF RESULTS FROM AIR TRAPS AND SUBMERGED TRAP

The total pollen catch in the air traps at 50m and 150m offshore averaged 5400 grains cm^{-2} for the year May 1978 to May 1979, a value only one-third to one-half of that calculated by Beales (1980) for the current annual rate of pollen sedimentation at the mud surface. Previous work (Bonny 1978) suggests that the Tauber trap probably underestimates by about 50% the actual pollen flux to the water surface at wind speeds around those recorded near Crose Mere ($2cm\,s^{-1}$ on average), so it would seem that aerial input to mid-lake could account for most of the pollen being deposited at the central mud surface. However, the comparatively high input of pollen to the submerged trap (about 77 000 grains cm^{-2} during the year) suggests that there is considerable circulation of pollen within the water column before sedimentation occurs. The pollen increment which comes permanently to rest on the central mud surface, therefore, is not necessarily composed only of pollen supplied by air to the central water surface: indeed, the composition of the catches in the mid-lake air traps and in the submerged trap differed in a way which suggested that much of the pollen in the water column at the centre of the lake could have been circulated from the littoral region.

Results from the air-trap transect show that input of *Alnus* pollen, for example, fell to background level within 25m of the lake shore, and yet substantial numbers of *Alnus* grains were caught by the submerged trap during most sampling periods (fig. 9.5). Underwater catches were particularly high during and just after the *Alnus* flowering season in spring, suggesting that there was immediate circulation of pollen by water from the inshore zone where deposition from local alders was very high at this

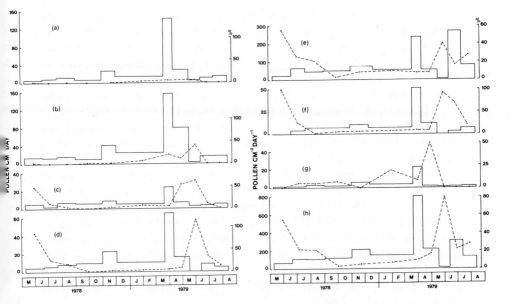

Fig. 9.5   Input of selected pollen taxa to the air traps and submerged trap in mid-lake. Histogram blocks (left-hand scale) show the numbers of 'extra' pollen caught in the submerged trap. The dashed line (right-hand scale) indicates the proportion of the catch in the submerged trap which could be accounted for by apparent aerial input of pollen to mid-lake (calculated as the mean of pollen catches in the air traps at 50m and c. 150m offshore, multiplied by two to compensate approximately for underestimation by the Tauber trap of the actual pollen flux to the water surface – see text.) Key to taxa: (a) *Corylus*, (b) *Alnus*, (c) *Betula*, (d) *Quercus*, (e) Gramineae, (f) *Pinus*, (g) Filicales, (h) total determinable pollen.

time of year. However, underwater catches of pollen taxa not present in marginal vegetation (e.g. *Corylus*, *Quercus*, *Pinus*) followed a similar pattern despite the fact that the aerial input of these pollen types was no higher near the shore than in mid-lake (fig. 9.5). In addition, the spring peak for underwater catches of some of these taxa (e.g. *Pinus*, *Quercus*) preceded their flowering periods in May–June, so their availability in March–April should have been at a minimum. These inconsistencies suggest that the supply of pollen to the submerged trap was not only being controlled by the circulation of freshly-produced pollen within the water column, but also by the resuspension of out-of-season pollen from the sediment surface. This explanation is supported by the fact that, outside the main flowering season, peak catches in the submerged trap coincided with periods when water turbulence was at a maximum and most likely to disturb the mud surface.

High input to the submerged trap in November 1978, for example, can best be ascribed to the resuspension of pollen during the autumn overturn. The succeeding winter was relatively severe in Britain and Crose Mere was frozen over for several weeks early in 1979. It may be, therefore, that the peak of pollen input to the submerged trap in March–April was due partly to resuspension brought about by a spring overturn of the type more usually associated with continental climates (cf. Davis 1968). In addition, there was a period of high winds just after the ice melted in early March which would have added to any turbulence induced already by thermal instability in the water column. The size of the underwater pollen catch must have reflected trap design to some extent since open-topped cylinders of the type used are known to over-estimate the rate of particle sedimentation in moving water, but provide a good estimate of the actual downward flux of particles through calm water (Pennington 1974; Gardner 1980a, 1980b). Somewhat higher pollen input to the submerged trap would be expected to occur, therefore, at times of greatest water turbulence in Crose Mere (at the autumn overturn and in early spring) than during calmer periods (under ice cover and throughout the period of partial or complete thermal stratification in late spring and summer).

The effects of trap design are likely to have been only partly responsible, however, for the very high catch of pollen underwater in March–April 1979. This must have been due also to a real increase in the rate of sedimentation as much of the circulating particulate matter in the lake settled out at the end of the spring diatom maximum in April. The observations of Reynolds (1973b) show that a peak in the populations of *Asterionella formosa* Hass. and *Fragilaria crotonensis* Kitton occurs in Crose Mere some time between February and May, the exact timing depending each year on several environmental factors. The high concentration of these diatoms in the water column is depleted rapidly by sinking, however, as soon as turbulence is reduced at the onset of stratification, and thereafter other algae usually predominate. In 1971, for example, *Asterionella* cells in the water column diminished from $3000ml^{-1}$ to $30ml^{-1}$ between mid-March and mid-April as stratification became established (Reynolds 1973b: fig. 5).

It was clear from examination of the raw seston collected by the submerged trap during the spring of 1978 that the usual algal succession was taking place. Seston caught during the period 13 December 1978 to 21 March 1979 was olive-brown in colour and contained a mixture of diatoms with the taxa *Asterionella*, *Fragilaria*, and *Stephanodiscus* predominating. Many of the cells had partly-decayed chloroplasts, indicating that some of the material caught was not fresh but had been resuspended during the autumn and winter – as the high proportion of out-of-season pollen also suggested. Seston trapped between 21 March and 11 April was green in colour and contained an abundance of fresh-looking *Asterionella* and *Fragilaria*, with some *Stephanodiscus* and *Cyclotella* types. By the next visit, 16 May, diatoms no longer predominated in the plankton and the water of the mere was turbid with a bloom of blue–green algae. Stratification was established by this time with the metalimnion at a depth of about 2m.

In a highly productive lake like Crose Mere, it is probable that pollen grains do not, in the main, sink through the water as separate entities, but as components of aggregations of particles. Microscopic examination of floccules of seston from the submerged trap in March–April, for example, showed that particles of diverse origin (e.g. mineral, algal, fungal and insect fragments and pollen) had become coagulated and sedimented in a sticky matrix of colloid produced by bacteria and lysing algal cells. The sedimentation rate of these floccules is likely to have been higher than that of individual pollen grains because of the incorporation into the floccules of materials of high density such as mineral particles and diatoms with a heavy siliceous ornament. It seems, therefore, that water movements in this lake will influence the seasonal dynamics of pollen sedimentation in two ways: directly by resuspending bottom sediment containing pollen, which is circulated and subsequently settles out under calm water conditions, and indirectly through the effects of turbulence and stratification on the growth and maintenance of algal populations which, when senescent, entrap grains as they descend through the water column so that pollen sedimentation is enhanced.

RESULTS OF ANALYSES OF SURFACE SEDIMENT

Distribution of pollen percentages. The spatial distribution of major pollen taxa over the mud surface was assessed by comparing the percentage pollen composition of sediment samples from 32 locations in the mere (fig. 9.6). Two independent multivariate numerical methods were used in this comparison: principal components analysis (PCA) and minimum-variance cluster analysis (MINVAR) (see Birks *et al.* 1975; Bonny 1976). Only taxa which occurred with a frequency of at least 2% in one or more samples were included in the analyses. The original counts of these pollen and spore taxa were recalculated as percentages of the total included taxa before the analyses were carried out.

Results obtained by MINVAR are displayed as a dendrogram (fig. 9.7) which shows the samples linked in an hierarchy according to their similarity in composition. Most are linked at a low level into one of three groups (A, B, D), but two samples (C and E) are not linked until a relatively high level. Inspection of the recalculated pollen percentages of the samples (plotted on fig. 9.7 in the order indicated by the analysis)

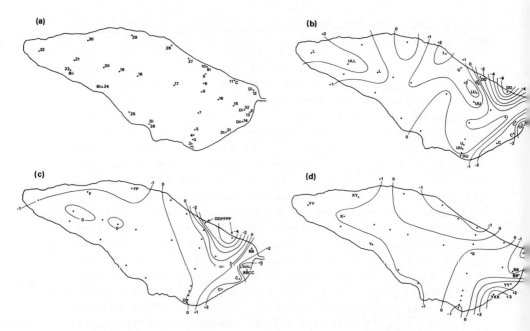

Fig. 9.6   (a) Locations of surface sediment samples in Crose Mere. Letters Bi–E
indicate groups to which samples were allocated by MINVAR analysis; all
unlettered samples belong to group A. (b)–(d) Maps of principal component scores
of individual samples on the first (b), second (c), and third (d) principal
components. The letters denote samples which have percentages of certain pollen
taxa more than one (single letter) or two (double letter) standard deviations greater
than the mean percentage for all samples. Key to taxa: B, *Betula*; C, *Corylus*; D,
Degraded; F, Filicales; G, Gramineae; L, *Plantago*; P, *Pinus*; U, *Urtica*; X, *Fraxinus*;
Y, Cyperaceae.

suggests some reasons for this grouping. The 22 samples in groups A and
B are all fairly similar in composition, but group B samples clearly have
lower percentages of Gramineae pollen and higher percentages of *Pinus*
and degraded pollen (Bi) or *Alnus* pollen (Bii) than group A samples.
Sample C differs from all others in having the minimum percentage of
Gramineae pollen and the maximum percentages of *Pinus* and degraded
pollen for the set. Sample C and group D samples all have higher
percentages of *Corylus* pollen than groups A and B, combined with higher
percentages of *Betula* and/or *Salix* and lower percentages of Gramineae
(Di), or a minimum value for *Alnus* pollen (Dii). Sample E has the
maximum percentage of *Corylus* pollen for the set. Fig. 9.6 illustrates the
spatial distribution of these groups of samples over the mud surface of
Crose Mere. It is evident from this map that proportions of the taxa
included in the analysis are fairly similar over much of the mud surface

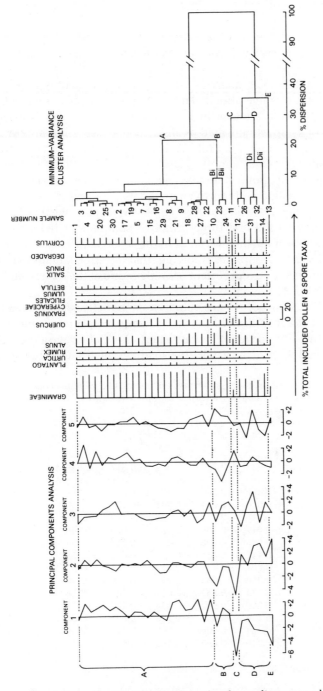

Fig. 9.7   A comparison of multivariate analyses of surface sediment sample data from Crose Mere. PCA sample scores on the first five principal components are shown on the left of the diagram, the samples being arranged in the order suggested by the MINVAR dendrogram on the right; pollen percentages of the 15 taxa used in the analyses are shown in the centre of the diagram.

(i.e. group A samples), but that sediment of markedly different composition characterizes certain littoral areas, especially at the eastern end of the lake.

Results of the PCA analysis, shown as component scores for the samples on the first five principal components, are also plotted on fig. 9.7. At first sight there appears to be little or no agreement between the groupings of samples indicated by the two analyses, since almost all the sample groups delimited by MINVAR include, within each group, both positive and negative sample scores on each of the five principal components. Only group B displays some consistency in that all sample scores on the second and fourth components are negative, and scores on the fifth component are all positive. However, maps of the distribution of sample scores on the first three principal components (which together account for 55% of the total variance in the set of data) agree with the map of the MINVAR groups in showing that steep gradients in pollen composition over the mud surface are only found at the eastern end of the mere (fig. 9.6).

Correlation of the occurrence of individual pollen taxa in the samples is provided by the loadings of the taxa on the principal components (table 9.3). On the first principal component the pollen taxa Gramineae, *Plantago*, and *Urtica* have positive loadings of 0.3 or more, whereas

Table 9.3. Loadings of major pollen taxa on the first five principal components of a principal components analysis of surface sediment sample data from Crose Mere.

	Component				
	1	2	3	4	5
Alnus	0.238	−0.127	0.069	−0.317	0.549
Betula	−0.185	0.410	−0.414	−0.052	0.018
Corylus	−0.382	0.304	0.084	0.082	−0.070
Cyperaceae	0.009	−0.004	0.698	−0.171	−0.238
Degraded	−0.315	−0.371	−0.126	0.149	0.082
Filicales	−0.017	−0.489	0.030	0.026	−0.048
Fraxinus	0.089	0.258	0.405	0.282	0.303
Gramineae	0.411	0.031	−0.088	0.034	−0.158
Pinus	−0.291	−0.411	−0.017	0.117	0.081
Plantago	0.358	0.083	0.050	0.110	−0.315
Quercus	0.174	−0.287	−0.011	0.120	−0.408
Rumex	0.269	−0.002	−0.234	0.427	0.170
Salix	−0.195	0.097	−0.126	−0.032	−0.456
Ulmus	−0.184	0.106	0.198	0.724	0.038
Urtica	0.316	0.005	−0.168	0.097	−0.029
% total variance	27.2	17.9	9.9	8.1	7.7
Cumulative % of total variance	27.2	45.1	55.1	63.2	70.9

Corylus and degraded pollen have negative loadings of 0.3 or more. Hence these two groups of taxa vary inversely between samples: where percentages of Gramineae, *Plantago*, and *Urtica* are high, those of *Corylus* and degraded pollen are low, and *vice versa*. Similarly, on the second principal component there is covariance between *Corylus* and *Betula*, which both have positive loadings of 0.3 or more, and *Pinus*, degraded pollen, and Filicales spores, which all have negative loadings of 0.3 or more. On the third principal component Cyperaceae and *Fraxinus* pollen have positive loadings of 0.3 or more, and so vary inversely with *Betula*, which has a loading of −0.4.

Maps of the sample scores on the first three principal components show clearly the influence of the loadings of individual taxa (fig. 9.6). The patterns of contours mapped cannot be explained entirely on the basis of current aerial pollen input to the mere, although localized input from reedswamp sedges probably accounts for the high proportions of Cyperaceae pollen which influence sample scores on the third principal component. Similarly, the high incidence of *Urtica* pollen along the south–north transect, evident in the sample scores on the first principal component, must reflect local dispersal by nettles in the nature reserve from which, as the trap results show, pollen is borne out over the lake by winds from the south and south-west. Prevailing winds from this quarter also prevent the accumulation of fine sediment along the north-west shore. Here, high percentages of degraded pollen are found, presumably as a result of constant abrasion of grains by mineral particles in the zone of wave action (fig. 9.6). The high proportion of Filicales spores in this area is probably a consequence of the destruction of pollen grains by mechanical attrition, to which the spores are more resistant. *Pinus* pollen percentages are also higher in this region than elsewhere, but this is probably because the grains tend to float and so may be driven towards the north-west by the prevailing winds – possibly while entrapped in scum and decaying algal blooms which become concentrated along this part of the shore by wind-driven surface currents.

The high percentages of *Corylus*, *Betula*, and *Fraxinus* pollen which influence strongly the scores of samples from the eastern end of the mere (fig. 9.6) cannot be due to modern local input of these pollen taxa as no parent plants are present in the adjacent fen vegetation. Birch and ash grow on the drier soils of Lloyd's Wood behind the fen (fig. 9.1), but the nearest sources of *Corylus* pollen are occasional hazel bushes along roadsides well away from the mere. The sediment exposed at the eastern end of Crose Mere is probably not contemporary, therefore, but was deposited at a time when *Corylus*, *Betula*, and *Fraxinus* were still growing locally, perhaps in woodland which has since been cleared. To test this hypothesis, two short cores of sediment were obtained from stations 13 and 32 and analysed to determine the underlying pollen stratigraphy.

Analyses of sediment cores. Pollen diagrams from the short cores are shown in fig. 9.8. Spectra from the base of the longer profile (50cm depth) are dominated by tree pollen, indicating that sediment at this level in the core was deposited when the vegetation around Crose Mere was chiefly woodland. The subsequent increase in the proportion of NAP, and the appearance of Linaceae and *Humulus/Cannabis* pollen types at 40cm depth, suggests the gradual clearance of woodland and increasing arable use of the fertile soils around the mere. Percentages of *Pinus* pollen, which increase towards the top of the profile, may reflect the establishment of local conifer plantations in recent centuries. Pollen spectra from the shorter profile indicate a similar sequence of events.

The spectra from the upper part of these short cores show a resemblance to those from Zone 10c (50–11cm depth) of a long core from the deepest point in Crose Mere, analysed by Beales (1980). There are similarities in respect of percentages of NAP taxa and in the increasing *Pinus* frequencies towards the surface, although proportions of most tree pollen taxa and of *Corylus* are higher in the minicores. Historical evidence suggests that these spectra post-date AD 1600, when the final period of forest clearance and drainage associated with cattle rearing commenced in north Shropshire, and planting of conifers began. Since that date, marked by the increase in *Pinus* pollen percentages, it seems that 50cm of sediment have accumulated in the deepest part of Crose Mere, whereas only 20cm have been deposited at the eastern end of the lake where the minicores were taken. This may reflect a slowing or cessation of sediment accumulation in recent centuries in this part of the littoral. Alternatively, it may be that lowering of lake level (by 2–3m) in the nineteenth century was sufficient to bring about some erosion of shallow-water sediments and exposure of older material characterized by higher proportions of AP and *Corylus* pollen than are now found in surface sediment elsewhere in the lake. Beales (1980) suggests that transfer to deep water of reworked littoral material could be partly responsible for the – apparently – unduly old ^{14}C dates obtained from his long core. While this is possible, it seems from the present distribution of pollen percentages over the mud surface that only localized erosion of sediment is likely to have taken place, chiefly in the area of shallow water at the eastern end of the lake.

Discussion

The results from air traps at Crose Mere provide new information regarding the spatial pattern of pollen input to the inshore water surface

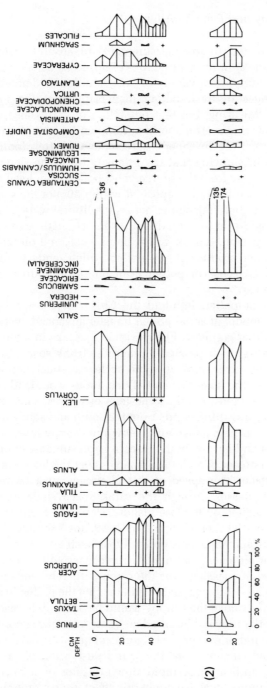

Fig. 9.8 Pollen diagrams (selected taxa only) from two minicores taken in Crose Mere at (1) station 32, and (2) station 13. The percentages plotted are of total arboreal pollen excluding *Alnus* and *Corylus*.

of a lake, and add to the measurements of absolute rates of pollen deposition made at or near the centres of lakes elsewhere. Evidence from the traps also supports a variety of other studies by demonstrating that seasonal changes in fresh pollen dispersal reflect the intensity and phenology of flowering, but that reflotation of grains also occurs, both during the flowering season and in autumn and winter. Results from the sampling transect are in general agreement with the findings of land-based studies in showing that the intensity of pollen deposition diminishes rapidly away from the source: most pollen from local plants in the nature reserve is apparently deposited within 30m of the southern shore of Crose Mere. Pollen trajectories across the littoral water surface appear to depend, in addition to wind direction and strength, upon such factors as height of emission, type of anther dehiscence, and on whether or not the parent plants produce sticky pollen which tends to form clumps. Deposition of all types of pollen from non-local sources is relatively uniform, however.

The pattern of pollen input to the Crose Mere traps lends further support to the model of aerial pollen transfer advanced by Tauber (1965, 1977), namely that grains will be recruited to a lake in a forested area in three main ways: by air passing through the trunk space, by air flowing over the canopy layer, and by more or less vertical rainout from the atmosphere over the lake. The belt of deciduous woodland in the nature reserve at Crose Mere is only some tens of metres wide, but is dense enough to split prevailing winds from the south and south-west into two airstreams, which pass above and below the canopy layer. Local pollen from the herbs and shrubs in the reserve is presumably emitted into the trunk-space airstream, which then emerges over the lake as a layer above the water surface. The speed of this airstream must be reduced considerably as it passes through the trunk space since most pollen is apparently deposited quite close inshore. However, there are differences between taxa – sticky entomophilous NAP, for example, is not transported as far as anemophilous pollen types such as *Urtica*.

Slow-moving air emerging from the trunk space will exercise drag upon the unimpeded air passing over the canopy, thereby enhancing its tendency to descend. Tauber (1965) calculated that the pollen in air deflected downwards from above the canopy would be unlikely to reach the water surface in less than several hundred metres from a lake shore. Currier and Kapp (1974) found, however, that significant quantities of regional pollen were deposited inshore on the windward side of a small lake (80m × 100m), indicating a rapid down-turning of a horizontal drift current from canopy level. No doubt this effect is most pronounced where a small lake in dense forest forms an opening in the vegetation in which an air vortex can develop, but Raynor *et al.* (1974) have observed a similar phenomenon in non-forested areas, namely that pollen accumulates preferentially in the cavity region downwind of such minor vegetative barriers as hedges, irrespective of whether these are dense or relatively porous to the wind. These results suggest that deposition from air currents flowing over, rather than through, the vegetation around lakes – even quite small lakes – may be more important than has been thought hitherto. Certainly, the results from Crose Mere are consistent with the idea that air passing through the trunk space delivers a high flux of mainly local pollen to the littoral region of a lake, the water surface outside this narrow belt (~30m) receiving pollen at relatively low intensity chiefly from (1) air currents which flow over the surrounding vegetation and are deflected downwards during their passage across the lake, plus (2) a component washed from the atmosphere by rain.

Restriction of most local pollen input to the inshore zone could explain why results from traps at the centres of European lakes of very different sizes are so similar and, furthermore, why the average rates of pollen deposition recorded, so far, in mid-lake are very low in comparison with rates in traps sited in vegetation on land (see Bonny 1980: table 8). However, the relative importance of pollen components supplied by rainout and by the air flowing through and over the vegetation around lakes is bound to vary with such factors as lake size, surrounding topography, local vegetation type and its density, meteorological conditions, and so on: further investigation of the intensity and spatial distribution of the pollen flux to a variety of lakes is clearly needed so that generalized models of aerial pollen input can be refined.

Investigation in other lakes is also required of the extent to which the initial pattern of pollen input by air is modified by water movements. A comparison between results from the air traps and the submerged trap in Crose Mere supports the conclusions of the small number of existing studies, however, in showing that there is considerable circulation of pollen within the lake, both of fresh grains which enter the water column during the flowering season and of those which sink to the mud surface and are resuspended with bottom sediment during periods of strong turbulence. The redistribution of pollen grains by lake water must tend to homogenize pollen percentages over the mud surface, although some residual trends which can be related to input from local vegetation may remain – as is evident, for example, from the distribution of percentages of certain pollen taxa over the mud surface of Crose Mere. Quantitatively, water circulation is important in transferring to the central area of a lake some of the huge amount of pollen supplied by air to the littoral region. While this process may counteract the uneven input of local pollen by air, it may also result in unduly high rates of pollen accumulation in the deepest and least-distributed parts of a lake basin (Davis 1968). Along exposed shores, vigorous wave action may prevent the accumulation of any fine sediment in shallow water, cause selective destruction of the more delicate pollen taxa, or remove completely material deposited previously, as may have occurred at the eastern end of Crose Mere.

The gross effects of some limnological processes have now been demonstrated in a few lakes, but more examples are needed to survey the range of variation likely to exist between, for example, shallow lakes in which the whole water column is persistently disturbed by wind, and deep meromictic basins in which both thermal and wind-induced turbulence are fairly restricted. It is to be hoped that advances in trapping technology will facilitate such investigations (see Reynolds et al. 1980). More specific studies are also required in relatively neglected areas concerned with the water transport of pollen – for example, on the behaviour of grains at the air/water interface and in minor density currents in the water column (cf. Bradley 1969), on the relative sinking rates of different pollen taxa in water, on the possibly enhanced sedimentation of grains coagulated by algae and colloids in the water of productive lakes, and on the effects which residence in lake water and in sediment undergoing diagenesis may have upon the physical structure of pollen grains. In addition, it must be emphasized that aspects of the transport of pollen to lakes by inflows deserve further study, although this is obviously not relevant to enclosed lakes such as Crose Mere. However, the streamborne pollen component may be of great quantitative significance, especially in areas of high relief, and of importance qualitatively, since streams can recruit to a lake a variety of non-local pollen taxa which are dispersed relatively poorly by air (Pennington 1964; Birks 1972; Peck 1973; Bonny 1978).

In a recent paper, Jacobson and Bradshaw (1981) discuss the criteria which investigators should bear in mind when choosing a lake sediment or peat profile suitable for the elucidation of a particular palaeoecological problem. While it is desirable that care should be exercised in the selection of all sites, it may be that, as in Britain, most deposits available for any kind of pollen-analytical study are concentrated in certain geographical areas, outside which investigators have to make the best of any profiles available. Where such constraints are imposed – and, indeed, perhaps also where they are not – it is important that the manner in which the pollen accumulated should be considered, since transport processes and, at lake sites, the subsequent action of limnological variables, may have had a profound effect upon the pollen assemblages finally incorporated in a deposit. It is to be hoped that palaeoecological interpretations will be aided by a more thorough understanding of the factors affecting pollen recruitment, gained from present-day observations. Results from Crose Mere add to the information currently available, but more investigations are needed at other sites.

Acknowledgments

It is a pleasure to acknowledge the great extent to which this investigation has depended upon the earlier work of Professor Winifred Tutin, F.R.S., with sediment traps, and upon her foresight, enterprise, and encouragement. Acknowledgment is also due to the Nature Conservancy Council and to the owners, the Mercantile and General Insurance Co. Ltd, for permission to work on Crose Mere, and to Mr C. Marsh, farmer, for allowing access to the southern shore. We are indebted to the staff of Preston Montford Field Centre for meteorological data and some assistance with sampling. Help in the field was also given by members of Liverpool Sub-Aqua Club, under the guidance of Dr G.H. Evans (Liverpool Polytechnic), and by Dr Elizabeth Y. Haworth, Dr D.J.J. Kinsman, and Dr C.S. Reynolds (Freshwater Biological Association). We also thank Professor H. Smith (Botany Department, University of Leicester) for providing laboratory facilities, and Dr H.J.B. Birks (University of Cambridge) for carrying out the multivariate analyses of Crose Mere pollen counts.

References

Andersen, S.T., 1970. The relative pollen productivity and pollen representation of north European trees, and correction factors for the pollen spectra determined by surface pollen analyses from forests. *Danm. Geol. Unders.*, Series II, *96*: 1–99.

Andersen, S.T., 1974. Wind conditions and pollen deposition in a mixed deciduous forest: II. Seasonal and annual pollen deposition, 1967–1972. *Grana palynol. 14*: 64–77.

Beales, P.W., 1980. The Late Devensian and Flandrian vegetational history of Crose Mere, Shropshire. *New Phytol. 85*: 133–61.

Birks, H.H., 1972. Studies in the vegetational history of Scotland II. Two pollen diagrams from the Galloway Hills, Kirkcudbrightshire. *J. Ecol. 60*: 183–217.

Birks, H.J.B., Webb, T. and Berti, A.A., 1975. Numerical analysis of pollen samples from central Canada: a comparison of methods. *Rev. Palaeobot. Palynol. 20*: 133–69.

Bonny, A.P., 1972. A method for determining absolute pollen frequencies in lake sediments. *New Phytol. 71*: 393–405.

Bonny, A.P., 1976. Recruitment of pollen to the seston and sediment of some Lake District lakes. *J. Ecol. 64*: 859–87.

Bonny, A.P., 1978. The effect of pollen recruitment processes on pollen distribution over the sediment surface of a small lake in Cumbria. *J. Ecol.* 66: 385–416.

Bonny, A.P., 1980. Seasonal and annual variation over 5 years in contemporary airborne pollen trapped at a Cumbrian lake. *J. Ecol.* 68: 421–41.

Bonny, A.P. and Allen, P.V., 1983. Comparison of pollen data from Tauber traps paired in the field with simple cylindrical collectors. *Grana palynol.* 22: 51–8.

Bradley, W.H., 1969. Vertical density currents – 2. *Limnol. Oceanogr.* 14: 1–3.

Currier, P.J. and Kapp, R.O., 1974. Local and regional pollen rain components at Davis Lake, Montcalm County, Michigan. *Mich. Acad.* 7: 211–25.

Davis, M.B., 1968. Pollen grains in lake sediments: Redeposition caused by seasonal water circulation. *Science, N.Y.* 162: 796–9.

Gardner, W.D., 1980a. Sediment trap dynamics and calibration: a laboratory evaluation, *J. Mar. Res.* 38: 17–39.

Gardner, W.D., 1980b. Field assessment of sediment traps. *J. Mar. Res.* 38: 41–52.

Gorham, E., 1957. The chemical composition of some waters from lowland lakes in Shropshire, England. *Tellus* 9: 174–9.

Grosse-Brauckmann, G., 1978. Absolute jährliche Pollenniederschlags-mengen an verschiedenen Beobachtungsorten in der Bundesrepublik Deutschland. *Flora* 167: 209–47.

Hardy, E.M., 1939. Studies of the Post-glacial history of British vegetation. V. The Shropshire and Flint Maelor Mosses. *New Phytol.* 38: 364–96.

Jacobson, G.L., and Bradshaw, R.H.W., 1981. The selection of sites for palaeovegetational studies. *Quaternary Res.* 16: 80–96.

Krzywinski, K., 1977. The Tauber pollen trap, a discussion of its usefulness in pollen deposition studies. *Grana palynol.* 16: 147–8.

Mackereth, F.J.H., 1969. A short core sampler for sub-aqueous deposits. *Limnol. Oceanogr.* 14: 145–51.

Peck, R.M., 1973. Pollen budget studies in a small Yorkshire catchment. In *Quaternary Plant Ecology*, ed. H.J.B. Birks and R.G. West, 43–60.

Pennington, W., 1964. Pollen analyses from the deposits of six upland tarns in the Lake District. *Phil. Trans. R. Soc. B, 248:* 205–44.

Pennington, W., 1974. Seston and sediment formation in five Lake District lakes. *J. Ecol.* 62: 215–51.

Pennington, W., Haworth, E.Y., Bonny, A.P. and Lishman, J.P., 1972. Lake sediments in northern Scotland. *Phil. Trans. R. Soc. B, 264:* 191–294.

Raynor, G.S., Ogden, E.C. and Hayes, J.V., 1974. Enhancement of particulate concentrations downwind of vegetative barriers. *J. Agric. Met.* 13: 181–8.

Reynolds, C.S., 1973a. The phytoplankton of Crose Mere, Shropshire. *Br. Phycol. J.* 8: 153–62.

Reynolds, C.S., 1973b. The seasonal periodicity of planktonic diatoms in a shallow eutrophic lake. *Freshwater Biol.* 3: 89–110.

Reynolds, C.S., 1979. The limnology of the eutrophic meres of the Shropshire–Cheshire plain: a review. *Fld Stud.* 5: 93–173.

Reynolds, C.S., Wiseman, S.W. and Gardner, W.D., 1980. An annotated bibliography of aquatic sediment traps and trapping methods, *Occ. Publ. Freshwat. biol. Assoc. 11.*

Sinker, C.A., 1962. The North Shropshire meres and mosses, a background for ecologists, *Fld Stud. 1:* 101–38.

Tauber, H., 1965. Differential pollen dispersion and the interpretation of pollen diagrams. *Danm. Geol. Unders.* Series II, *89:* 1–69.

Tauber, H., 1974. A static non-overload pollen collector. *New Phytol. 73:* 359–69.

Tauber, H., 1977. Investigations of aerial pollen transport in a forested area. *Dansk Bot. Ark. 32:* 1–121.

10 Sediment focusing and pollen influx

Margaret Bryan Davis, Robert E. Moeller and Jesse Ford

Introduction

SEDIMENT FOCUSING is differential deposition that results in accumulation of greater amounts of sediment in the deeper parts of lake basins (Likens and Davis 1975). Sediment focusing is an important phenomenon to palynologists who measure pollen influx (grains accumulated cm^{-2} yr^{-1}) in the hope that pollen influx in a sediment core is representative of inputs to the lake as a whole. The variable of real interest is the whole-lake input, because palaeoecologists seek to measure changes in regional pollen production that reflect changes in the abundance of source plants on the landscape. In this way fossil pollen might function as a rough estimate of population densities over many thousands of years.

Sedimentary influences on the spatial patterns of pollen accumulation are important because they confuse the record of real changes in pollen input. In a hypothetical vertical-sided lake basin with sediment laid down in even layers, the accumulation rate at the centre would be the same as accumulation rates everywhere in the basin. However, in conical basins, which are more typical of real lakes, sediment is generally focused toward the centre. This would not present major difficulties if the same proportion of the total were always deposited at the coring site. Recent papers recognize, however, that as a lake basin fills in, younger sediment may be spread over a larger area of the lake bottom than was the older sediment, thus accumulating in thinner layers. Under these circumstances the sediment core can register a misleading decline in influx, even though inputs to the lake as a whole remain constant (Likens and Davis 1975, Davis and Ford 1982). Under different conditions (e.g. Frains Lake, Michigan: Kerfoot 1974) focusing can become more intense as the lake fills in, resulting in increasing influx rates at the central coring site. Hypothetical patterns of infilling are discussed by Lehman (1975) who uses geometric models to calculate ways in which constant inputs to lakes might be expressed as changing influx rates in a central core.

It is the purpose of this paper (1) to describe sediment focusing at

Mirror Lake, New Hampshire, where a large number of cores and probes have been studied (Davis and Ford 1982, 1984, Moeller 1984); and (2) to discuss the relevance of the results to the interpretation of pollen-influx diagrams. Pollen-influx data from a number of nearby lakes help to distinguish local from regional events. (3) A third purpose has been to investigate the depositional processes that cause focusing of pollen and sediment. Sediment traps and surficial cores were studied for this purpose in Mirror Lake.

Methods

Mirror Lake is located at about 200m elevation in the foothills of the White Mountains in northern New Hampshire, U.S.A. (fig. 10.1). The lake is oligotrophic, dimictic, 15ha in area, and has a maximum depth of 11m. The deciduous forest of the adjacent watersheds has been the subject of an integrated ecosystem study (Likens et al. 1977, Bormann and Likens 1979) and the modern lake is described in a monograph edited by Likens (1984).

The history of the lake and its watershed has been investigated through study of a sediment core by a number of cooperating investigators. This core was collected near the centre of the basin and is referred to as the 'Central core'. The results are summarized in Likens (1984). Palaeoecological studies of other lakes in the vicinity provide useful comparisons (Davis et al. 1980, Spear 1981, Davis 1984, Ford 1984). Our work on sedimentary infilling of the basin (Davis and Ford 1982, 1984) was initiated to answer questions about sediment focusing raised in an initial report (Likens and Davis 1975) and by Lehman (1975). Knowledge of the pattern of sediment deposition was also essential to situate sedimentary fluxes within the nutrient budget of the lake (Moeller and Likens 1978, Moeller 1984).

The shape and size of the sediment body was deduced from 22 cores and 27 probes taken in various parts of the basin. Several samples were collected in shallow water by SCUBA divers using augers. Four of the cores selected for detailed study were collected with a Livingstone corer 5cm in diameter; the Central core was collected with a Swedish foil sampler (Likens and Davis 1975, Davis and Ford 1982). Cores of surficial sediment were collected in plastic tubes with pistons; the tubes were kept upright until extruded in the laboratory.

As part of our study of sedimentary processes in Mirror Lake, sediment traps were maintained at three sites (fig. 10.2) over a period of three full annual cycles. The traps were wide-mouth jars, 9cm diameter, suspended 1m above the sediment surface in water 7m, 10m and 7m deep (Moeller

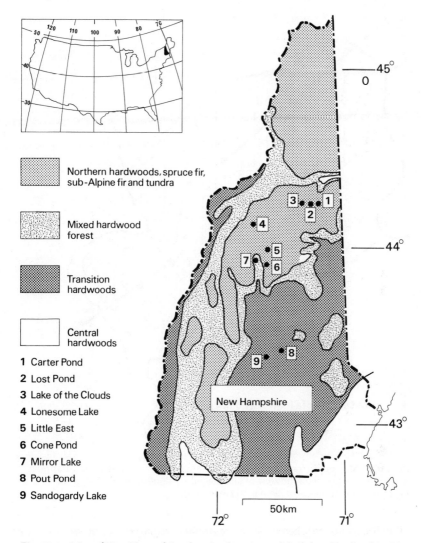

Fig. 10.1   Map of New Hampshire showing locations of the lakes discussed in the text.

and Likens 1978). Samples were retrieved from the sediment traps at intervals to record the seasonality of deposition and to provide quantitative insights into the importance of sediment resuspension within the basin. Pollen concentrations and percentages were calculated in subsamples of sediment from the traps at the Central station. Pollen was also analysed at 2cm intervals throughout the post-settlement layer of surficial cores collected near each trapping site.

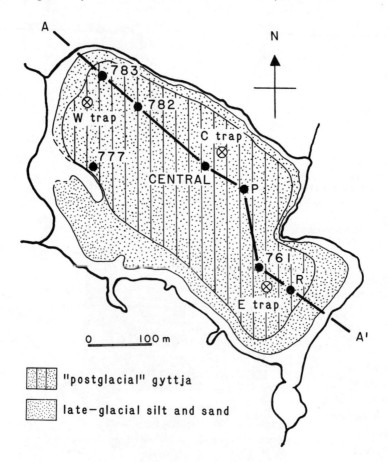

Fig. 10.2   Map of Mirror Lake, showing locations of cores and sediment-trapping sites. Superimposed on the zone of late-glacial silt deposition (dotted) is the maximal limit of pre-settlement gÿttja (hatched).

The locations of lakes used for regional comparison of pollen influx are shown in fig. 10.1. Lake characteristics are summarized in table 10.1. Coring of these lakes was done with a Livingstone corer from a raft or through the ice in the wintertime.

Five cores from Mirror Lake were studied in detail. Three of these cores (777, 783, and 761) are from intermediate depths just below the thermocline; core 782 and the Central core are from deeper water (fig. 10.2). Sediments from ten levels in the Central core were dated by the radiocarbon method (Likens and Davis 1975); two dates were also obtained from core 783, and four from the deepest levels in core 782, which are older than the deepest levels recovered in the Central core. Age

Table 10.1.   Characteristics of the lakes used for regional comparison of pollen influx.

Lake	Latitude/ longitude	Elevation (m)	Area (ha)	Maximum depth (m)	Surrounding vegetation	Watershed area[1] (ha)	Watershed: lake ratio
Carter	44°15′30″/71°11′45″	1055	1.6[3]	4.3	spruce-fir-birch	39[2]	13.4
Cone	43°54′30″/71°37′	469	4.5[3]	8.5	northern hard- woods/spruce-fir	74[2]	16.4
Little East	44°01′15″/71°35′	791	4.3[2]	1.4	spruce-fir/birch	193[2]	44.9
Lonesome	44°08′30″/71°42′	835	16.4[2]	2.7	spruce-fir/birch	468[2]	28.5
Lost	44°15′/71°15′	610	1.6[3]	0.9	northern hard- woods/spruce-fir	70[2]	7.8
Mirror	43°56′30″/71°42′	211	15.0	11.0	northern hard- woods	107[4]	7.1
Pout	43°26′15″/71°29′30″	149	8.6[2]	21.4	white pine – hemlock	90[2]	10.5
Sandogardy	43°23′/71°37′	121	20.0[2]	4.6	white pine – hemlock	1285[2]	64.3

1. Excluding lake area.
2. Estimated from topographic maps.
3. As reported by New Hampshire Fish and Game Dept Survey Report No. 8a.
4. Likens and Davis (1975).

determinations of samples from other cores were made by cross-correlating the pollen assemblages with those in the radiocarbon-dated Central core, and with those in the radiocarbon-dated basal segement of core 782 (Davis and Ford 1982). Those few levels that could not be dated easily by cross-correlation of pollen were radiocarbon-dated. Methods of sample preparation and pollen counting are described elsewhere (Davis and Ford 1982).

We are defining *pollen concentration* as the number of pollen grains from terrestrial plants per cm^3 of sediment (units are grains cm^{-3}). *Pollen influx* is the number of grains accumulated per cm^2 per year on the sediment surface at a coring site (units are grains cm^{-2} yr^{-1}). Influx is calculated by dividing the pollen concentration in a sediment core by the number of years necessary for the deposition of 1cm thickness of sediment. The latter variable, the *deposition time* (yr cm^{-1}), varies in different parts of the core. It is estimated from age determinations made at numerous levels in the core. Total *pollen inputs* are all the pollen grains entering the lake from outside the lake basin each year, divided by the area of the lake, i.e. grains per cm^2 of lake surface per year. Therefore, although pollen inputs are expressed in the same units as pollen influx, the two terms are not synonymous. *Pollen assemblage* refers to the relative abundances or percentages of all pollen types in a sediment sample.

Sediment focusing at Mirror Lake

SEDIMENT ACCUMULATION

The original lake basin extended 23m below the present lake surface. It formed by the melting of glacial ice buried beneath outwash deposits (Winter 1984) and was characterized by steep slopes which extended unbroken from the shores to the center of the lake. Fig. 10.3 presents a cross-section of the lake basin and its sediments along the transect A–A'. Infilling during the last 14,000 years has occurred in three stages.

Stage 1 (14,000–11,000 BP) During Stage 1 a grey silty sediment was deposited over 85% of the area of the lake basin. Near shore this sediment

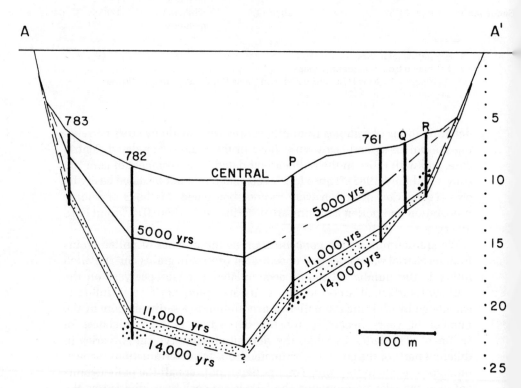

Fig. 10.3   Cross-section of Mirror Lake basin showing sediment accumulated along the transect A–A' (fig. 10.2).

can be quite sandy. The silts and sands are inorganic (2–10% loss on ignition), and one or two metres thick (fig. 10.3); they contain pollen and macrofossils from a tundra-type vegetation and were deposited at a time when the export rate of particulate inorganics from the lake's watershed was more than 30 times higher than export from modern, forested watersheds (Likens and Davis 1975, Davis and Ford 1982, 1984). The export rate fell rapidly to levels typical of modern forested watersheds between 12,000 and 10,000 years ago, when spruce trees began to grow on the watershed.

Stage 2 (11,000–140 BP). Deposition of gÿttja was initiated 10,000–11,000 years ago and continued to the time of settlement. The gÿttja is flocculent, greenish-brown in colour, and more organic than the underlying sediment. Percentage loss on ignition was about 10% 10,000 years ago and rose gradually to 30–40% by 140 years ago. Most of the inorganic material in the gÿttja is diatomaceous silica, rather than mineral silt (Davis and Ford 1984). In contrast to the grey silt deposited earlier in the lake's history, the organic gÿttja was strongly focused into the central part of the basin (figs. 10.2 and 10.3). No gÿttja at all accumulated in water that today is shallower than 6m, meaning that only 50% of the present area of the lake received sediment during a long time interval. Accumulation was not uniform even within the central area of the lake. At depths 6–8m below the present water surface, 2–6m of sediment accumulated, while sediments 11–12m thick accumulated in the deepest parts of the basin, where water depths are now 8–11m (fig. 10.3). Over short time intervals the differences between sites are larger than those implied by the total thickness accumulated, because the area of most rapid accumulation shifted from one part of the basin to another from one millennium to the next (Davis and Ford 1982, 1984). The intensity of focusing, i.e. the differential between central and peripheral sites, has also changed several times.

Stage 3 (140–0 BP). The third stage of infilling has occurred within the last 140 years, during which time human disturbance of the lake and its watershed has caused changes in both the nature of the sediment and its distribution within the basin (Moeller 1984). A dam established in the mid-nineteenth century raised the level of the lake 1–2m. The midsummer thermocline now begins 4–5m below the lake surface; presumably the pre-settlement thermocline did also. The damming therefore had the effect of raising the thermocline 1–2m higher above the sediment surface; as a result, organic gÿttja began to accumulate at depths of 5–6m where gÿttja had not accumulated before, increasing the area of gÿttja deposition from 7.5ha (50% of the present lake area) to 10.2ha (68% of the present lake area; see fig. 10.4). In some areas this was the first episode of net accumulation of sediment since the late-glacial period; consequently in

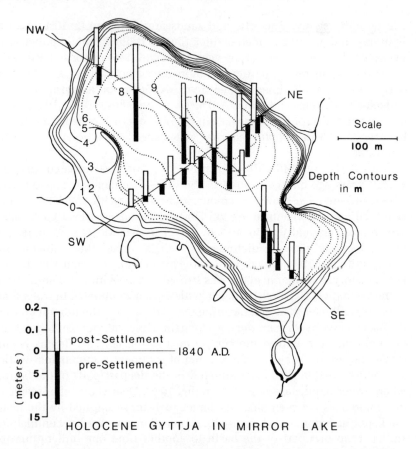

Fig. 10.4 Depositional pattern of gÿttja during the pre-settlement Holocene (solid histogram) and post-settlement period (open histogram) along two transects across Mirror Lake. The thickness of pre-settlement gÿttja is based on probes that may include some of the late-glacial silt. The thickness of post-settlement gÿttja is based on the 1840 horizon, corrected 2cm for downward mixing of the horizon at water depths less than 8m, and 4cm at depths greater than 8m. Note that sites of surface cores and probings do not always coincide.

these areas post-settlement sediment directly overlies late-glacial silt. Within the zone of accumulation focusing has been relatively weak (fig. 10.4), in contrast to the earlier pattern. Presumably the relaxation of focusing was caused by the abrupt increase in lake level (Moeller 1984).

For almost a century prior to 1925, a major stream was diverted into Mirror Lake (Likens 1972), increasing the source area for streamborne pollen. The inwashing of soil organic material, including pollen, may account for an increase in the proportion of poorly preserved pollen in post-settlement sediment (Pennington 1979; see discussion below).

In contrast to the apparently continuous deposition occurring in four of the five cores, core 777 contains a sharp stratigraphic unconformity at 2.2m. No net accumulation of sediment occurred between about 11,500 and 7,500 years ago. Even after sediment began to accumulate again, the site apparently continued to be subject to currents that winnowed away organic matter. The sediments in this core are consistently less organic than at the other coring sites in the lake.

POLLEN ACCUMULATION

Percentages. Pollen-percentage diagrams from the five cores are shown in fig. 10.5. The diagrams are almost identical. All the cores begin with a herb zone; this is poorly developed in the Central core where the bottom of the core was lost and the oldest sediment retrieved was only 11,400 years old. The herb zone is overlain by a spruce pollen zone deposited between 9,500 and 11,500 years ago. Pine and oak pollen dominate early Holocene sediments at all sites except 777, where sediment of this age as well as from the underlying spruce zone is missing. Cores 761, 782, and the Central core all show the clear minimum in hemlock percentages in mid-profile that is a regional feature of pollen diagrams from the eastern U.S. and Canada (Davis 1981). The hemlock minimum does not appear in pollen diagrams from cores 777 and 783, apparently because samples are widely spaced in time (one per 10^3 years average).

Concentrations. The total terrestrial pollen concentrations in the sediment of the five cores are plotted against the estimated age of the sediment in fig. 10.6a. Core 761 appears to be the richest in pollen; the Central core has a lower concentration of pollen than the others in sediment 8500 years old. Otherwise the values are the same in sediment of similar age.

Influx. In contrast to pollen percentages and concentrations, the stratigraphic patterns of total pollen influx differ greatly from site to site (fig. 10.6b). Core 761, despite its slightly higher concentration of pollen, accumulated less sediment and consequently has a relatively small pollen influx 9000–5000 years BP. Cores 783 and 777 display low rates of pollen influx throughout, due to low sediment-accumulation rates.

These data show that the sediment accumulating at all five sites is homogeneous, well-mixed sediment, with similar pollen concentrations and pollen percentages. However, the amount of sediment that accumulates in different parts of the lake varies. Sediment is focused into the various parts of the deep basin, where it consequently accumulates more rapidly. This process has occurred continuously for the last 11,000 years, even though the increment of sediment accumulated at each deep-water site has varied from century to century.

270

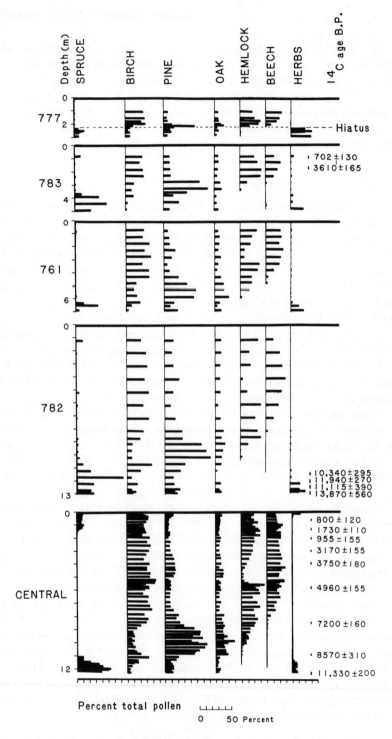

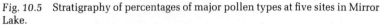

Fig. 10.5 Stratigraphy of percentages of major pollen types at five sites in Mirror Lake.

The pattern of sediment deposition has changed since settlement, but there is still a near uniformity of pollen composition across the zone of gÿttja. Pollen percentages were calculated in post-settlement gÿttja from cores collected near the sediment trap stations (fig. 10.2). The relative abundances within the sum of terrestrial pollen at 7m water depth at sites E and W near the margins of the gÿttja deposit differ in only a few cases from those at the 10m site (fig. 10.7). These pollen assemblages were compiled by combining counts from samples throughout the cultural layer, thereby assuring an identical time interval for deposition at all sites. The uniformity is greater than we had anticipated. If the two inlet streams at the western end of the lake add a floristically distinct, streamborne component to the lake (cf. Bonny 1976, 1978), the distinctiveness of that source is obliterated by mixing within the water column before permanent deposition.

Lakewide pollen inputs

MIRROR LAKE

The annual pollen input to a lake can be estimated by averaging influx rates measured in cores and extrapolating this rate to the area of the lake receiving sediment at that time. For Mirror Lake these estimates are probably most accurate for the time prior to 11,000 years ago, when sedimentation was fairly uniform from site to site. The subsequent irregular accumulation of gÿttja means that average data approximate true inputs less precisely.

Stage One (14,000–11,000 BP). Estimates of lakewide pollen inputs have been made for two periods within the first stage of sedimentation. Thirteen thousand years ago, when tundra surrounded Mirror Lake, the average influx at the five coring sites was 500 grains $cm^{-2} yr^{-1}$ (fig. 10.6b). Extrapolating this to 12.8ha (85% of the present-day lake area) results in an annual lakewide pollen input of 6.4×10^{11} grains, or 425 grains cm^{-2} of modern lake surface.

Pollen influx measured in the five cores had increased to an average of 9000 grains $cm^{-2}yr^{-1}$ by 11,000 years ago, when spruce began to grow on the local landscape. The area of the lake receiving sediment at that time is not known with certainty, but if we assume that it was about 80% of the present-day lake area, annual pollen inputs were 7200 grains cm^{-2} of modern lake surface, a 17-fold increase.

Stage Two (11,000–140 BP). Lakewide pollen inputs rose during the second stage of sedimentation. About 9100 years ago, when pollen influx at the Central site was in excess of 80,000 grains $cm^{-2}yr^{-1}$, average influx at all five sites was only 35,400. This is true despite the fact that four of

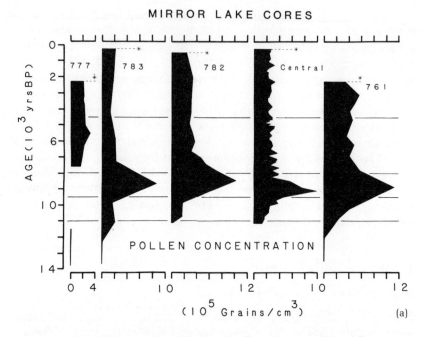

Fig. 10.6(a)   Concentration of total terrestrial pollen in five cores from different sites in Mirror Lake.

the five coring sites attained maximal pollen influx sometime during the period 8000–9600 BP, because (1) the actual time of maximum influx differed from core to core, and (2) site 777 was accumulating no sediment during this period. This site may have served as one of the source areas for the pollen and sediment focused into other parts of the basin (fig. 10.6b). Applying the average influx value of 35,400 grains $cm^{-2}yr^{-1}$ to the portion of the lake basin receiving sediment during this period (about 40% of the present-day lake area) results in annual pollen inputs of 14,200 grains cm^{-3} of modern lake surface. This figure is only twice as high as the 11,000BP estimate of 7200 grains cm^{-2}, in contrast to the ten-fold increase implied by influx values. Clearly, then, corrections must be made for the changing proportion of the basin receiving sediment in order to assess realistically changes in pollen inputs.

Over the Holocene as a whole, pollen influx averaged about 24,700 grains $cm^{-2}yr^{-1}$. Extrapolating this to the 50% of the present-day lake basin receiving sediment during most of the Holocene, we find that the average annual pollen input to the lake was 12,400 grains cm^{-2} of modern lake surface.

Stage Three (140–0 BP). During the post-settlement stage of sedimenta-

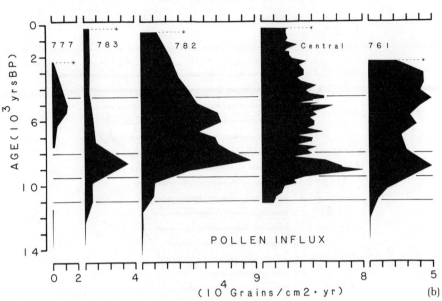

Fig. 10.6(b)   Pollen influx in the same cores.

tion the annual influx to the sediment at the three trap sites has averaged 19,400 grains $cm^{-2}yr^{-1}$, based on accumulations measured in cores at those sites. Extrapolating this value to the increased area of the lake basin receiving sediment (about 68% of the present-day lake area), average annual pollen inputs to the lake have been 13,200 grains cm^{-2} of modern lake surface. This rate is very close to the mean Holocene input of 12,400 grains cm^{-2}, although stream diversion into the lake during this period is believed to have significantly increased total pollen inputs (Moeller 1984).

In summary, our analysis of lakewide pollen inputs indicates that total inputs of pollen were low at the time the landscape supported tundra vegetation, and rose in the expected manner as spruce trees colonized the region. Total pollen production was apparently not as variable over the Holocene as estimates based on core-specific influx values alone might suggest. At Mirror Lake, focusing was extremely strong during the Holocene, resulting in influx rates of 35,000–80,000 grains $cm^{-2}yr^{-1}$. Total lakewide pollen inputs, by contrast, are more modest, varying between 10,000–15,000 grains $cm^{-2}yr^{-1}$. These lower rates appear to be reasonable estimates of the pollen rain produced by a mixed deciduous–

PERCENT OF POLLEN

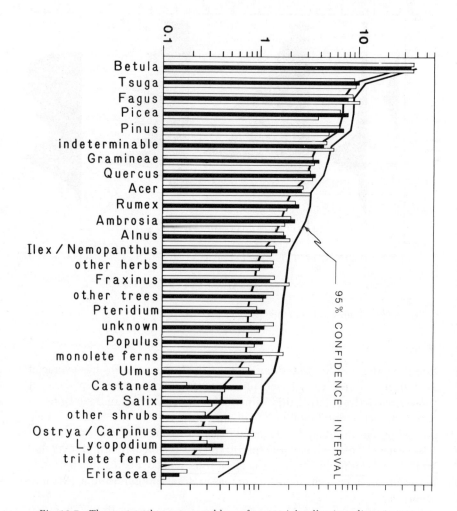

Fig. 10.7   The post-settlement assemblage of terrestrial pollen in sediment cores from the West and East stations in Mirror Lake (fig. 10.2) (upper and lower open histograms, respectively) compared to the ranked assemblage in a core from the Central station (dark histogram). A 95% confidence interval is calculated around the Central assemblage (pollen sum = 2500).

coniferous forest, and should be substituted for the higher values based on influx reported by M.B. Davis *et al.* (1973). However, they are still almost twice as high as measured in forest regions in Europe (Pennington 1979).

Other New Hampshire lakes

We have compared pollen concentration and influx from six lakes in the vicinity of Mirror Lake (fig. 10.8) in order to distinguish between regional changes in pollen inputs through time, presumably related to changes in pollen productivity, and changes due to sedimentary processes. Lost Pond and Cone Pond are surrounded by deciduous forest with some conifers. These might be expected to produce quantities of pollen similar to those produced by the forests around Mirror Lake. Little East Pond, Lonesome Lake, and Carter Pond are surrounded by forests of spruce, fir, and paper birch. Spruce and fir produce little pollen, but birch is a prolific pollen producer. Sandogardy Pond, which is further south, is surrounded by hemlock and white pine trees that produce moderate to abundant amounts of pollen (Davis and Goodlett 1960). Inputs to individual lakes may also be influenced by the fetch of the lake, which affects the ratio of local to regional pollen deposited on the lake surface (Tauber 1977, Berglund 1973, Jacobson and Bradshaw 1981) and by exposure, which may affect wind velocities and pollen inputs. Inflowing streams also affect pollen inputs; these will be discussed below.

The Mirror Lake cores show the following four features:

1. Low pollen concentration and influx prior to 11,000 years ago.
2. Increasing pollen concentration and influx 11,000–9000 years ago.
3. Peak concentrations and influx in the early Holocene between 9600 and 8000 years ago, with variation from site to site within the lake.
4. Constant pollen concentrations from 7000 years ago to the present, but variable influx. In several cores influx reaches a secondary maximum 6000–4000 years ago and then declines in the late Holocene.

Features 1 and 2 appear at all the regional sites. The consistent changes in all three variables – pollen percentages, concentrations, and influxes – appear to reflect real changes in pollen inputs as the vegetation changed from tundra to spruce woodland. However, the magnitude of the change in individual cores is exaggerated by increased intensity of sediment focusing as sediments became more organic. This is true in Mirror Lake, and presumably at the other sites as well (Davis and Ford 1982, 1984).

Feature 3, the early Holocene peak in pollen influx, was clearly the result of sediment focusing at Mirror Lake. Total input of pollen to the lake was not significantly higher during this period than later in the Holocene, as discussed earlier. Equivalent early Holocene peaks occur at Carter Pond and Lonesome Lake, and perhaps at Sandogardy Pond. However, Lost Pond, Little East Pond and Cone Pond fail to show this effect.

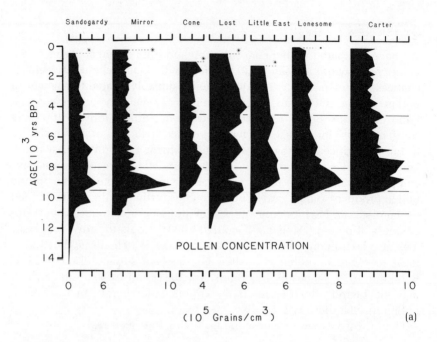

Fig. 10.8(a)   Concentration of total terrestrial pollen in cores collected at the deepest site in seven lakes in New Hampshire.

The stratigraphic records from Rogers Lake, Connecticut (Davis 1969), Moulton Pond, Maine (Davis R.B. *et al.* 1975) and Berry Pond, Massachusetts (Whitehead *et al.* 1973), also show peak influx rates in the early Holocene. Whitehead *et al.* (1973) interpret the apparently simultaneous maxima for organic matter and microfossil influx as an early productivity maximum. Davis R.B. *et al.* (1975), on the other hand, believe that the unusually high pollen influx during this period may be exaggerated due to sediment focusing. Our studies demonstrate that sediment focusing at Mirror Lake indeed exaggerated influx of pollen, organic matter, and inorganic matter to deep-water sites during this time period (Davis and Ford 1982). The possibility that sediment focusing contributed to high early Holocene influx at Berry Pond needs to be investigated before a strong case can be made for the existence of an early Holocene productivity maximum.

Feature 4 is a late-Holocene decline in pollen influx following an apparent maximum 4000–6000 years ago (Davis and Ford 1982). The decline in influx at Mirror Lake is matched by a late-Holocene decline at Lost Pond, Cone Pond, and possibly Sandogardy Pond. Only Lost Pond, however, shows a decline in concentration as well. A decline in influx

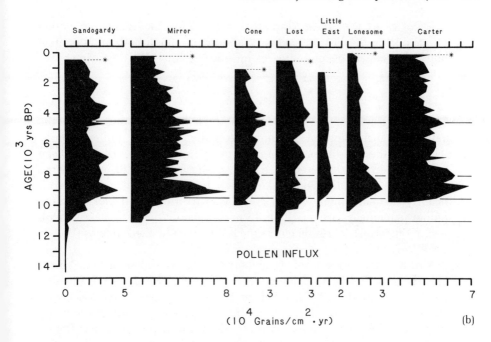

Fig. 10.8(b)   Pollen influx in the same cores.

does not occur in Lonesome Lake, Carter Pond, or Little East Pond. Because the record is inconsistent and unrelated to changes in concentration, it seems unlikely that Feature 4 was caused by regional changes in pollen productivity.

Late-Holocene declines in pollen influx have been noted at other New England sites (Davis R.B. *et al.* 1975, Whitehead 1979) and in Labrador (Lamb 1980). Davis R.B. *et al.* (1975) and Whitehead (1979) suggest that the late-Holocene declines at Rogers, Moulton, and Berry Pond may be related to changing sedimentation patterns. Although our data from Mirror Lake do not demonstrate this conclusively (we have no coring sites where influx increased in the late-Holocene) the alternative explanation, that the decline is due to a regional decline in pollen productivity, is not supported when all our New Hampshire sites are taken into consideration. Lamb's (1980) assertion based on single cores that the forests of south-eastern Labrador underwent a decline in productivity due to unfavourable climate needs to be re-evaluated in this context.

Late-Holocene declines in influx are not universal features of pollen diagrams in North America. Frains Lake, Michigan, in fact, shows a three-fold increase in pollen influx during the second half of the Holocene

(Kerfoot 1974). Rutz Lake, Minnesota, shows a five-fold increase in pollen influx during the last 500 years (Waddington 1969). Both authors attributed the late Holocene increases in influx to intensified focusing.

STREAM INPUTS

The comparison of the New Hampshire lakes is complicated by the varying contribution of streamborne pollen to total pollen inputs. Although the streamborne component in Mirror Lake has not been measured directly, inferential evidence suggests that the streamborne component is small. Whole-lake pollen inputs did not increase significantly after settlement when a stream was diverted into the lake, and the pollen percentages are nearly uniform throughout the basin even though half of the terrestrial runoff enters the lake as channelized streams (Winter 1984).

A hint to the presence of streamborne pollen in sediment from Mirror Lake is provided by the abundance of 'deteriorated' gains (cf. Cushing 1967). Crumpled and corroded pollen exines have been associated stratigraphically with erosion of drainage basin soils (Tolonen 1980) and statistically, with the non-airborne component of pollen influx (Bonny 1978). When counting surface cores from Mirror and two other New Hampshire lakes, we scored the condition of all deteriorated Betulaceous (>95% *Betula*) grains encountered. *Betula* pollen grains predominate both above and below the settlement horizon; conditions of their preservation provide a graphic index to the preservation of pollen as a whole.

The major increase in herb pollen at Mirror Lake, dating to approximately AD 1840 (Moeller 1984), coincided with a doubling in the proportion of deteriorated grains within the Betulaceae, from 15–25% to 30–45% (fig. 10.9). We surmise that forest clearance affected the hydrology of the watershed, increasing streamflow and inputs of pollen eroded from the humic horizon of the soil, much in the manner described for English lakes (Pennington 1979). The partial diversion of Hubbard Brook into Mirror Lake also substantially increased the watershed area, although the whole-lake input of pollen appears to have increased only slightly. At Pout Pond, New Hampshire, the proportion of deteriorated grains increased almost two-fold at the settlement horizon, but this increase was not sustained throughout the settlement period. At Little East Pond, the change in preservation associated with disturbance was small and shortlived. Forests near this lake were logged, but never cleared for agriculture (fig. 10.9).

The differences in the presettlement proportion of deteriorated pollen (Little East >Mirror>Pout) can be related to the ratio of watershed area to lake area and to the importance of surficial streams. Little East's ratio of

DETERIORATED   BETULA   POLLEN   (% OF TOTAL BETULA)

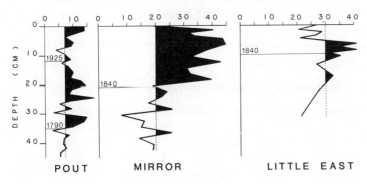

POUT   MIRROR   LITTLE EAST

Fig. 10.9   Recent stratigraphic changes in the preservation of *Betula* pollen in three New Hampshire lakes. The increase in deteriorated pollen above the pre-settlement background level is represented by the black part of each curve. The dates of 1790 and 1840 represent the major increase in herb pollen at the cultural horizon, that of 1925 corresponds to the abrupt decrease in *Castanea* pollen at one site.

about 45 : 1, combined with a substantial channelized runoff from its steep watershed, may explain the greater input of deteriorated pollen at Little East Pond than at Mirror Lake, which has substantial channelized flow but a watershed : lake ratio of only 7 : 1. The watershed : lake ratio of Little East Pond resembles Blelham Tarn in the English Lake District (Bonny 1978). Pout Pond is similar to Mirror Lake, with a ratio of watershed area to lake surface of about 10.5 : 1, but most inflow arrives as seepage through the sandy outwash in which the lake was formed.

At Mirror Lake we have observed deteriorated pollen in debris collected from inlet pools, and severely deteriorated pollen, mainly Betulaceous, in the humus layer of local forest soil. The interpretation of deteriorated pollen as a soil residue is weakened, however, by the presently untested possibility that deterioration also takes place within the lakes. The ordering of deterioration (Little East>Mirror>Pout) is the inverse of the ordering based on maximum depth (Pout>Mirror>Little East: table 10.1).

Streamborne inputs have been shown to provide an important part of the pollen supply to some lakes. At Blelham Tarn, which is about two-thirds the size of Mirror Lake but 14.5m deep, Bonny (1978) found that pollen deposition within large, midlake enclosures was only 15% of the annual pollen influx to sediment traps outside the enclosures. The floristic composition of the pollen trapped outside the tubes, combined with a correlation between deposition and rainfall, strongly supported earlier suggestions that streams were an important source of pollen to English reservoirs and lakes (Peck 1973, Bonny 1976). Some undetermined portion of the extra pollen deposited outside the traps during autumn and winter presumably had been resuspended from the lake bottom. During the millennia before major forest clearance, pollen influx to Blelham Tarn sediments was twice the influx measured in a core from a nearby seepage lake (Pennington 1979), a result compatible with streamflow as a source of at least half of the pollen to Blelham Tarn when the landscape was forested. The importance of this source increased when the landscape was cleared and inwash of soils began. In near-surface sediment at Blelham Tarn, influx was 8–9 times higher than in the seepage lake. Blelham Tarn has a watershed : lake ratio of 39, and is located in a region of relatively heavy rainfall (mean annual precipitation 175–200cm (Macan 1970) v 125–130cm for Mirror Lake (Bormann and Likens 1979)). Several years ago Pennington (1973) pointed out that differences in pollen concentration and influx in lakes with different hydrologic characteristics suggested real differences in pollen inputs. In her discussion she distinguished carefully between variations in influx resulting from focusing, and those due to differences in pollen recruitment *via* inflowing streams. A factor that we have not considered, but which was discussed by Pennington (1973), is the loss of pollen from lakes with rapid throughput of water.

Mechanisms of sediment focusing in Mirror Lake

Two processes can be identified in Mirror Lake that cause sediment focusing. Mechanism 1 is episodic, and mechanism 2 is continuous.

Mechanism 1 depends on unusual weather with strong winds that create exceptionally strong currents (cf. Johnson 1980). These water movements erode inorganic sediment from the shallower parts of the basin and move it in density currents to the deeper parts of the lake. Deposits of homogenous sediment a few millimetres to several centimetres thick may subsequently be laid down on top of the previous sediment surface. Although the actual frequency of these events is not known, we have found several sand layers in cores collected as much as 100m from shore. Prior to 11,000 years ago sand was occasionally deposited near the lake centre, but during the Holocene sand layers have been confined to areas where the present water depth is 6–8m.

Gÿttja can be moved by lower water velocities than sands and silts because of its lesser density. There is direct evidence that episodic movement of large volumes of gÿttja occurred in Mirror Lake shortly after the land surrounding the lake was settled by farmers in the last century. A layer of gÿttja 6–8cm thick was apparently deposited near the centre of the lake during a period of at most a few years. This layer, which lies at

24–30cm depth in a core from the deepest part of the lake, was detected by its uniform ^{210}Pb content (Von Damm *et al.* 1979). Although we had not suspected any such depositional anomaly on the basis of pollen percentages, the pollen assemblage throughout this layer is in fact sufficiently uniform to be indistinguishable from a mixed layer (Davis 1984). The deep-water post-glacial sediments in Mirror Lake are so homogenous in appearance that it would be difficult to identify a layer deposited in this manner from appearance alone. Most pollen samples from older sediment in the Central core were collected 10–20cm apart. If some of them are from layers like this they were not distinguishable from others.

Sediment-trap data from the late autumn of 1973 revealed another episode of rapid deposition near the centre of the lake (Moeller and Likens 1978). As a result of a prolonged rainstorm, the inlet streams carried large amounts of fine inorganic mineral silt which was deposited in the Central traps suspended at 9m. Little silt was deposited in traps suspended at 6m at all three sites, suggesting that the silt was carried along the lake bottom as a density current. At the Central site, inorganic matter equivalent to 1–2 years' deposition (at the mean annual trap rate) arrived during this single exposure period. The silt was probably eroded from highway embankments constructed on the drainage basin in 1969 (Likens 1972). This mechanism of focusing does not involve resuspension of pre-existing deposits, and contributes little to the focusing of fine organic matter, including pollen, into deeper water. Episodes of severe redeposition of organic gÿttja within the basin may operate analogously, but none were detected during the sediment-trap studies.

The second mechanism for sediment focusing is a more regular, more or less continuous, process of sediment resuspension and redeposition. Every autumn, during the period of isothermal conditions within the water column, flocculent surficial sediment is eroded from the shallower portion of the lake bottom. Much of this material eventually settles onto deeper sediments, and was included in the annual flux of sediment to the gÿttja as measured by traps.

The pollen composition of sediment in the traps demonstrates that resuspended material is included in sediment deposited during the autumn and winter (see also Davis 1968). In fig. 10.10 we examine the concentrations of tree pollen and ragweed (*Ambrosia*) pollen in organic matter deposited within sediment traps. Because tree pollen is produced only from late April to mid-June, its presence in trap samples from autumn and winter demonstrates resuspension of sediment. The background level of tree pollen in the summer (mid-June to mid-September) can be attributed both to inwash of pollen in streams and to some redeposition of pollen initially retained within the littoral zone. Ragweed pollen is produced during late summer and early autumn. The paucity of

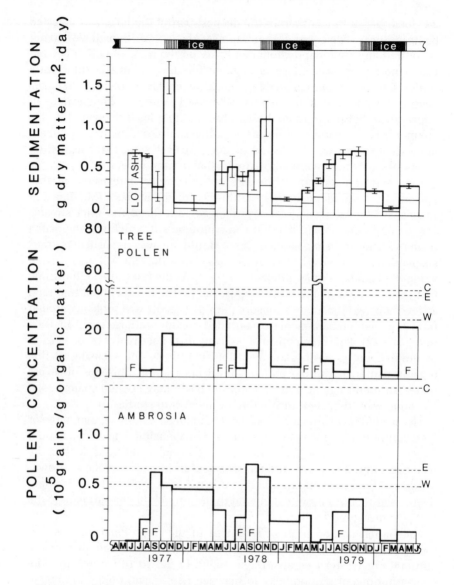

Fig. 10.10   Sedimentation into collecting traps at the Central site in Mirror Lake.
The top diagram presents the deposition rate of dry sediment, with 95% confidence
interval (n = 4 traps). Deposition is divided into loss-on-ignition (lower part of
histogram) and ash (upper part). The lower diagrams present the concentration of
pollen in the organic component. The symbol F designates exposures during which
trees or *Ambrosia* were flowering. Dotted lines indicate the pollen concentrations in
the surficial sediment at the collecting sites.

ragweed pollen in sediment deposited in early summer indicates the absence of significant resuspension during the latter portion of the period when trees are in flower (late May and June). Ragweed pollen is present in sediment deposited during April and early May. However, ice persists on Mirror Lake well into the season of warmer weather. As a result spring overturn is frequently incomplete or absent. Some resuspension may occur during the brief time before thermal stratification develops, but ragweed and other pollen types also might be washed into the lake during the high runoff associated with snowmelt in April.

Judging from the amount of dry matter and tree pollen collected in traps during autumn and winter, resuspended matter appears to make up about 40% of the total annual sediment influx to the lake bottom at a depth of 10m. This resuspended material has presumably been eroded from shallower parts of the lake bottom. During summer, sediment traps at the marginal 7m sites collect the same amount of sediment as the Central traps. Deposition is more variable during the autumn-winter period, both from year to year (fig. 10.10) and from site to site (table 10.2). Resuspension of pollen must be very low in winter. The 1979–80 results (fig. 10.10) show this clearly; in other years autumnal and winter deposition were included within a single sampling period. Thus the differential deposition that is observed among trapping sites apparently occurs during the autumn, when resuspension occurs. The annual influx of sediment measured by traps exceeds long-term core accumulation rates by the same proportion at all three sites (table 10.2). This constant proportionality is

Table 10.2.   A comparison of differential sedimentation revealed by surface cores and sediment trap collections at three sites within the gÿttja zone of Mirror Lake. Dry matter collected in sediment traps has been allocated approximately to that deposited during spring and summer (average 209 days), and during autumn and winter (156 days) when resuspension predominates.

	West (7m)	Central (10m)	East (7m)
Sediment trap rate[1]			
Dry matter (mg cm^{-2} yr^{-1})			
Spring & summer	12 ± 2	10 ± 1	10 ± 1
Autumn & winter	9 ± 5	8 ± 2	5 ± 2
Total	21 ± 6	18 ± 2	15 ± 2
Pollen (10^3 grains cm^{-2} yr^{-1})		17 ± 3	
Core rate[2]			
Dry matter (mg cm^{-2} yr^{-1})	14	12	11
Pollen (10^3 grains cm^{-2} yr^{-1})	17	24	17

1. Mean of 2 (East spring & summer) or 3 years, with standard deviation.
2. Calculated assuming the main herb rise dates to AD 1840, and is corrected for downward mixing of the herb rise (2cm at West and East, 4cm at Central).

evidence that the gÿttja at 7m is not subjected to more severe autumnal erosion than gÿttja at 10m. Apparently resuspension is confined to parts of the lake shallower than 5–6m, where sediment fails to accumulate.

The annual rate of dry matter deposition measured in traps was about 1.4 times the post-settlement rates of accumulation measured in cores, a difference that reflects the presence in trapped sediment of organic matter that is destined to be decomposed within a few years time (Moeller 1984). The annual rate of dry matter deposition in both cores and traps was highest at the West site. In contrast, pollen is differentially focused to the deepest site: the influx of pollen to post-settlement gÿttja has been greater $(24 \times 10^3 \text{grains cm}^{-2}\text{yr}^{-1})$ at the Central site (10m) than at both marginal sites $(17 \times 10^3 \text{grains cm}^{-2}\text{yr}^{-1})$ during the last 140 years. Evidently variations in pollen concentration similar to those observed in the Holocene cores (fig. 10.6a) occur also in post-settlement sediment, even though pollen percentages are nearly uniform.

The magnitude of autumnal resuspension suggests that in Mirror Lake erosion during most years is restricted to littoral substrates and the marginal reaches of the gÿttja shallower than 7m. Not coincidentally, the amount of resuspended tree pollen in the annual influx to the Central site (38% of the annual tree-pollen influx) is similar in magnitude to the portion of the lake bottom where little if any gÿttja has accumulated since settlement (32% of the total area). Because little if any resuspension appears to occur in deep water, the resuspended sediment and pollen detected in the traps is properly counted as part of the net accumulation at these sites.

Blelham Tarn (11ha) and Ennerdale Water (291ha) are additional examples of lakes where a significant amount of resuspended pollen contributes to the total pollen influx to profundal sites; pollen influx measured in traps in the profundal zone is equivalent to the long-term accumulation determined from dated surficial cores. Unlike Mirror Lake, both lakes accumulate sediment in the littoral zone. Blelham Tarn apparently has a significant amount of autumnal resuspension (Pennington 1974), but its actual magnitude is obscured by the large amount of streamborne pollen entering the lake throughout the year (Bonny 1978). Focusing at Blelham Tarn produces two-fold variation in pollen influx within the profundal zone (Pennington et al. 1976).

Frains Lake is smaller (7ha) and shallower (10m) than Blelham Tarn and Mirror Lake. Extensive resuspension of sediment occurs in autumn and spring within both the littoral and profundal zones, and resuspended pollen makes up 80% of pollen deposited in traps near the lake centre. Pollen deposition in traps is 2–4 times the long-term accumulation on the lake bottom, except in a very small depression in the centre of the basin. In that depression, the gross pollen influx to traps approximately equals the net long-term accumulation, illustrating how a small portion of the profundal region still functions as a permanent sediment trap (Kerfoot 1974, Davis 1976). Over most of the rest of the basin, including the littoral zone, pollen influx is much lower and relatively uniform (2–3 fold variation) (Davis and Brubaker 1973). At nearby Sayles Lake, which unlike Frains Lake has no profundal zone at all, resuspended pollen represents 90% of the gross influx, and this influx exaggerates long-term accumulation by large but variable proportions (Davis 1973). The irregular deposition in Sayles Lake illustrates the pattern found by Lundqvist (1927) for shallow, sediment-choked basins: resuspension occurs continuously, and produces an irregular, temporally changing stratigraphy.

Mirror Lake exemplifies Lundqvist's 'profundal' type of sedimentation, in which there is no accumulation in the littoral zone. In lakes with a 'profundal' type of deposition, sediment accumulates across the entire lake bottom below the sedimentation limit, with the focus of greatest influx lying near the deepest part of the basin. This pattern resembles the hyperbolic, sinusoid, and ellipsoid models of infilling presented by Lehman (1975), except that the lateral sedimentation limit is in deeper water rather than at the edge of the lake. Because the sedimentation limit moves up and down with changes in water level, the continuity of deposition at the margin of the accumulation zone depends on constancy of the water level. The hiatus in core 777, which was taken near the margin of the gÿttja zone, and the sand layers extending into the gÿttja from the exposed south and east shores, could have been caused by rare erosional events (Davis and Ford 1984). It is possible that the effect of these events was augmented by lower water levels, at least at certain seasons of the year, that enabled wind-generated deep-water currents to reach the sediment surface.

At Mirror Lake resuspension occurs mainly during the autumn, but even during summer stratification epilimnial currents reduce the accumulation of settling seston and inwashed debris on littoral substrates. Instead this material is added to the hypolimnial water column. Ohle (1962) found that sediment traps suspended at increasing depths within the hypolimnia of several relatively deep lakes (22–45m) in northern Germany collected progressively more sediment than the traps above them. Ohle envisaged sediment kept in suspension as it decomposed, while settling through progressively smaller cross-sections of the hypolimnion until it finally reached the deepest part of the basin. Ohle referred to this pattern of sediment concentration as sediment 'funnelling' ('Trichtereffekt'). Our results from Mirror Lake, however, imply that periodic resuspension of sediment is probably a more significant process than funnelling for accomplishing sediment focusing.

Resuspension is frequently detected in studies employing sediment traps; it shows up as peaks of deposition that are out of synchrony with inputs of particulate matter to the lake (Lastein 1976, Serruya 1977, Mothes 1981) and that are correlated with windy weather or isothermal conditions in the water column (White and Wetzel 1975, Lastein 1976, Koidsumi and Sakurai 1968). Resuspension has also been deduced from a chemical similarity between trapped material and surficial sediment (Pennington 1974, Gasith 1976, Serruya 1977, Ulén 1978), from the presence in trapped material of 'out-of-season' pollen (Davis 1968, 1973) or littoral organisms (Mueller 1964, Simola 1981), and from the excess of annual sedimentation computed from traps over net annual influx measured in surficial cores (Davis 1968, 1973, Pennington 1974, Serruya 1977).

We agree with the emphasis Lundqvist placed in his classic 1927 paper on the interaction between wind and lake morphology in determining the pattern of sediment deposition. Unfortunately in the 55 years since his paper was published little additional information has been added to our understanding of the hydrodynamic and sedimentological processes that control sediment deposition in small lakes. The distribution and intensity of currents in small lakes have not been measured, nor have the benthic shear-stresses that govern deposition and resuspension of the flocculent, organic-rich sediment with which pollen is often associated.

Conclusions

Our results from Mirror Lake show that pollen influx varies from one site to another within the lake. The differences are large even though all five cores were collected in relatively deep water, below the thermocline. During the Holocene, influx at some sites was five times higher than at others. In the early Holocene, there was a period of exceptionally intense focusing, when unusually high pollen influx occurred at some sites (although peak influxes were not synchronous from core to core), while no sediment at all accumulated in other areas. Even variations as large as we have observed would not make interpretation difficult, however, if the focusing pattern had remained constant – that is, if the sediment deposited at one site always represented the same proportion of inputs to the lake. Our data show, however, that the spatial patterns of deposition have changed through time. Each coring site has experienced a different history of changes in pollen influx. No single core provides a record of pollen inputs to the lake.

There are several factors that control pollen influx. The first, and most basic, is the pollen supply to the lake. Pollen production by vegetation is the ultimate factor affecting pollen supply. It is the parameter that palaeoecologists seek to measure, because it relates directly to the population densities of particular species of plants. However, a number of proximate factors affect pollen supply, as well. Bonny (1978), Peck (1973), and Pennington (1979) have shown that streamborne inputs are significant in British lakes and reservoirs and Pennington (1973) has shown that throughput of water can decrease the pollen supply. Although we have not made direct measurements of this factor, inferential data suggest a less important role in the lakes we have studied. Lakes having larger watershed : lake ratios and channelized inlets are expected to resemble the British lakes more strongly.

A second proximate factor affecting pollen supply is topographic exposure of a lake. Tauber (1977), Berglund (1973), and others have shown from pollen assemblages that large lakes with a long fetch receive a larger proportion of pollen from regional, rather than local sources (Jacobson and Bradshaw 1981). It is not clear, however, that this results from greater trapping efficiency for regional pollen. Large lakes may have trapping efficiencies similar to smaller lakes, but relatively lower inputs of local pollen, since the ratio of inputs from local v. regional vegetation should equal the ratio of shoreline to lake area (Tauber 1977). For this reason inputs per unit surface area might be expected to be higher in small lakes surrounded by forest than in larger lakes, although this has not been demonstrated convincingly. Low influx rates in sediments have been attributed to properties of the lake that cause it to 'collect and preserve pollen' less effectively than other lakes (Brubaker 1975). However, we suspect that a lack of sediment focusing is a more likely cause for consistently low influx rates.

Spurious variations in pollen-influx rates in sediment cores may be introduced by short-term changes in matrix deposition. These will result in changes in pollen concentration within the matrix. In principle, most short-term changes in the deposition of sediment matrix should be resolvable by frequent radiocarbon dates. If one were able to correct for all changes in matrix deposition, pollen influx could be measured accurately. However, in practice, the numbers of dates obtained are frequently too few to detect short-term changes. Furthermore, investigators must make a subjective decision: to smooth out the age-depth relationship, assuming a relatively large error for each age determination, or to accept each date literally, thereby creating – in most cases – abrupt changes in calculated sedimentation rate at each dated level. The number of dated levels should be sufficient to make no difference which approach is used, but in practice there are usually too few dates. For this reason the confidence interval surrounding pollen-influx measurement is much larger than the counting errors of either pollen or radiocarbon.

Variations in sedimentation rate due to changes in the intensity of sediment focusing are more serious, because they are more difficult to detect and to compensate for. Changes in focusing can be recognized by comparing dated cores collected at different sites within the lake basin. To calculate lake-wide inputs and thus correct for changes in focusing, several cores must be studied and the spatial and temporal patterns of deposition must be determined.

Mechanisms that affect pollen deposition have been studied in detail in only three lakes, Blelham Tarn, Frains Lake, and Mirror Lake, and these three are so different from one another that we are hesitant to advise palynologists about the kind of lake most suitable for pollen-influx measurement. Pennington (1973) and Bonny (1976, 1978) have emphasized the importance of the size of the watershed. Pennington's (1979) comparisons demonstrate that enclosed basins provide pollen-influx measurements more closely related to vegetation than lakes with major inflowing streams. Our study of sediment focusing in Mirror Lake suggests, in addition, that lakes without littoral sediment are not the most suitable for pollen-influx measurement, because they are likely to have strong, temporally variable patterns of sediment focusing. We believe that this factor accounts for the extremely high pollen influx found at Lake of the Clouds, New Hampshire, an alpine lake without littoral sediment in which influx reaches 100,000 grains $cm^{-2}yr^{-1}$ (Spear 1981). On the other hand, the presence of sediment in the littoral zone is no guarantee that strong focusing has not occurred at some time in the past.

Palynologists tend to collect cores from the deepest spot in lake basins. We feel that this is probably the best strategy, even though the deepest part of the lake may be the centre for modern sediment focusing. It may or may not have been the focal point in the past. The deepest part of the lake currently may be experiencing the most rapid deposition, and the expanded stratigraphy can be a useful advantage in detailed palaeoecological study. The record should at least be complete, without obvious sedimentary unconformities (such as occurred in Mirror Lake core 777), and without small missing segments which may be very difficult to detect. Such a core may, however, include layers deposited rapidly and irregularly as density currents, such as the layer at 24–30cm depth in the surficial core from Mirror Lake. The influx may be irregular through time, but in a lake like Mirror Lake the influx is irregular at all sites; none of the sites we investigated was preferable to the Central site in this regard (fig. 10.6B).

Deep lakes with broad, flat bottoms *may* collect sediments in a regular way, without strong focusing toward the centre. However, the present lake bottom gives little clue to past morphometry.

It is difficult to know in advance which lakes will have the fewest problems with pollen influx, and in many regions there is little or no luxury of choice. However, data presentation can help with interpretation. Pollen concentration, pollen influx, and pollen percentages are all intercorrelated variables, but each gives a different, and useful, kind of information about pollen inputs to the lake. All three kinds of information should be presented in pollen-influx studies. Pennington (1979) presents a good example of interpretation based on all three parameters, pointing out that pollen changes in sediment cores that appear as changes in all three are most likely to represent real changes in pollen inputs to the lake. Such changes can be distinguished from those resulting from focusing, which show up as changes in pollen influx without a correlative change in concentration (Pennington 1979). Over short segments of the core the pollen concentration can be very helpful in interpretation; it can be used, for example, to detect increased pollen input due to the local arrival of a plant species.

In an earlier paper (Davis *et al.* 1973) we assessed the variability of pollen-influx measurements. We attempted at that time to calibrate pollen influx with population densities of source trees. We estimated that pollen-influx rates commonly varied by a factor of three from one site to another within a lake, although still larger, 10-fold variations might be found in exceptional cases. These estimates were too conservative. Variations by a factor of 5 are common, even when one considers only sediment from the deep basin of the lake. Another way to assess variability is to inspect pollen-influx curves through time at individual lakes. During the Holocene, when pollen productivity was probably fairly constant, deep-water cores from the lakes we have studied in New Hampshire show total pollen influx values that vary as follows:

Sandogardy Pond	3-fold
Mirror Lake Central core	4-fold
core 782	5-fold
Cone Pond	4-fold
Lost Pond	2-fold
Little East Pond	3-fold
Lonesome Lake	5-fold
Carter Pond	3-fold

Despite the lack of precision of pollen-influx measurements inherent in the methodology, influx measurements are fairly consistent for the late-glacial portion of sediment cores, at least in New England. This is fortunate, as late-glacial sediments are those for which pollen-influx measurements are most needed: late-glacial pollen percentages are often misleading and difficult to interpret because the pollen productivity of the vegetation changed greatly as tundra was replaced by spruce forest (Davis and Deevey 1964). It seems likely that the low variance of late-glacial influx is a function of the nature of the sediment, which is silt or silty sand at most sites in northeastern United States. These sediments are not strongly focused at Mirror Lake, nor, presumably, at other sites. Pennington (1973) has also shown consistent influx values for small and large lakes during the warmer interstadial phase of the British late-glacial, but warns that large lakes may show exceptionally low influx during colder stadial phases when increased throughput of glacial meltwater decreased the pollen supply.

Pennington (1973: 80, 79) anticipated our study 10 years ago when she pointed out that 'the very uneven distribution of sediment on the bottom of some lakes . . . raise(s) doubts about the meaning of deposition rates per unit area per year'. She suggested then that it would be 'necessary . . . to consider for each site what local influences, apart from pollen production by the vegetation represented, . . . may affect pollen deposition rates'. Our studies have extended her investigations of British lakes to New Hampshire, corroborating her insights on the processes affecting sediment and pollen accumulation in lakes.

Acknowledgments

This work has been supported by the National Science Foundation. This paper is a contribution to the Hubbard Brook Ecosystem study.

References

Berglund, B.E., 1973. Pollen dispersal and deposition in an area of Southeastern Sweden – some preliminary results. In *Quaternary Plant Ecology*, ed. H.J.B. Birks and R.G. West, 117–130.

Bonny, A.P., 1976. Recruitment of pollen to the seston and sediment of some Lake District lakes. *J. Ecol.* 64: 859–87.

Bonny, A.P., 1978. The effect of pollen recruitment processes on pollen distribution over the sediment surface of a small lake in Cumbria. *J. Ecol.* 66: 385–416.

Bormann, F.H. and Likens, G.E., 1979. *Pattern and Process in a Forested Ecosystem* (New York).

Brubaker, L.B., 1975. Postglacial forest patterns associated with till and outwash in northcentral Upper Michigan. *Quaternary Res.* 5: 499–527.

Cushing, E.J., 1967. Evidence for differential pollen preservation in late Quaternary sediments in Minnesota. *Rev. Paleobot. Palynol.* 4: 87–101.

Davis, M.B., 1968. Pollen grains in lake sediments: Redeposition caused by seasonal water circulation. *Science* 162: 796–9.

Davis, M.B., 1969. Climatic changes in southern Connecticut recorded by pollen deposition at Rogers Lake. *Ecology* 50: 409–22.

Davis, M.B., 1973. Redeposition of pollen grains in lake sediment. *Limnol. Oceanogr. 18:* 44–52.

Davis, M.B., 1976. Erosion rates and land-use history in southern Michigan. *Environ. Conserv. 3:* 139–48.

Davis, M.B., 1981. Outbreaks of forest pathogens in Quaternary history. *Proc. IV int. palynol. Conf. Lucknow (1976–77), 3:* 216–27.

Davis, M.B., 1984. History of the vegetation on the watershed. In *An Ecosystem Approach to Limnology: Mirror Lake and its Watershed*, ed. G.E. Likens (New York).

Davis, M.B. and Brubaker, L.B., 1973. Differential sedimentation of pollen grains in lakes. *Limnol. Oceanogr. 18:* 638–46.

Davis, M.B., Brubaker, L.B. and Webb, T., III., 1973. Calibration of absolute pollen influx. In *Quaternary Plant Ecology*, ed. H.J.B. Birks and R.G. West, 9–25.

Davis, M.B. and Deevey, E.S., Jr, 1964. Pollen accumulation rates: estimates from late-glacial sediment of Rogers Lake. *Science 145:* 1293–5.

Davis, M.B. and Ford, M.S. (J.), 1982. Sediment focusing in Mirror Lake, New Hampshire. *Limnol. Oceanogr. 27:* 147–50.

Davis, M.B. and Ford, J., 1984. Late-glacial and Holocene sedimentation. In *An Ecosystem Approach to Limnology: Mirror Lake and its Watershed*, ed. G.E. Likens (New York).

Davis, M.B. and Goodlett, J.C., 1960. Comparison of present vegetation with pollen spectra in surface samples from Brownington Pond, Vermont. *Ecology 41:* 346–57.

Davis, M.B., Spear, R.W. and Shane, L.C.K., 1980. Holocene climate of New England. *Quaternary Res. 14:* 240–50.

Davis, R.B., Bradstreet, T.E., Stuckenrath, R. and Borns, H.W., Jr, 1975. Vegetation and associated environments during the past 14,000 years near Moulton Pond, Maine. *Quaternary Res. 5:* 435–65.

Ford, J., 1984. The influence of lithology on ecosystem development in New England: A comparative palaeological study. Ph.D. thesis, University of Minnesota.

Gasith, A., 1976. Seston dynamics and tripton sedimentation in the pelagic zone of a shallow eutrophic lake. *Hydrobiologia 51:* 225–31.

Jacobson, G.L., Jr and Bradshaw, R.H.W., 1981. The selection of sites for paleovegetational studies. *Quaternary Res. 16:* 80–96.

Johnson, T.C., 1980. Sediment redistribution by waves in lakes, reservoirs and embayments. In *Proceedings of the Symposium on Surface Water Impoundments*, ed. H. Stefan, 1307–17 (American Society of Civil Engineers: June 2–5 1980, Minneapolis).

Kerfoot, W.C., 1974. Net accumulation rates and the history of cladoceran communities. *Ecology 55:* 51–61.

Koidsumi, K. and Sakurai, Y., 1968. Precipitating substances of lake Suwa (Materials for the limnology of Lake Suwa, IV). (Japanese, English summary). *Jap. J. Ecol. 18:* 212–17.

Lamb, H.F., 1980. Late Quaternary vegetational history of southeastern Labrador. *Arctic Alpine Res. 12:* 117–35.

Lastein, E., 1976. Recent sedimentation and resuspension of organic matter in eutrophic Lake Esrom, Denmark. *Oikos 27:* 44–9.

Lehman, J.T., 1975. Reconstructing the rate of accumulation of lake sediment: The effect of sediment focusing. *Quaternary Res.* 5: 541–50.

Likens, G.E., 1972. Mirror Lake: its past, present, and future? *Appalachia 39:* 23–41.

Likens, G.E., 1984. *An Ecosystem Approach to Limnology: Mirror Lake and its Watershed* (New York).

Likens, G.E. and Davis, M.B., 1975. Post-glacial history of Mirror Lake and its watershed in New Hampshire U.S.A.: an initial report. *Verh. int. Verein. theor. angew. Limnol. 19:* 982–93.

Likens, G.E., Bormann, F.H., Pierce, R.S., Eaton, J.S. and Johnson, N.M., 1977. *Biogeochemistry of a Forested Ecosystem (New York).*

Lundqvist, G., 1927. Bodenablagerungen und Entwicklungstypen der Seen. In *Binnengewässer 2,* ed. A. Thienemann (Stuttgart).

Macan, T.T., 1970. *Biological Studies of the English Lakes* (New York).

Moeller, R.E., 1984. Contemporary sedimentation. In *An Ecosystem Approach to Limnology: Mirror Lake and its Watershed,* ed. G.E. Likens (New York).

Moeller, R.E. and Likens, G.E., 1978. Seston sedimentation in Mirror Lake, New Hampshire, and its relationship to long-term sediment accumulation. *Verh. int. Verein. theor. angew. Limnol. 20:* 525–30.

Mothes, G., 1981. Sedimentation und Stoffbilanzen in Seen des Stechlinseegebiets. *Limnologica (Berlin) 13:* 147–94.

Mueller, W.P., 1964. The distribution of cladoceran remains in surficial sediments from three Northern Indiana lakes. *Invest. Indiana Lakes Streams 6:* 1–63.

Ohle, W., 1962. Der Stoffhaushalt der Seen als Grundlage einer allgemeinen Stoffwechseldynamik der Gewässer. *Kieler Meeresforsch. 18:* 107–20.

Peck, R.M., 1973. Pollen budget studies in a small Yorkshire catchment. In *Quaternary Plant Ecology,* ed. H.J.B. Birks and R.G. West, 43–60.

Pennington, W., 1973. Absolute pollen frequencies in the sediments of lakes of different morphometry. In *Quaternary Plant Ecology,* ed. H.J.B. Birks and R.G. West, 79–104.

Pennington, W., 1974. Seston and sediment formation in five Lake District lakes. *J. Ecol. 62:* 215–51.

Pennington, W., 1979. The origin of pollen in lake sediments: an enclosed lake compared with one receiving inflow streams. *New Phytol. 83:* 189–213.

Pennington, W. (Mrs T.G. Tutin), Cambray, R.S., Eakins, J.D. and Harkness, D.D., 1976. Radionuclide dating of the recent sediments of Blelham Tarn. *Freshwater Biol. 6:* 317–31.

Serruya, C., 1977. Rates of sedimentation and resuspension in Lake Kinneret. In *Interactions Between Sediments and Fresh Water,* ed. H.L. Golterman, 48–56, Proceedings of International Symposium, Amsterdam, The Netherlands, 6–10 Sept. 1976. (The Hague).

Simola, H. 1981. Sedimentation in a eutrophic stratified lake in S. Finland. *Ann. Bot. Fennici 18:* 23–36.

Spear, R.W., 1981. The history of high-elevation vegetation in the White Mountains of New Hampshire (Ph.D. thesis, University of Minnesota).

Tauber, H., 1977. Investigations of aerial pollen transport in a forested area. *Dansk Bot. Ark. 32:* 1–121.

Tolonen, M., 1980. Degradation analysis of pollen in sediments of Lake Lamminjärvi, S. Finland. *Ann. Bot. Fennici 17:* 11–14.

Ulen, B., 1978. Seston and Sediment in Lake Norrviken I. Seston composition and sedimentation. *Schweiz. Z. Hydrol. 40:* 262–86.

Von Damm, K.L., Benninger, L.K. and Turekian, K.K., 1979. The ^{210}Pb chronology of a core from Mirror Lake, New Hampshire. *Limnol. Oceanogr. 24:* 434–9.

Waddington, J.C.B., 1969. A stratigraphic record of the pollen influx to a lake in the Big Woods of Minnesota. *Geol. Soc. Am. Spec. Pap. 123:* 263–82.

White, W.S. and Wetzel, R.G., 1975. Nitrogen, phosphorus, particulate and colloidal carbon content of sedimenting seston of a hardwater lake. *Verh. int. Verein. theor. angew. Limnol. 19:* 330–9.

Whitehead, D.R., 1979. Late-glacial and post-glacial vegetational history of the Berkshires, Western Massachusetts. *Quaternary Res. 12:* 333–57.

Whitehead, D.R., Rochester, H., Jr, Rissing, S.W., Douglass, C.B. and Sheehan, M.C., 1973. Late-glacial and post-glacial productivity changes in a New England pond. *Science 181:* 744–7.

Winter, T.C., 1984. Physiographic setting and origin of Mirror Lake. In *An Ecosystem Approach to Limnology: Mirror Lake and its Watershed,* ed. G.E. Likens (New York).

11   Stages in soil development reconstructed by evidence from hypha fragments, pollen, and humus contents in soil profiles

Svend T. Andersen

Introduction

IN AN EARLIER WORK (Andersen 1979), it was noticed that dark-coloured fungal hyphae preserved in various acid forest soils in Denmark had been comminuted into fragments of varying length, and it was assumed that the fragmentation was due to the activity of various soil animals.

Fungal hyphae are present in the leaves when they fall to the ground in the autumn, but increase rapidly in frequency after litter fall due to invasion by soil-inhabiting species (Hering 1965, Minderman and Daniëls 1967, Jensen 1974). Nagel-de Boois and Jansen (1967) found that mycelial growth was slightly higher in calcareous soil than in acid soil; the breakdown rate of mycelium, however, was significantly higher in the calcareous soil than in the acid soil (Nagel-de Boois and Jansen 1971, cf. Waid 1960).

Pigmentation by melanin and other substances protects fungal hyphae in soils against breakdown (Gray and Williams 1975). Pigmented hyphae occur in some Basidiomycotina and Deuteromycotina (Pugh 1974). According to Waid (1960), the proportion of dark-coloured hyphae in soils increases with age, and these hyphae may form 80–90% of the mycelial population after one year. The abundance of pigmented hyphae in acid soils compared with neutral soils noticed by Müller (1878) and other authors may thus, to some degree, be due to the slower breakdown and longer life-span of the hyphae in the acid soils.

In a study of a podzol profile from a *Quercus petraea-Fagus sylvatica* woodland, Nagel-de Boois and Jansen (1971) found that the highest amount of fungal hyphae (per gram organic matter) was in the humus layer and the A_1 horizon of the mineral soil; live mycelia, however, occurred mainly in the leaf layer and were already decreasing in frequency in the leaf fragmentation layer. 95% of the hyphae incorporated in the soil were dead.

In the soils examined by the present author, unbroken hypha strands occurred only in the litter layer, whereas the dark-coloured hyphae preserved in the soil were fragmented. Many kinds of soil animals feed on live mycelia, but it appears that the thick-walled dark hyphae must be fragmented by animals equipped with specialized cutting mouth-parts. Among soil-inhabiting macroarthropods, such mouth-parts are found in Isopoda (woodlice), Diplopoda (millipedes), and the larvae of Diptera (flies). These animals are important for the initial comminution of leaf litter (Heath *et al.* 1965, Edwards 1974) and may bite fungal hyphae into long fragments (Wallwork 1970). Several microarthropods feed on fungal hyphae (Farahat 1966). Acarina (mites), particularly Cryptostigmata (oribatids), have cutting mouth-parts, whereas Collembola (springtails) tend to crush their food (Harding and Stuttard 1974). Oribatids bite plant tissue into small fragments and may produce faecal pellets with abundant short hypha fragments (Forsslund 1938, Schuster 1956, Wallwork 1970).

Macroarthropods such as isopods and diplopods feed mainly in the litter layer and the topmost centimetres of the soil (Bocock and Heath 1967). Oribatids occur to considerable depth levels in brown earth (Sall *et al.* 1949; Haarløv 1960), whereas animal life is concentrated within the topmost centimetres of acid humus, and the bleached sand beneath the humus is nearly void of animals (Weis-Fogh 1947–8; Murphy 1955). Hypha fragments in brown earth are thus comminuted by animals present in the soil, whereas the fragmentation of hyphae in podzols and beneath the topmost centimetres of humus layers is due to former animal communities, whose activity has now ceased. Andersen (1979) found characteristic changes in the occurrence of the hypha-fragment length-classes in a brown-earth profile and a podzol profile.

Measurements of hypha fragments in faecal pellets from various soil animals are compared with measurements of hypha fragments in forest soils in the present work. Hypha fragments in samples from a number of soil profiles have been measured, and the relationship between changes in soil animal communities, soil and vegetation will be discussed.

Site and vegetational history

Eldrup Forest, Djursland, in eastern Jutland (fig. 11.1), and its vegetational history, have been described by Andersen (1978, 1979). A protected research area is dominated by *Fagus sylvatica* with scattered *Quercus petraea* (fig. 11.2). The substrate is sandy till, poor in lime and covered in some places by meltwater sand. The till and the sand are covered by an

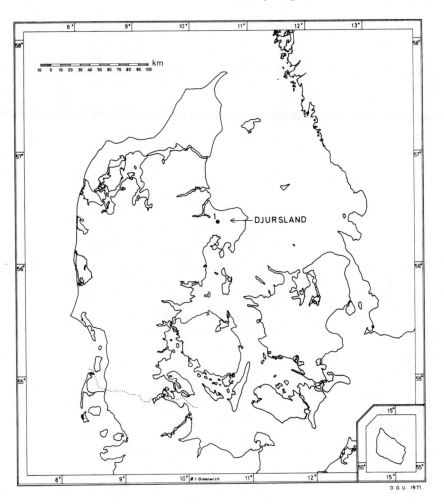

Fig. 11.1   1 = Location of Eldrup Forest in Denmark.

acid humus layer of varying thickness and are podzolized to varying depth levels. Oligotrophic brown earth occurs in a few patches. The ground flora is poor.

The Holocene vegetational history of the forest was studied by pollen analyses from two small hollows with gyttja and peat layers (fig. 11.2). They show a succession of *Betula, Corylus-Pinus-Betula-Populus, Tilia-Quercus-Corylus,* and *Fagus* forest, with restricted human influence in late Sub-boreal, early Sub-atlantic, the medieval period, and the eighteenth century.

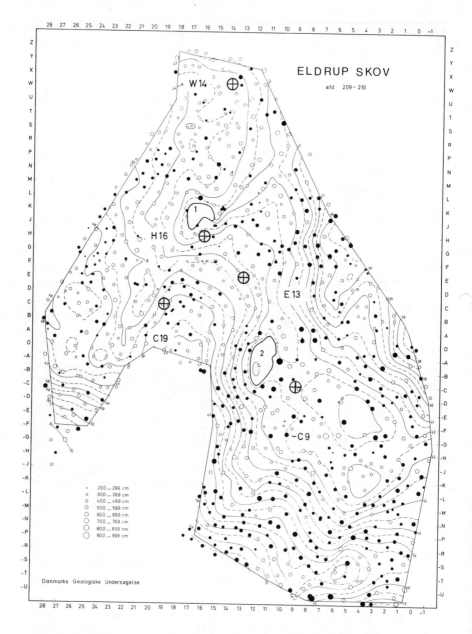

Fig. 11.2 The fenced research area in Eldrup Forest with the soil sections W 14, C 19, E 13, H 16 and –C 9. Coordinate intervals 10m, surface contour intervals 0.5m. ● *Fagus sylvatica*, ○ *Quercus petraea*. 1 and 2 indicate the hollows from which the pollen diagrams in Andersen 1978 were worked out.

Materials

Faecal pellets

Hypha fragments were measured in faecal pellets of soil-inhabiting animals. The pellets were kindly provided by Professor N. Haarløv, the Danish Agricultural University, Copenhagen. Pellets from the following animals were examined:

Glomeris marginata, Julus scandinavicus (Diplopoda), Oniscus asellus, Philoscia muscorum (Isopoda), and Tipulidae larvae (Diptera). Pellets were derived from animals which had fed on agar cultures of mycelia isolated from beech forest soil.
Platynothrus peltifer (Cryptostigmata). Pellets were derived from animals fed on forest soil (1942 and 1950).
Belba sp. (Cryptostigmata). Pellets were attached to an animal collected in Eldrup Forest.
Nothrus silvestris and N. palustris (Cryptostigmata). Pellets were derived from animals extracted from a soil sample from Eldrup Forest.

Soil samples

Soil samples were derived from five soil sections in Eldrup Forest. Four sections were described by Andersen (1979). The sections H16 and C19 are podzols on sandy till, E13 is a podzol on meltwater sand, and W14 is an oligotrophic brown earth developed on sandy till (fig. 11.2). Section C9 is also a podzol developed on sandy till (table 11.1, fig. 11.2).

Table 11.1 Soil section – C9 in Eldrup Forest. The surface of the mineral soil occurs at 27cm depth, according to the loss on ignition determination. This limit was not visible at the soil section.

A_0	0–4cm	Brown, brittle humus
A_0	4–22cm	Blackish-brown, greasy humus
A_{1+2}	22–44cm	Blackish-grey sandy till
A_2	44– cm	Rust-coloured hard sandy till

Andersen (1979) reported measurements of hypha fragments in samples from the sections C19 and W14. Hypha fragments were measured in some additional samples from these sections and in samples from the other sections above.

Methods

The faecal pellets from *Platynothrus peltifer* were mounted in slides with Faure's medium. The other pellets were soaked in benzene, mounted in slides with silicone oil, and crushed by pressure on the coverslip.

Hypha fragments in soil samples were measured in samples prepared for pollen analysis. The samples were boiled in potassium hydroxide, hydrofluoric acid, and acetolysis mixture, and were mounted in silicone oil after washing with alcohol and benzene. The hyphae were undamaged by the chemical treatment. The hypha fragments were drawn on paper by means of a Leitz drawing apparatus and measured with a ruler. The class unit was 1mm, which corresponds to 2μm. If curved, the total length of the curved fragment was measured. Two to three hundred fragments were measured per sample.

Iversen (1964) estimated the frequency of hyphae in soils in relation to pollen numbers. Hypha frequencies were not measured in the present investigation. Pollen percentages were calculated as percentages of all pollen and spores and as percentages of the sum of tree pollen after numerical correction with the correction factors found by Andersen (1970).

Pollen corrosion was measured by noting the number of grains with corrosion scars in each taxon and calculated as a percentage of all grains of that taxon in cases where ten or more grains were counted. The average corrosion percentages shown in the diagrams were calculated as averages of the corrosion percentages found for each tree taxon in a sample (excluding *Pinus* and including *Corylus*). This type of corrosion is due to biological activity (Havinga 1971, Elsik 1971).

Mineral and organic content of the soil samples were determined by ignition at 550°C. Values for ignition residue, measured as per cent of dry weight of the soil, were drawn on logarithmic scales. These curves are likely to illustrate changes in volumetric composition (on a linear scale, cp. Andersen 1979).

Hypha fragments in faecal pellets compared with soil samples

Size-frequency distributions of hypha fragments in the faecal pellets are compared with selected soil samples in figs. 11.3 and 11.4. Macroarthropods may devour fragments already bitten by smaller animals when they feed on soil material. Hence, only fragments in pellets derived from animals which had fed on intact mycelia were measured in these cases. The various oribatid species are the smallest animals living in the soils, and it is assumed that the size-frequency distribution of fragments found in pellets produced by animals which have fed on soil material are characteristic of the species.

The size-frequency distributions for fragments produced by the macro-arthropods *Glomeris marginata*, *Oniscus asellus*, *Julus scandinavicus*, and *Philoscia muscorum* have pronounced modes at 30–45μm and are characteristically skewed to the right (fig. 11.3). The modal value for *Glomeris* is smallest (32μm) and that for *Philoscia* largest (45μm). The modes for *Oniscus* and *Julus* are at 34 and 35μm respectively. The modal value represents the most common fragment length and presumably corresponds to the distance between the mandibles of the animal. Fragments shorter than this value probably derive from cases where the end of a hypha thread was bitten off, whereas the longer fragments are likely to derive from cases where a hypha thread was oriented obliquely to the mandibles. The very long fragments probably passed lengthwise into the oesophagus of the animal.

The fragments produced by Tipulidae larvae show a mode at 40μm (fig. 11.3); the size-frequency distribution curve is much flatter than those for

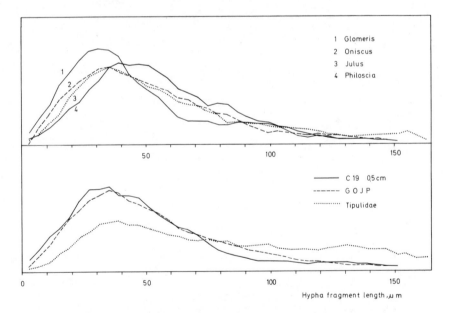

Fig. 11.3   Frequency distribution of the length of hypha fragments in faecal pellets from macroarthropods (*Glomeris marginata*, *Oniscus asellus*, *Julus scandinavicus*, *Philoscia muscorum*), and Tipulidae larvae fed on mycelial cultures, and in a soil sample (section C 19, 0.5cm depth). GOJP = average curve for pellets of *Glomeris*, *Oniscus*, *Julus*, and *Philoscia*. The distribution curves were smoothed by running averages.

the other macroarthropods. These larvae apparently tend to tear the hypha threads apart and produce fragments of greatly varying length.

A size-distribution curve for hypha fragments from the topmost sample in profile C19 has a modal value at about 30μm and is also characteristically skewed to the right (fig. 11.3). The curve differs essentially from that for the Tipulidae larvae and is very similar to an average curve for *Glomeris, Oniscus, Julus,* and *Philoscia.* The hypha fragments in this soil sample thus were apparently produced by an assemblage of macroarthropods. Isopods and diplopods are important for the initial comminution of leaf litter and prefer *Fagus* and *Quercus* leaves in a weathered state (Edwards 1974). Weathered leaves are infected with dark fungal hyphae, which apparently were comminuted as the animals fed on the litter.

The hypha fragments in faecal pellets from oribatids are considerably smaller than those measured in the macroarthropod pellets. The fragments in the two collections produced by *Platynothrus peltifer* (fig. 11.4) differ somewhat as to modal values (16 and 20μm). Hence different animals of the same species may produce fragments of slightly different

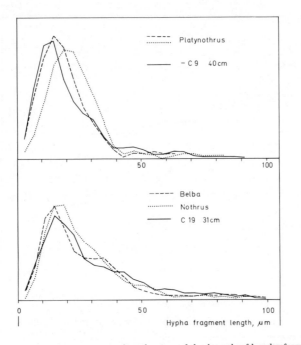

Fig. 11.4   Frequency distribution of the length of hypha fragments in faecal pellets from oribatids (*Platynothrus peltifer, Belba* sp., and *Nothrus silvestris* + *N. palustris*) fed on forest soil material, and in two soil samples (sections −C 9, 40cm, and C 19, 31cm depth). The distribution curves were smoothed by running averages.

length. The two frequency distribution curves are, however, similar in being nearly symmetrical with a steep decrease of the frequencies for fragments up to 40μm long and a right-hand tail of low frequencies for fragments 40–90μm long. The curves for *Belba* sp. and the *Nothrus* species have similar modes (at 15 and 17μm), but the frequencies for the longer fragments decrease more gradually than in the *Platynothrus* pellets. It is not possible to decide whether this difference is accidental or due to different feeding habits. The curves indicate that the oribatids also produce hypha fragments of a characteristic length and accidently swallow longer fragments up to 100μm long.

Two size-frequency curves from soil samples (–C9, 40cm and C19, 31cm depth) match the *Platynothrus* or the *Belba-Nothrus* pellets respectively, and it is indicated that the organic material in these samples has been comminuted thoroughly by oribatids.

Andersen (1979) showed that the hypha fragments in samples from the brown earth profile (W14) could be grouped in the size classes 1–21μm, 21–35μm, 35–61μm, and longer than 61μm, according to their correlation with depth. The frequencies of the four size-classes of hypha fragments in the faecal pellets and soil samples from figs. 11.3 and 11.4 are shown in fig. 11.5. Fragments shorter than 21μm dominate, and fragments longer

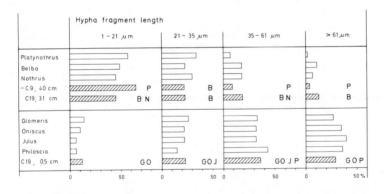

Fig. 11.5 Frequencies of hypha fragment length classes in the faecal pellets of oribatids and macroarthropods, and in the soil samples shown in figs. 11.3–11.4. The letters indicate that the frequency of the size class did not differ significantly (P<0.05) from the frequency of the same size class in the faecal pellets from the animal indicated (P = *Platynothrus*, B = *Belba*, N = *Nothrus*, G = *Glomeris*, O = *Oniscus*, J = *Julus*, P = *Philoscia*). Significance tests by the chi-squared method.

than 35μm are rare in the oribatid pellets. The fragments in the two soil samples match the *Platynothrus* sample (–C9) and the *Belba* and *Nothrus* samples (C19, 31cm). The macroarthropod pellets predominantly contain fragments longer than 35μm, and fragments shorter than 21μm are rare, whereas the frequencies for fragments 21–35μm long are similar to those found in the oribatid pellets. The soil sample C19, 0.5cm, matches the pellets of the four macroarthropods.

The frequencies for hypha fragments 21–35μm in all the samples measured are uncorrelated, whereas the fragments longer than 35μm are strongly negatively correlated with the frequencies for fragments shorter than 21μm (fig. 11.6). Oribatids thus bite hyphae longer than 21μm and

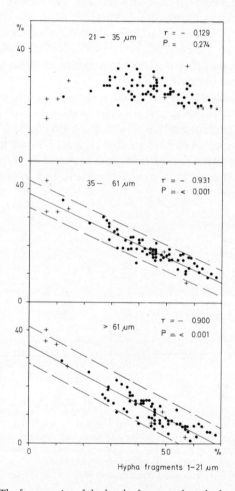

Fig. 11.6 The frequencies of the hypha fragment length classes 21–35μm, 35–61μm, and >61μm compared with the frequencies of the length class 1–21μm in faecal pellets from macroarthropods and oribatids (+) and in all the soil samples measured (●). r = least squares correlation. Calculated regression lines with one standard deviation are shown.

produce fragments shorter than 35μm, with the result that short fragments increase, intermediate fragments increase and decrease equally, and long fragments decrease in frequency.

Several of the soil samples in fig. 11.6 are similar to the oribatid pellets, and samples with less than 45% short fragments are intermediate between the oribatid and the macroarthropod pellets. The latter samples were apparently comminuted by oribatids to varying degrees. The intensity of oribatid activity can thus be measured by the frequency of short fragments. If the frequency of short fragments is about 10%, then the samples have been treated by macroarthropods alone, and samples with 45% or more short fragments have been completely comminuted by oribatids. The hypha fragments in samples with intermediate frequencies (10–45%) of short fragments have been more or less completely treated by oribatids.

Vertical variation of hypha fragment length, organic content, and pollen corrosion in soil samples

The vertical distribution of short hypha fragments in the five soil sections examined is compared with mineral and organic content and with pollen corrosion (fig. 11.7).

MINERAL SOIL

Section W14. Organic content is high in the uppermost 30cm of the brown earth (fig. 11.7) indicating intensive downward transportation by burrowing earthworms and incomplete mineralization. Short hypha fragments are infrequent in the topmost samples. The fungal hyphae are thus comminuted mainly by macroarthropods in the litter layer and are transported downward into the soil by earthworms before they have been treated thoroughly by oribatids. The frequency of the short hypha fragments increases below 7cm depth, indicating that the organic material becomes further comminuted by oribatids living in the soil.

Pollen grains present in the topmost 30cm were transported downwards mainly by earthworms (Andersen 1979). Corroded pollen is scarce (less than 10%) throughout the section. The pH-level is about 5–6 in the topmost 30cm. The organisms which attack pollen grains are, thus, nearly absent at these pH-levels.

Section C19. The organic content is very low below 6cm depth in the mineral soil of the podzol at section C19 (fig. 11.7). The organic matter was apparently strongly mineralized, and only resistant particles such as pollen grains and hypha fragments are preserved. Organic content increases in the topmost 5cm, which constitute the A_1 horizon. The

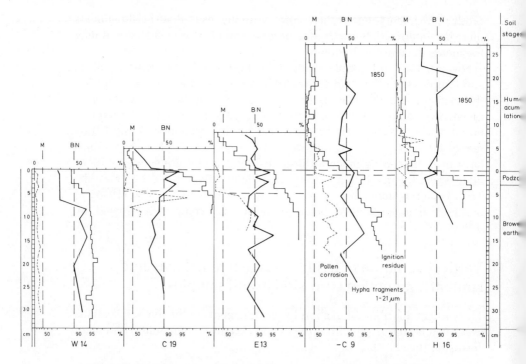

Fig. 11.7 The frequencies of hypha fragments 1–21μm long, average pollen corrosion (uppermost percentage scale), and ignition residue (lowermost percentage scale) in humus layers above, and in mineral soil below, the mineral surface (at 0cm) in the five soil sections examined. Humus accumulation began at C 19 and E 13 about AD 1850. The AD 1850 level is indicated for –C 9 and H 16. M indicates the frequency of short hypha fragments in macroarthropod pellets (10%), and BN the lower limit for oribatid pellets (45%, *Belba* and *Nothrus*).

frequencies of the short hypha fragments found in the mineral soil differ from those found in the brown earth. Short fragments are very frequent in the topmost 5cm; they are less frequent at 5–20cm depth, and increase again below 20cm. Andersen (1979) found that the increase of short fragments below 20cm depth indicates a former brown-earth stage, where the hypha fragments were transported into the soil by burrowing earthworms and were comminuted by oribatids. It appears, however, that the oribatid activity was lower than in the brown earth at W14, because the short fragments increase at a greater depth level (20cm) than at W14 (7cm). The high frequencies of short hypha fragments in the topmost 6cm at C19 indicate according to Andersen (1979) that the brown-earth stage was replaced by a podzoloid stage with a soil fauna dominated by macroarthropods and oribatids. The downward transport of organic matter was restricted to the uppermost part of the mineral soil due to the disappearance of the burrowing earthworms, and hypha fragments became thoroughly comminuted by oribatids. Perel *et al.* (1966) thus have shown that organic matter becomes mixed into the soil only to a shallow depth, if earthworms are excluded.

Pollen originally deposited at the surface was transported downwards during the brown-earth and podzoloid stages (Andersen 1979). Pollen corrosion is moderate to high (20–70%) below 4cm depth, indicating that the pollen transported downward in the brown-earth stage was exposed to a high biological activity. pH accordingly was higher than 5.0 during the brown-earth stage. The pollen buried in the podzoloid stage is slightly corroded (less than 5%) and pH had accordingly decreased to about 5.0 or less.

Section E13. A development similar to that described for section C19 was found in the mineral soil of the podzol profile E13. Short hypha fragments are frequent at the lowermost levels (below 10cm); they have low values at 6–10cm, and increase again in the topmost 6cm, at which level organic content is also high. Hence, a change from a brown-earth stage with earthworm and oribatid activity to a podzoloid stage with an arthropod community is recorded. It appears that the oribatid activity during the brown-earth stage at E13 was stronger than at C19, because short fragments dominate up to a higher level (10cm) than at C19 (20cm).

Pollen corrosion is 30–40% below 5cm depth, indicating a pH-level higher than five during the brown-earth stage, and is lower than 5% in the podzoloid stage indicating increased acidity similar to section C19.

Section –C9. Organic content in the mineral soil at the podzol profile –C9 is higher than at C19 and E13 probably because larger amounts of colloidal humus were transported down from the thick humus layer above. There is no distinctive increase in organic matter in the topmost centimetres. Hypha fragments are scarce compared with the other sections, indicating that the production of dark hyphae in the litter layer was low. The frequencies of the short hypha fragments are generally high, and only a slight increase is recorded in the topmost sample. Former brown-earth and podzoloid stages are thus only slightly differentiated at this section.

Pollen corrosion is moderate (20–40%) below 1cm depth, indicating a pH-level above five in the brown-earth stage. The corrosion percentage is somewhat lower in the topmost sample (10%), indicating increased acidity.

Section H16. The soil profile at H16 is shallower than at the other sections due to a high groundwater level (Andersen 1979). The content of organic matter increases in the topmost centimetres of the mineral soil. Short hypha fragments are frequent below 5cm depth; there is a minimum at 2–4cm, and increased frequency in the uppermost sample. It appears that there occurred a brown-earth stage, where earthworm penetration was hampered by groundwater and oribatids were frequent. This stage was replaced by a podzoloid stage which lasted for only a short time.

Pollen corrosion is low (c. 10%), indicating a fairly low pH-level (slightly above 5.0), and decreases only slightly in the topmost sample of the mineral soil.

HUMUS LAYERS

A stage where humus accumulated on top of the mineral soil followed the podzoloid stage. The podzolization of the mineral soil was completed during this stage. The onset of humus accumulation varied somewhat in the four podzol profiles in fig. 11.8. Humus began to accumulate at about AD 1850 at the sections C19 and E13 (Andersen 1979), at about 2000 BC at −C9 (see below), and at about AD 1500 at H16 (Andersen 1979). Since AD 1850, a level which is easily recognized in the pollen diagrams, 4.5cm humus has accumulated at C19, 8cm at E13, 6cm at −C9, and 10cm at H16. Humus accumulation thus is highest at H16 and lowest at C19. The humus-accumulation rate depends on the amount of leaf litter fall and, hence, the exposure to wind. Section C19 is at a less sheltered position than the other sites.

Hypha fragments and pollen grains are not displaced vertically in the humus, and the degrees of comminution and corrosion thus reflect biological activity at the time of deposition or shortly afterwards.

Short hypha fragments are scarce in the humus layer at C19 and their frequency decreases towards the surface. Hence, oribatids were scarce and are nearly absent at the surface. Oribatid activity was much higher in the humus layers at E13, −C9, and H16, as shown by the high frequencies of short hypha fragments. The topmost samples have in some cases (E13, H16) a lower content of short hypha fragments than the samples just below. This presumably indicates that oribatids living in the topmost centimetres of the humus have not yet comminuted all the long fragments produced by the macroarthropods.

Corroded pollen grains are nearly absent in the humus layers at the sections C19 and E13. The pH-level is about 4.0 in the topmost samples, indicating that the organisms which attack pollen grains are absent at pH less than 5.0. The pH-level thus decreased from about 5.0 to about 4.0 at the onset of humus accumulation. At the sections −C9 and H16 there are low frequencies (c. 10%) of corroded grains in the lowermost 6–8cm of the humus layer. Hence a slight biological breakdown of pollen grains occurred, and the humus was apparently only mildly acid (pH at 5.0 or slightly higher). Later, acidity increased further, and only a few corroded grains (less than 5%) occur.

Soil development in the podzol profiles

A change from a former brown-earth stage with an earthworm and oribatid community and moderately attacked pollen grains, to a podzoloid stage with an arthropod community and with low biological breakdown of the pollen grains, is recorded in the mineral soil of the four podzol profiles studied. The podzol developed after the onset of humus accumulation. The level of oribatid activity in the brown-earth stages and the distinction of the podzoloid stages differ somewhat in the four sections.

At C19 and E13 there are distinctive minima in oribatid influence (at 5–20cm and at 5–10cm), indicating that the comminution of the hyphae during the transportation downward in the brown-earth stage was fairly slow, whereas biological attack on the pollen grains was moderate, indicating a mildly acid soil (pH higher than 5.0). There was a distinctive podzoloid stage with strong comminution of hyphae, increased organic content, and only slight breakdown of pollen grains, indicating low biological activity and increased acidity (pH about 5.0). Acidity was high at the onset of the humus accumulation (pH about 4.0). The level of the oribatid activity was distinctly lower at section C19 than at section E13. It appears that the differences in oribatid activity at the two sections was due to differences in exposure to wind and hence desiccation.

The productivity of dark fungal hyphae was low during the brown-earth stage at −C9, and their comminution by oribatids was strong. Oribatid activity was intensive at section H16, probably because of a high moisture level. Pollen corrosion is moderate at section −C9 and fairly low at H16, indicating pH-levels above or slightly above 5.0. Podzoloid stages are only slightly differentiated, indicating more rapid changes from the brown-earth stages to the stages with humus accumulation than at the sections C19 and E13. There was low biological attack on pollen grains indicating moderately acid conditions (pH at 5.0 or slightly higher) during the earliest phase of humus accumulation. The acidity later increased and biological breakdown of the pollen grains ceased.

Two types of soil development thus can be distinguished: the former with a distinctive podzoloid stage (C19, E13) and the latter with a more rapid change from a brown-earth stage to a stage with humus accumulation (−C9, H16).

Vegetation and soil development

Pollen grains are destroyed in neutral soils, whereas they are preserved in acid soils and are transported downwards into the mineral soil by soil fauna (Andersen 1979). Pollen diagrams from podzols thus reflect changes in vegetation during the stages of soil development immediately preceding the onset of podzolization. Pollen assemblages buried in acid brown earth become strongly mixed in the vertical direction during the transportation downward and the pollen curves are smoothed out, whereas the pollen assemblages are better differentiated in the topmost part of podzolized mineral soil due to decreased mixing activity (Andersen 1979). The pollen assemblages preserved in acid humus layers are well-differentiated.

Pollen diagrams from the sections W14, C19, E13, and H16 show a succession of *Fagus* forest (seventeenth century), grazed *Fagus* forest (eighteenth century), and rejuvenated *Fagus–Quercus* forest (nineteenth century), whereas pollen from the former *Tilia*-dominated forest stage was destroyed (Andersen 1979).

The former oligotrophic brown-earth stage at the sections C19 and E13 belongs to the *Fagus*-forest stage. The change to the podzoloid stage took place during the grazing stage, and humus accumulation began during the reforestation after grazing was abandoned (Andersen 1979).

The change from the brown-earth stage to the stage with humus accumulation at section H16 took place during the *Fagus*-forest stage. Section H16 is in a shallow depression where leaves accumulate. Hence the change from the brown-earth stage to the humus-accumulation stage cannot have been due to desiccation. The pH-level was only slightly above five during the brown-earth stage. The heavy litter accumulation thus probably caused quicker acidification of the soil than at the neighbouring sites, C19 and E13. Humus accumulated continuously during the grazing stage and the shallow depression was apparently not trampled by the cattle.

A pollen diagram from section −C9 is shown in fig. 11.7. The pollen analyses from the mineral soil show a forest stage with dominant *Tilia* and some *Quercus* and *Corylus*. Plants indicative of acid soil (*Gymnocarpium, Melampyrum*) were frequent. Pollen corrosion is moderate and it can be suggested that the *Tilia*-forest grew on slightly acid brown earth. A similar *Tilia*-forest stage, dating from Atlantic and early Sub-boreal time, is recorded in the pollen diagram from a wet hollow 30m from section −C9 (site 2, fig. 11.2, Andersen 1978). Pollen from the preceding *Betula*- and *Corylus*-forest stages is not preserved in the soil at −C9, and it appears that the brown earth at the site became sufficiently acid for pollen preservation during the *Tilia*-forest stage. *Tilia* leaves disintegrate more rapidly than *Fagus* leaves in woodland soil (Heath *et al.* 1965). The composition of the leaf litter thus explains the low productivity of dark fungal hyphae noticed in the brown-earth stage at this section.

Tilia decreases somewhat, and *Quercus* and *Betula* increase in the

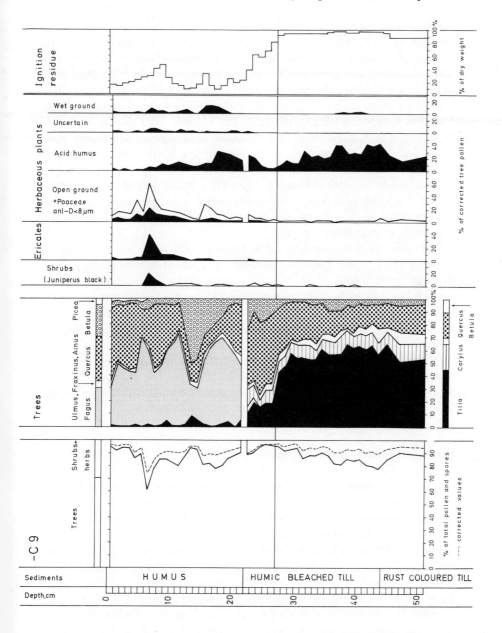

Fig. 11.8 Pollen diagram from section –C 9. The frequencies for tree pollen and non-tree pollen to the right were calculated as percentages of the tree pollen after numerical corrections with the following factors: *Quercus, Betula, Alnus, Corylus* × 0.25; *Ulmus, Picea* × 0.5; *Fagus* × 1; *Tilia, Fraxinus* × 2. The heavy line indicates the surface of the mineral soil.

topmost centimetres of the mineral soil at −C9 (27–29cm below the surface, fig. 11.8), and *Quercus* is dominant in the lowermost centimetres of the humus layer (22–27cm, fig. 11.8). A similar abrupt replacement of the *Tilia*-forest by *Quercus–Betula*-forest is recorded in the pollen diagram at site 2. This change was interpreted by Andersen (1978) as the onset of a stage with anthropogenic influence and was dated to middle Sub-boreal time (at about 2000 BC). The change from the brown-earth stage to the podzoloid stage, and the onset of humus accumulation, thus coincided with a change in the composition of the leaf litter from dominant *Tilia* leaves to dominant *Quercus* leaves. *Quercus* leaves are less palatable to earthworms than are *Tilia* leaves (Heath *et al.* 1965). Hence the disappearance of the earthworms and the onset of humus accumulation seem to have been due to changes in the forest community caused by human disturbance.

The *Quercus*-dominated forest at section −C9 is replaced by *Fagus*-forest at 22cm depth. This change occurred at about AD 500 in the pollen diagram from site 2. A *Betula*-maximum at 12–18cm depth can be ascribed to human activity in medieval times; and grazing activity with subsequent forest regeneration in the eighteenth and nineteenth centuries is reflected by maxima for open-ground herbs, Ericales (mainly *Calluna*), *Juniperus*, and *Quercus*.

It thus appears that the well-developed podzoloid stage at sections C19 and E13 coincided with a grazing stage, whereas the change from the brown-earth stage to the stage with humus accumulation at H16 and −C9 took place during forest stages without grazing influence.

Summary and discussion

Dark-coloured fungal hyphae develop in slowly decomposing leaf litter and become fragmented by the soil fauna. Macroarthropods such as diplopods and isopods are responsible for the initial comminution, and microarthropods, particularly oribatids, bite the hyphae into still smaller fragments. It has been shown that the level of oribatid influence in soil samples can be measured by the frequency of fragments shorter than 21μm. Hypha fragments produced by macroarthropods contain about 10% short fragments, whereas the fragments produced by oribatids contain 45% or more short fragments. Intermediate frequencies of the short fragments indicate various degrees of oribatid influence.

Dark-coloured hypha fragments produced at the surface of acid mineral forest soil are transported downwards by the fauna and comminuted by oribatids living in the soil. The frequency of short fragments increases with depth in oligotrophic brown earth. A former brown-earth stage in podzols is therefore revealed by increasing frequencies of short fragments with depth. Moderate pollen corrosion indicates moderate biological activity and moderate acidity (pH higher than 5.0) during the brown-earth stage.

Podzoloid soil stages, where burrowing earthworms were absent, are characterized by high frequencies of short hypha fragments, low pollen corrosion and a concentration of organic matter in the topmost part of the mineral soil. A change from a former brown-earth stage to a podzoloid stage can thus be traced in podzols by an increase of short hypha fragments, a decrease in pollen corrosion, and an increase in organic content in the topmost part of the mineral soil. Two types of soil development can be distinguished according to the length of the podzoloid stage.

Various levels of oribatid influence can be detected in the humus layers accumulated on top of the mineral soil, probably depending on variations in moisture.

The changes in soil conditions indicated by hypha measurements can be related to changes in vegetation.

The influence of oribatids was very high during the brown-earth stage at section −C9, probably because of a rapid disintegration of the dominant *Tilia* leaves in the litter and a low production of dark fungal hyphae. A short podzoloid stage, and the transition to humus accumulation, coincided with a change to *Quercus*-dominated forest due to human activity at about 2000 BC and, hence, a change in the composition of the leaf litter to leaves less palatable to earthworms.

Tilia–Quercus forest persisted until nearly AD 1000 near the sites C19, E13, and H16 (Andersen 1978), but no traces of this forest stage are preserved in the soil profiles. *Tilia* probably disappeared due to human activities during the Middle Ages, from which time agricultural activity is indicated (Andersen 1979). *Fagus* forest growing on brown earth can be traced in the podzol profiles. The brown earth at section H16 was rather acid. It changed to a short podzoloid stage, and humus accumulation began during the *Fagus*-forest stage. These soil changes were not provoked by human influence, but were rather due to locally heavy litter accumulation.

The well-developed podzoloid stage at the sections C19 and E13 coincided with a period of cattle grazing in the eighteenth century. The change to the podzoloid stage was rather abrupt and the podzoloid stage persisted for some time. Humus accumulation began in the nineteenth century during reforestation. The hypha measurements indicate higher oribatid activity at the sections E13 and H16 than at C19, probably due to a higher soil moisture. Brown earth persisted in a few patches until today. The section W14 is from such a patch of brown earth.

As mentioned above, traces of the former *Tilia*-dominated forest are absent in the mineral soil at the sections W14, C19, E13, and H16, and traces of the *Betula* and *Corylus* stages are absent at section −C9. It appears that pollen buried in brown earth is eventually destroyed, and that only pollen buried immediately before the cessation of earthworm activity is preserved.

The initially raw soils in Eldrup Forest seem to have developed into oligotrophic brown earth by a progressive spontaneous process (retrogressive succession *sensu* Iversen 1964). Oligotrophic brown earth had already developed in the *Tilia*-forest stage of the early Sub-boreal period and has persisted until today in some small areas. The change to podzoloid soil and podzol was provoked by human activity in most cases. The podzoloid stage was short in the cases where the change from brown earth was due to a change in the composition of the tree leaf litter (−C9) or a high leaf litter fall (H16), whereas the podzoloid stage provoked by grazing activities persisted for some time until humus accumulation began during reforestation (C19, H13).

Andersen (1979) showed that the genesis of soils can be illuminated by a study of the pollen and hypha fragments preserved in the soils. The present study shows that differences in soil evolution can be traced even within a small area due to variations in the timing and nature of human disturbance, and due to local variations in exposure and soil moisture. The change from brown earth to podzol was spontaneous in one case.

Dark-coloured fungal hyphae resistant to decomposition form in an acid terrestrial environment. The presence of eroded organic material of terrestrial origin in lake sediments can be detected by an estimation of the hypha content (Cushing 1964, Birks 1970, 1973, Jóhansen 1977). Tutin (1969, cf. Pennington 1975, Jóhansen 1978), has shown that such eroded material can cause disturbance of radiocarbon dates. Hence, studies of hypha remains in lake sediments are also important.

Acknowledgments

The research area in Eldrup Forest was established in agreement with the Løvenholm Foundation and with financial support from the Carlsberg Foundation. The field work was carried out together with H. Bahnson, The Geological Survey of Denmark. Special thanks are due to Professors N. Haarløv and V. Jensen, The Danish Agricultural University, for providing faecal pellets from arthropods and for useful discussions; to forester E. Due, Supervisor of the Løvenholm Foundation, for stimulating discussions in the field; and to S. Gravesen and A. Kiøller for information about fungi. B. Stavngaard carefully performed the numerous hypha measurements, F. Wienberg made the drawings, and O. Cole Collin revised the manuscript. Dr H.J.B. Birks, University of Cambridge, was kind enough to revise the English.

References

Andersen, S.T., 1970. The relative pollen productivity and pollen representation of North European trees, and correction factors for tree pollen spectra. *Danm. geol. Unders. II: 96.*

Andersen, S.T., 1978. Local and regional vegetational development in eastern Denmark in the Holocene. *Danm. geol. Unders., Årbog 1976: 5–27.*

Andersen, S.T., 1979. Brown earth and podzol: soil genesis illuminated by microfossil analysis. *Boreas 8: 59–73.*

Birks, H.J.B., 1970. Inwashed pollen spectra at Loch Fada, Isle of Skye. *New Phytol. 69: 807–20.*

Birks, H.J.B., 1973. *Past and Present Vegetation of the Isle of Skye. A Palaeoecological Study.*

Bocock, K.L. and Heath, J., 1967. Feeding activity of the millipede *Glomeris marginata* (Villers) in relation to its vertical distribution in the soil. In *Progress in Soil Biology,* ed. O. Graff and J.E. Satchess, 233–40. (Braunschweig).

Cushing, E.J., 1964. Redeposited pollen in late-Wisconsin pollen spectra from east-central Minnesota. *Am. J. Sci. 262: 1075–88.*

Edwards, C.A., 1974. Macroarthropods. In *Biology of Plant Litter Decomposition 2:* ed. C.H. Dickinson and G.J.F. Pugh, 533–44.

Elsik, W.C., 1971. Microbiological degradation of sporopollenin. In *Sporopollenin,* ed. J. Brooks, P.R. Grant, M. Muir, P. van Gijzel and G. Shaw, 480–511.

Farahat, A.Z., 1966. Studies on the influence of some fungi on Collembola and Acari. *Pedobiologia 6: 258–68.*

Forsslund, K.H., 1938. Bidrag til kännedomen om djurlivets i marken inverkan på markomvandlingen. *Meddr. Statens Skogsförsöksanst. 31: 87–107.*

Gray, T.R.G. and Williams, S.T., 1975. *Soil Micro-organisms.*

Haarløv, N. 1960. Microarthropods from Danish soils, *Oikos, Suppl. 3.*

Harding, D.J.L. and Stuttard, R.A., 1974. Microarthropods. In *Biology of Plant Litter Decomposition 2* ed. C.H. Dickinson and G.J.F. Pugh, 489–532.

Havinga, A.J., 1971. An experimental investigation into the decay of pollen and spores in various soil types. In *Sporopollenin,* ed. J. Brooks, P.R. Grant, M. Muir, P. van Gijzel and G. Shaw, 446–79.

Heath, G.W., Arnold, M.K. and Edwards, C.A., 1965. Studies in leaf litter breakdown: I. Breakdown rates of leaves of different species. *Pedobiologia 6: 1–12.*

Hering, T.F., 1965. Succession of fungi in the litter of a Lake District oakwood. *Trans. Br. mycol. Soc. 48: 3, 391–408.*

Iversen, J., 1964. Retrogressive vegetational succession in the Post-glacial. *J. Ecol. 52 (Suppl.): 59–70.*

Jensen, V., 1974. Decomposition of Angiosperm tree leaf litter. In *Biology of Plant Litter Decomposition 1*, ed. C.H. Dickinson and G.J.F. Pugh, 69–104.

Jóhansen, J., 1978. Outwash of terrestric soils into Lake Saksunarvatn, Faroe Islands. *Danm. Geol. Unders., Årbog 1978*: 31–7.

Minderman, G. and Daniëls, L., 1967. Colonization of newly fallen leaves by micro-organisms. In *Progress in Soil Biology*, ed. O. Graff and J.E. Satchell, 3–9 (Braunschweig).

Müller, P.E., 1878. Studier over Skovjord, som Bidrag til Skovdyrkningens Theori. I. *Tidsskr. Skovbrug 3*: 1–124.

Murphy, P.W., 1955. Ecology of the fauna of forest soils. In *Soil Zoology* ed. D.K.M. Kevan and Mc E. Keith, 99–124.

Nagel-de Boois, H.M. and Jansen, E., 1967. Hyphal activity in mull and mor of an oak forest. In *Progress in Soil Biology*, ed. O. Graff and J.E. Satchell, 27–36 (Braunschweig).

Nagel-de Boois, H.M. and Jansen, E., 1971. The growth of fungal mycelium in forest soil layers. *Revue Ecol. Biol. Sol. 8*: 109–520.

Pennington, W., 1975. The effect of Neolithic man on the environment in North-west England: the use of absolute pollen diagrams. In *The Effects of Man on the Landscape: The Highland Zone*, ed. J.G. Evans, S. Limbrey and H. Cleere (Council for British Archaeology Research Report No. 11).

Perel, T.S., Darpačevskij, L.O. and Jegorova, S.V., 1966. Experimente zur Untersuchung des Einflusses von Regenwürmern auf die Streuschicht und den Humushorizont von Waldböden. *Pedobiologia 6*: 269–76.

Pugh, G.J.F., 1974. Terrestrial fungi. In *Biology of Plant Litter Decomposition 2*, ed. C.H. Dickinson and G.J.F. Pugh, 303–36.

Sall, G., Hollick, F.S.J., Ran, F. and Brian, M.V., 1949. The arthropod population of pasture soil. *J. Anim. Ecol. 17*: 139–52.

Schuster, R., 1956. Der Anteil der Oribatiden an den Zersetzungsvorgängen im Boden. *Z. Morph. Ökol Tiere 45*: 1–33.

Tutin (née Pennington), W., 1969. The usefulness of pollen analysis in interpretation of stratigraphic horizons, both Late-glacial and Post-glacial. *Mitt. int. Verein. theor. angew. Limnol. 17*: 154–64.

Waid, J.S. 1960. The growth of fungi in soil. In *The Ecology of Soil Fungi, an International Symposium*, ed. D. Parkinson and J.S. Waid, 55–75.

Wallwork, H.A., 1970. *Ecology of Soil Animals.*

Weis-Fogh, T., 1947–8. Ecological investigations on mites and collemboles in the soil. *Natura Jutl. I*: 135–227.

12   Pollen diagrams from Cross Fell and their implication for former tree-lines

Judith Turner

Introduction

THE vegetational history of the northern Pennines has been reasonably well documented (Raistrick and Blackburn 1932, Precht 1953, Johnson and Dunham 1963, Bellamy *et al.* 1966, Squires 1971, Turner *et al.* 1973, Chambers 1978). In general terms it was similar to that of other upland areas of England. The open Late Devensian vegetation was replaced by forest during the early Flandrian according to the pattern described by Godwin (1975) for England and Wales, and in late Flandrian times the woodland gave way to grassland or blanket bog. Only on the north-east-facing slopes of the Derwent valley did the sequence of forest types differ, more closely resembling that of north-east Scotland (Turner and Hodgson 1981).

On the basis of their studies of the peats of the Moor House National Nature Reserve, Johnson and Dunham (1963) suggested that the forest did not extend above 760m during the Boreal, and several authors have referred to the treeless fell tops as possible refugia for the relict Teesdale flora (Pigott 1956, Valentine 1978). Macro-remains of *Salix sp.*, *Juniperus*, and *Betula* sp. have been found forming a basal layer in the peats up to, but not above, 760m and this has been taken to mean that trees did not grow above that altitude. My interest in the proposed former tree-line was first aroused while studying variations in the species composition of the early and mid-Flandrian forests of the region because many of the patterns found were inconsistent with the idea of treeless summits.

The data for the early Flandrian comes from 42 pollen sites scattered over an area from the Tyne to Stainmore (Turner and Hodgson 1979) and those for the mid-Flandrian from 38 sites covering the same area plus the hills bordering Swaledale and Wensleydale to the south. Most of the sites are topogenous mires which formerly were sufficiently small to have had forest or other non-mire vegetation growing within a few hundred metres and are thus likely to have been receiving a large proportion of their pollen from a discrete local catchment not overlapping with those of other sites. The remaining pollen was probably derived from further away and this regional component is likely to have been similar at sites not too far distant from each other.

Assuming a tree-line at 760m, then sites above it should have been receiving tree pollen only from the regional pollen rain which could have been coming from valleys well below the tree line as well as from the nearer fell slopes. Pollen assemblages from such sites should differ from those below the tree-line, not only with respect to the composition of their tree pollen, which would lack a local element, but also by having higher frequencies for pollen taxa such as Gramineae and Cyperaceae representing the local herbaceous vegetation above the tree line. Neither of these expectations was fulfilled by the data. Not only is the tree/non-tree pollen ratio similar at sites below, near, and above 760m but in addition the tree-pollen composition does not change abruptly at any particular altitude.

Because no palynological work had been done on Cross Fell, the summit of which is over 100m above the proposed tree-line, since Godwin and Clapham (1951) reported on eight samples taken from a peat deposit at 730m on its eastern flank, it was decided to prepare pollen diagrams from the very small peat deposits on the summit in the hope that they would provide more evidence concerning possible tree-lines.

Cross Fell

THE GEOLOGY AND SOILS

Cross Fell (893m OD, national grid reference NY690345) lies towards the northern end of the Pennine range and its western slope forms a steep scarp with commanding views over the Eden valley to the English Lake District. Its flattish summit, almost 1km² in area, like those of the neighbouring Little and Great Dun Fells, is largely composed of a massive bedded, medium to coarse-grained yellow sandstone, the Dun Fell Sandstone, which is about 18m thick (Johnson and Dunham 1963). There are a few small outcrops of overlying shales and flags. The beds dip only slightly to the north-east and the summit is almost completely surrounded by a narrow zone of steep scree which has formed at the junction with the underlying shales. These shales form the lower slopes of the fell summits including the col at Teeshead between Cross and Little Dun Fells (fig. 12.1).

No limestone outcrops on the summit. The Lower Fell Top Limestone

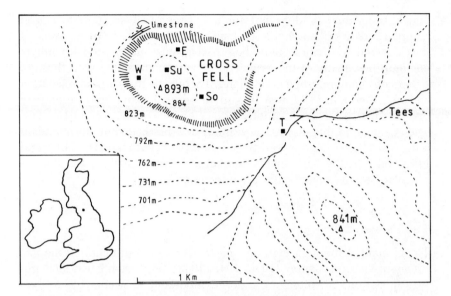

Fig. 12.1　Map of Cross Fell Summit showing the position of the sites studied. E East, W West, So South, Su Summit, T Teeshead. Location as inset.

shown on the Geological Survey maps as a band in the shales below the Dun Fell Sandstone has not been seen at the surface in pits dug for the purpose. Its position was inferred on general stratigraphic grounds. Johnson has mapped a small patch of limestone on the north-west of the fell (see fig. 12.1) on the basis of holes interpreted as sink holes, but again there is no outcrop.

The Dun Fell sandstone has given rise to sedentary humus iron podzols which are acid, sandy and base-poor (Hornung 1968). These soils contain numerous boulders lying at various angles and are over 1m deep in places. Hornung (personal communication) sees no reason why they should not have supported trees in the past, but knows of no positive evidence in the soil profiles he has examined that they did so. Polygonal patterns are well developed on the flat ground and there are stone stripes on the gentle slopes. Hornung is of the opinion that the surface has suffered intense solifluxion in the past, and may well be being affected by less intense activity today, especially during bad winters.

THE PRESENT-DAY VEGETATION

The vegetation of the nearby National Nature Reserve at Moor House, which unfortunately does not include Cross Fell summit, has been mapped and described phytosociologically by Eddy *et al.* (1968). Much of it is covered by ombrogenous bog, herbaceous, grassy, and bryophyte communities, many of the associations being comparable with those of either continental Europe or Scotland. There is no truly alpine vegetation and no woodland. The area is said to be well above the present-day tree-line which has been variously placed between 460m and 610m OD. The Festucetum that covers the summit of Cross Fell contains abundant sedges and lichens and is well grazed. Compared with the Tees valley and lower fells to the east, better known as Upper Teesdale, the flora is species-poor and botanically less interesting.

THE CLIMATE

The climate of the high Pennines has been studied in detail by the late Professor Gordon Manley (1936, 1942, 1943), who established meteorological stations at 560m on exposed moorland at Moor House, 4km east-south-east of Cross Fell, and at 834m near the summit of Great Dun Fell, only 3km south-east of Cross Fell. He compared the climate record from Moor House (1936) with that observed near sea-level in southern Iceland and summarized that at Dun Fell (1942) as follows:

> we therefore form a conception of an excessively windy and
> pervasively wet autumn, a very variable and stormy winter with long
> spells of snow-cover, high humidity and extremely bitter wind,
> alternating with brief periods of rain and thaw. April has a mean
> temperature little above the freezing-point and sunny days in May are
> offset by cold polar air; while the short and cloudy summer is not quite
> warm enough for the growth of trees. Throughout the year indeed the
> summits are frequently covered in cloud.

Manley (1942) considered that such an elevated sub-maritime region would be particularly sensitive to climatic change and that a slight increase in the frequency of warm anti-cyclones without any general rise in lowland temperatures would make a comparatively large difference, 1–2°F, to the mean temperature of the summits and that this would be almost sufficient to permit the growth of trees.

THE PEAT DEPOSITS

There is comparatively little peat on Cross Fell today. Johnson and Dunham (1963) mentioned the highest mound at a place which is 165m north-north-east and only 3m lower than the summit cairn (NY688344).

'Here 65cm of *Eriophorum–Sphagnum* peat occur in several residual mounds overlying well-developed stone polygons.' In 1978 there was even less peat than Johnson and Dunham had described, only one mound being anything like 65cm deep. Other similarly small and shallow remnants occur further to the north-north-east, to the west, and to the south-east of the summit, all at slightly lower altitudes but above the main screes. The deposit to the south-east is the most extensive. Only one of the mounds is associated with an obvious water seepage. There are also small patches of *Juncus squarrosus* L. on what appears to be the very last stages of eroding peat.

The fossil soils beneath the peat do not differ from the humus-iron podzols elsewhere on the fell top and Hornung (personal communication) considers that differential erosion of the once peat-covered summit has given rise to stone-pavement where the erosion continued to just below the soil peat interface, and to Festucetum where stabilization occurred at the soil surface. Thus peat, patches of *Juncus squarrosus*, Festucetum, and stone pavement all lie above the humus-iron podzol which formed earlier in the Flandrian.

There are also extensive areas of eroding blanket peat up to 2.5m deep at Teeshead and at comparable altitudes on the shelving eastern flank of Cross Fell.

The sites investigated and their pollen diagrams

Five deposits were studied. No detailed field mapping was carried out. Distances were estimated by pacing and a Thommen altimeter was used to give approximate altitudes, which are probably correct only to the nearest 10m (fig. 12.1 and table 12.1).

Table 12.1.   The sites investigated.

Site name	Altitude	Abbreviation
Teeshead	774m	T
Cross Fell East	863m	E
Cross Fell South	874m	So
Cross Fell West	875m	W
Cross Fell Summit	884m	Su

Apart from the deposit at Teeshead, which is 100m lower than the other sites, there is relatively little difference in altitude between them. There are, however, differences in aspect. Cross Fell East is sheltered from the prevailing westerlies and Cross Fell West is exposed to them and sheltered from the colder north-easterlies. Cross Fell Summit is the most exposed site and Cross Fell South is best positioned for maximum sunshine.

All pollen samples were prepared for counting by acetolysis and those which contained silica were also treated with hot HF (Faegri and Iversen 1964). They were mounted in glycerine jelly stained with safranin. Most pollen counts were terminated after 150 tree pollen grains (AP) excluding *Corylus* had been recorded. Pollen concentrations were not obtained. The diagrams show relative pollen frequencies. For each site the results are given in the form of:

1. A major tree and shrub pollen diagram with individual pollen taxa plotted as a percentage of AP. This includes a plot of the ratio between tree, shrub, and herb pollen.

2. A diagram showing the major herbaceous taxa, defined as those occurring at least one level with values of over 5% AP at any of the five sites.

3. A table showing the number of grains identified at each level for all other taxa, together with the tree pollen sum.

The diagrams have been zoned visually using horizons such as the 'elm decline' which are not always picked out by constrained cluster analyses, as well as those obtained by running the program ZONATION (Birks 1979) on the major tree taxa. The following local pollen-assemblage zones (p.a.z.) are recognized.

CF6   Herbaceous p.a.z.

$CF5_a^b$ *Quercus–Alnus* p.a.z.

CF4   *Quercus–Ulmus–Alnus* p.a.z.

CF3   *Betula–Quercus–Ulmus* p.a.z.

CF2   *Corylus–Betula–Ulmus–Quercus* p.a.z.

CF1   *Betula–Corylus* p.a.z.

CF5 has been divided on the basis of three features which occur at the same level on several of the diagrams, a rise in the herbaceous pollen frequency, a rise in the *Betula* pollen frequency and a peat stratigraphic horizon, a recurrence surface hereafter referred to as the *Grenzhorizont*. Diagrams from the other sites show two of these features and the 5a/b boundary has been placed accordingly. It is assumed in the absence of evidence to the contrary that the *Grenzhorizont* is a synchronous horizon over the short distances involved.

TEESHEAD

The deposit. The ombrogenous peat which has developed in Teeshead and on the neighbouring flanks of both Cross and Little Dun Fell is now deeply dissected and numerous small pieces of *Salix* wood are exposed near its base. The stratigraphy is typical of that described by Johnson and Dunham (1963) for the northern Pennines and on the sides of some of the mounds the *Grenzhorizont* can be seen.

The samples for pollen analysis were collected from one of the mounds of peat 50m north-east of the stream in the col and 50m east of the present (1980) Pennine Way. In June 1978 a core 150cm deep was taken by boring from a flat eroded surface of peat and the associated pollen diagram is labelled Teeshead 1 (figs. 12.2 and 12.3, table 12.2). The depths 110cm to 260cm refer to the top of the nearest mound. In August 1980 a second core of 190cm was taken from the surface of the most intact mound in order to obtain pollen samples from the most recent deposits. The associated diagram is labelled Teeshead 2 (figs. 12.4 and 12.5, table 12.3) and it overlaps stratigraphically with the top of Teeshead 1 by approximately 1m. On both occasions the cores were collected with a Russian-type borer, placed in plastic guttering, and wrapped in polythene for transport to the laboratory, where samples for pollen analysis were subsequently removed.

The pollen diagrams. The lower half of the Teeshead 1 diagram shows a forest history typical of the English Flandrian (West 1970), CF1, CF2, and CF3 being similar to zone FI, FII. CF4 resembles zone FII, but is unusual in having large fluctuations in the *Alnus* pollen frequencies. This is probably due to disturbance of the peat stratigraphy. Many diagrams from northern Pennine sites with a good depth of zone FI peat have a short zone FII profile and it is thought that such sites became too wet for continuous peat growth and were subjected to erosion during particularly wet periods. CF5 corresponds with zone FIII. The lowermost five samples of the Teeshead 2 diagram are a more typical zone FII assemblage and CF6 corresponds with zone FIII.

The CF4/5 boundary is equivalent to the VB III/IVa boundary on Chambers' (1978) pollen diagram from Valley Bog, Moor House, 7km to the east, where it has been dated to 4794 ± 55 BP (below) and 4596 ± 60 BP (above) and the CF5/6 boundary is equivalent to VB IVa/b, which has been dated to 2175 ± 45 BP (below) and 2212 ± 55 BP (above) (Chambers 1978). CF5 therefore is likely to have lasted about 2500 radiocarbon years and the *Grenzhorizont* which occurs about half way through the zone must date from sometime between about 4000 and 2500 BP.

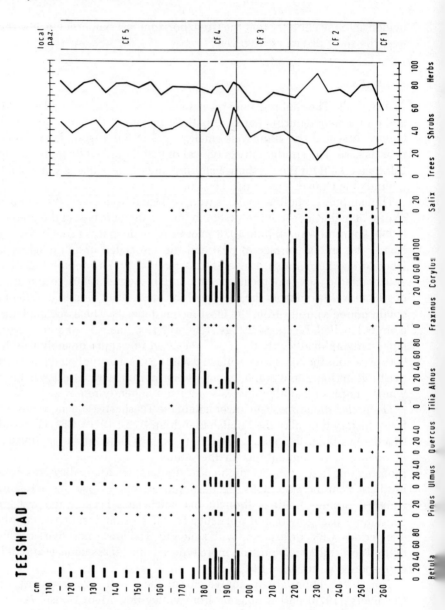

Figs. 12.2 & 12.3　Pollen diagram from Teeshead 1 with pollen frequencies expressed as a percentage of the tree pollen sum (excluding *Corylus* and *Salix*).

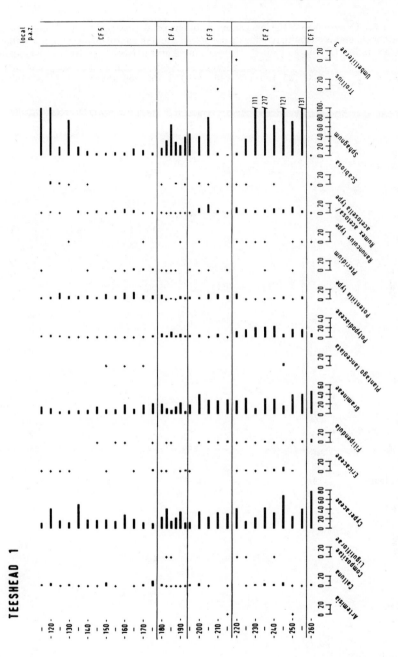

TEESHEAD 1

Table 12.2.   Occurrences of the less common pollen taxa at Teeshead 1. The figures represent the number of grains encountered during the pollen count of approximately 150 tree pollen grains (excluding *Corylus* and *Salix*). The exact number of tree pollen grains is also tabulated.

Sample depth in cm	115	120	125	130	135	140	145	150	155	160	165	170	175	180	182.5	185
Araliaceae																
Hedera helix																
Caprifoliaceae																
cf. Viburnum																
Caryophyllaceae																
Caryophyllaceae (un.)					1											
Lychnis type																
Chenopodiaceae						1										1
Compositae																
Anthemis type								1								
Bidens type								1	1	1				3	1	
Serratula type																
Corylaceae																
Carpinus																
Cruciferae																
Bunius type																
Cruciferae (un.)																
Cupressaceae																
Juniperus																
Equisetaceae																
Equisetum																1
Geraniaceae																
Geranium type																
Labiatae																
Lamium type																
Lycopodiaceae																
Lycopodium clavatum																
Plantaginaceae																
Plantago (un.)											1		1			
Polypodiaceae																
Polypodium vulgare	4	1	2				1				1	1	3	1		
Ranunculaceae																
Thalictrum type				1												
Rosaceae																
Dryas																
Prunus type																
Rosaceae (un.)									1		1					2
Salicaceae																
Populus													2			
Saxifragaceae																
Saxifraga stellaris														1		
Scrophulariaceae																
Melampyrum							2									
Sparganiaceae																
Typha angustifolia type																
Umbelliferae																
Umbelliferae 2		1													2	2
Umbelliferae (un.)												1		1		
Total tree pollen	163	176	163	153	160	155	155	162	152	154	150	156	155	158	178	15

Sample depth in cm	187.5	190	192.5	195	200	205	210	215	220	225	230	235	240	245	250	255	260
Araliaceae	1													1			
Hedera helix																	
Caprifoliaceae																	
cf. Viburnum									1								
Caryophyllaceae																	
Caryophyllaceae (un.)																	
Lychnis type																2	
Chenopodiaceae																	
Compositae																	
Anthemis type																	
Bidens type	2	5	2							1				1	1		
Serratula type																	1
Corylaceae																	
Carpinus				1													
Cruciferae																	
Bunius type										1							
Cruciferae (un.)							1		1								
Cupressaceae																	
Juniperus									2								
Equisetaceae																	
Equisetum																	
Geraniaceae																	
Geranium type																	1
Labiatae																	
Lamium type										1							
Lycopodiaceae																	
Lycopodium clavatum										1							
Plantaginaceae																	
Plantago (un.)																	
Polypodiaceae																	
Polypodium vulgare		2			1			1							1		
Ranunculaceae																	
Thalictrum type										1							
Rosaceae																	
Dryas															1		
Prunus type	2																
Rosaceae (un.)													1				
Salicaceae																	
Populus																	
Saxifragaceae																	
Saxifraga stellaris																	
Scrophulariaceae																	
Melampyrum					1		2	2	1	5		3	2				
Sparganiaceae																	
Typha angustifolia type													1				
Umbelliferae																	
Umbelliferae 2					2			1				1					
Umbelliferae (un.)				1		3		3							2		1
Total tree pollen	157	151	163	158	157	150	151	152	155	153	153	150	153	152	157	153	152

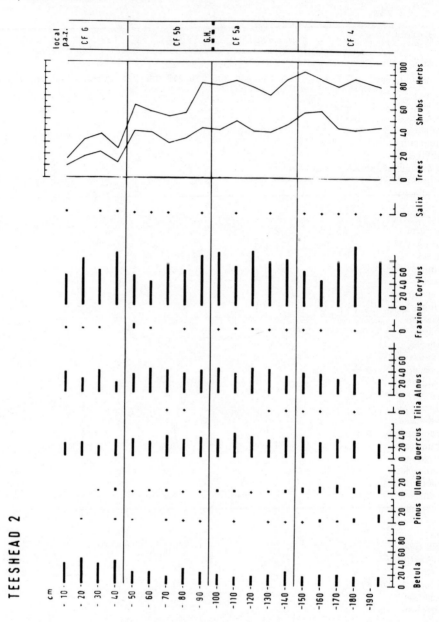

Figs. 12.4 & 12.5   Pollen diagram from Teeshead 2 with pollen frequencies expressed as a percentage of the tree pollen sum (excluding *Corylus* and *Salix*).

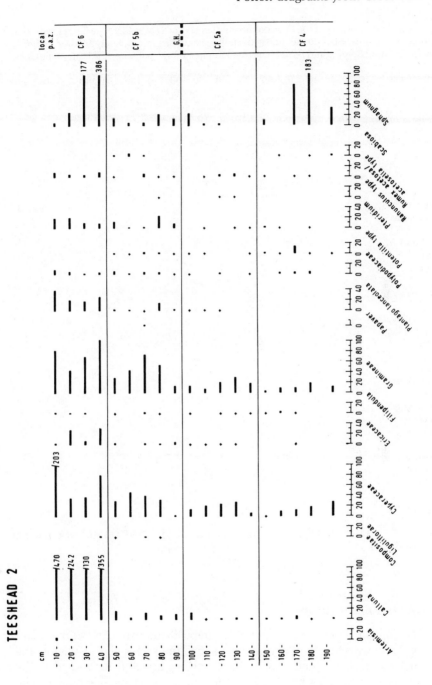

Table 12.3.   Occurrences of the less common pollen taxa at Teeshead 2. The figures represent the number of grains encountered during the pollen count of approximately 150 tree pollen grains (excluding *Corylus* and *Salix*). The exact number of tree pollen grains is also tabulated.

Sample depth in cm	10	20	30	40	50	60	70	80	90	100	110	120	130	140	150	160	170	180	19
Araliaceae																			
Hedera helix					1		1												
Caprifoliaceae																			
Sambucus						1													
Chenopodiaceae								1											
Cistaceae																			
Helianthemum				1															
Compositae																			
Bidens type	1	1						2		1			3	1		1	3	1	
Centaurea cyanus								1											
Cruciferae																			
Cruciferae (un.)						1													
Euphorbiaceae																			
Mercurialis								1											
Gramineae																			
cf. Cereal type			1																
Lycopodiaceae																			
Lycopodium clavatum				1															
Lycopodium selago							1												
Papaveraceae																			
Glaucium							1												
Plantaginaceae																			
Plantago (un.)								1											
Polypodiaceae																			
Polypodium vulgare					1	1	1	2				2				3	4	2	1
Ranunculaceae																			
Thalictrum type				1															
Rosaceae																			
Rosaceae (un.)								1			1								
Scrophulariaceae																			
Melampyrum				5						1									1
Umbelliferae																			
Umbelliferae (un.)						2													
Total tree pollen (excluding *Corylus*)	30	47	92	37	159	154	150	154	162	156	164	150	155	163	154	165	159	153	1

CROSS FELL SOUTH

The deposit. This is the largest of the deposits on Cross Fell Summit and lies to the west of the Pennine way above the screes and just below the final rise to the summit cairn. Stone stripes are well developed to the south of it. The peat bog is 90 by 40m and is being actively eroded. The depth varies but nowhere is it more than 180cm. Samples were collected directly into glass tubes from a peat profile exposed on the south-west side of the deposit. A clear *Grenzhorizont* was visible 70cm from the surface.

The pollen diagrams. The pollen diagrams (figs. 12.6 and 12.7, table 12.4) show that the peat at this site started forming during zone CF4 and accumulated during CF5 and CF6. The local assemblage zones are similar to those at Teeshead, but towards the top of CF6 the tree pollen frequency shows a rise that was not present at Teeshead. The *Grenzhorizont* again occurs half way through CF5, and both the *Betula* and the herbaceous pollen frequencies are slightly higher above than below it.

CROSS FELL WEST

The deposit. The ground to the south-west of the summit cairn slopes gently and then flattens to form a shelf above the main screes. On this shelf, there are several very shallow patches of peat, hardly worth the term mounds since they are so much shallower than those elsewhere. From the deepest, contiguous pollen samples were collected directly in tubes from a total depth of 31cm. The top 7.5cm was much less humified and formed a sharp boundary with the lower peat, resembling the *Grenzhorizont* at Teeshead and Cross Fell South.

The pollen diagrams. These diagrams (figs. 12.8 and 12.9, table 12.5) are less easy to equate with the others. The lowest sample is zoned as CF4 on the basis of the high *Ulmus* frequency. There is, however, no well-marked elm decline and whilst the CF4/5 boundary has been placed at 28cm it could also have been put at 18cm. CF6 is not represented. There is a distinct change in the tree pollen during CF5 at the level of the *Grenzhorizont*, allowing subdivision into subzones 5a and 5b with the *Betula* and *Fraxinus* and herbaceous frequencies rising and those of the other tree taxa falling.

CROSS FELL EAST

The deposit. This site lies 130m north-north-east of the Summit site at the foot of a gentle slope which carries grass-covered stone stripes. The sampled mound is small and about 60cm deep. Samples were collected directly into tubes from a cleaned face at the side of the mound.

The pollen diagrams. The zonation of the diagrams (figs. 12.10 and 12.11, table 12.6) was quite straightforward, CF4, 5a, and 5b being represented. A *Grenzhorizont* was not observed, although there may once have been one 20cm from the surface. The upper 20cm of peat differs in

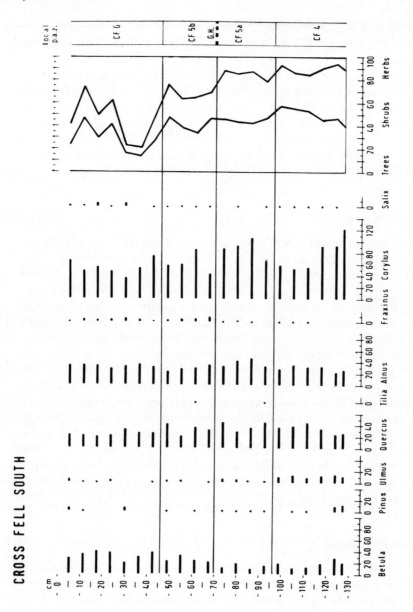

Figs. 12.6 & 12.7   Pollen diagram from Cross Fell South with pollen frequencies expressed as a percentage of the tree pollen sum (excluding *Corylus* and *Salix*).

CROSS FELL SOUTH

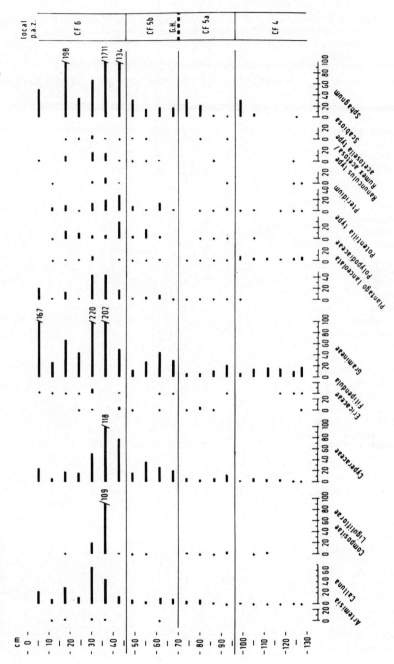

Table 12.4.   Occurrences of the less common pollen taxa at Cross Fell South. The figures represent the number of grains encountered during the pollen count of approximately 150 tree pollen grains (excluding *Corylus* and *Salix*). The exact number of tree pollen grains is also tabulated.

Sample depth in cm	5.0	11.5	17.5	23.5	30.0	33.0	36.5	39.5	42.5	48.5	55.0	61.5	67.5
Araliaceae													
Hedera helix													
Chenopodiaceae				1	1		1						
Compositae													
Anthemis type			1			1							
Bidens type	1			1	1		1			1			
Cruciferae				1									
Cupressaceae													
Juniperus				1									
Euphorbiaceae													
Mercurialis		2	1										1
Fagaceae													
Fagus			1								1		
Gramineae													
Cereal type				1			1						
Leguminosae													
Astragalus type				1									
Lycopodiaceae													
Lycopodium selago											2		
Plantaginaceae													
Plantago maritima type			1			1							
Plantago media/major type								1	1				
Plumbaginaceae													
Armeria type line B													
Polypodiaceae													
Polypodium vulgare			2			2						2	1
Ranunculaceae													
Anemone					1								
Rosaceae													
Prunus type								1					
Rubiaceae													
Galium type		3	1			2							
Saxifragaceae													
Saxifraga nivalis/													
stellaris type						1							
Scrophulariaceae													
cf. Digitalis type								1					
Melampyrum		1										1	
Rhinanthus type								1					
Selaginellaceae													
Selaginella													
Umbelliferae													
Umbelliferae (un.)			1	1		1							
Umbelliferae 2										1			
Urticaceae													
Urtica type							1		1				1
Total tree pollen (excluding *Corylus*)	52	154	90	154	48	112	44	82	100	155	153	150	157

Sample depth in cm	73.5	80.0	86.5	92.5	98.5	105.0	111.5	117.5	124.0	127.0
Araliaceae										
Hedera helix	1				1					
Chenopodiaceae										
Compositae										
Anthemis type										
Bidens type			4			1	3			8
Cruciferae										
Cupressaceae										
Juniperus										
Euphorbiaceae										
Mercurialis										
Fagaceae										
Fagus										
Gramineae										
Cereal type										
Leguminosae										
Astragalus type										
Lycopodiaceae										
Lycopodium selago			9	12						
Plantaginaceae										
Plantago maritima type										
Plantago media/major type										
Plumbaginaceae										
Armeria type line B		1								
Polypodiaceae										
Polypodium vulgare			1	3		1	2	2	2	1
Ranunculaceae										
Anemone										
Rosaceae										
Prunus type										
Rubiaceae										
Galium type										
Saxifragaceae										
Saxifraga nivalis/										
stellaris type										
Scrophulariaceae										
cf. Digitalis type										
Melampyrum										
Rhinanthus type										
Selaginellaceae										
Selaginella		3								
Umbelliferae										
Umbelliferae (un.)							1			1
Umbelliferae 2										
Urticaceae										
Urtica type										
Total tree pollen	152	158	153	156	156	151	152	156	166	161
(excluding *Corylus*)										

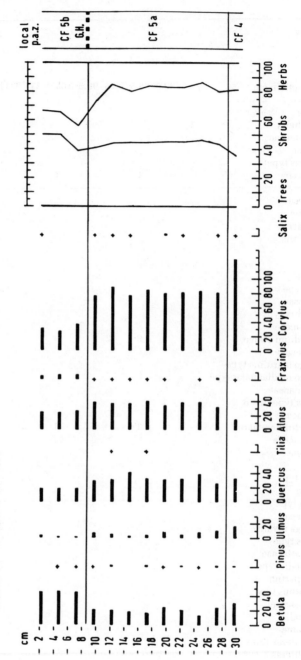

CROSS FELL WEST

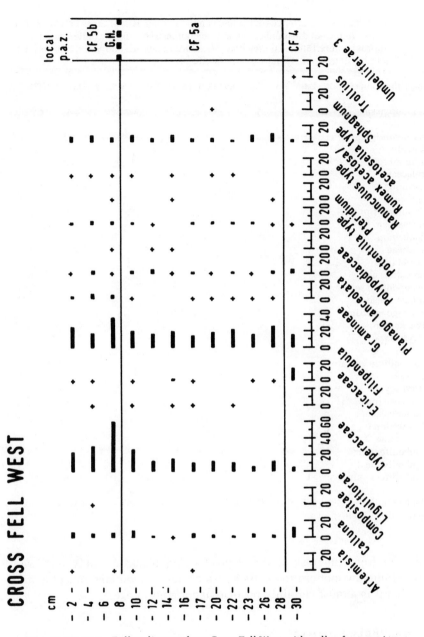

Figs. 12.8 & 12.9   Pollen diagram from Cross Fell West with pollen frequencies expressed as a percentage of the tree pollen sum (excluding *Corylus* and *Salix*).

Table 12.5.   Occurrences of the less common pollen taxa at Cross Fell West. The figures represent the number of grains encountered during the pollen count of approximately 150 tree pollen grains (excluding *Corylus* and *Salix*). The exact number of tree pollen grains is also tabulated.

Sample depth in cm	2.0	4.5	7.0	9.5	12.0	14.5	17.0	19.5	21.0	23.5	26.0	28.5
Chenopodiaceae		2	1									
Compositae												
Anthemis type		1	1	2								
Aster type											1	
Bidens type					1		1					
Corylaceae												
Carpinus		1										
Cruciferae												
Cruciferae (un.)	1											
Equisetaceae												
Equisetum					2							
Gramineae												
Cereal type	2											
Lycopodiaceae												
Lycopodium selago												1
Plantaginaceae												
Plantago maritima	1		1	2			1					
Polygonaceae												
Oxyria type		1	1									
Polypodiaceae												
Polypodium vulgare					1				1		1	
Rosaceae												
?Dryas type												1
Rosaceae (un.)			1									
Ranunculaceae												
Thalictrum type		1										
Umbelliferae												
Umbelliferae 1												3
Umbelliferae 2						1	1					
Umbelliferae (un.)	1								1	1		
Total tree pollen (excluding *Corylus*)	155	152	183	165	160	150	173	168	168	161	157	154

texture from the rest of the deposit being black, crumbly and well penetrated by modern roots. As it also contains larger amounts of *Betula* pollen it is almost certainly post-*Grenzhorizont* in age.

CROSS FELL SUMMIT

The deposit. This is the area originally described by Johnson and Dunham (1963) where several residual mounds of peat overlie reasonably distinct stone patterns, some 150m north-north-east of the summit cairn. It appears that a certain amount of erosion has taken place since then as the deepest mound is not as deep as when they described it. The ground surface around the mound is peaty indicating that the mound was formerly more extensive and may well have been linked with the neighbouring much-eroded patches of peat. Samples were collected directly from a cleaned face at the side of the intact mound. The *Grenzhorizont* could be seen 22cm below the surface.

The pollen diagrams. The pollen diagrams (figs. 12.12 and 12.13, table 12.7) indicate that the peat did not start forming at this site until sometime during CF5. CF5a and b are recognized. The most striking feature is the pollen spectrum at 32cm where there are large quantities of *Rumex acetosa/acetosella*-type pollen. *Rumex acetosella* colonizes burnt peat in the region today and this pollen spectrum is thought to indicate a temporary decrease in woodland cover not too far from the site. Small sooty fragments were observed on the pollen slide.

Also worthy of note are the large number of the less-abundant pollen taxa shown in table 12.7. There are an average of 2.1 for every CF5 sample, namely 38 taxa from 18 samples. This is much higher than the comparable ratios at the other sites, namely 1.5 at Teeshead 2, 1.0 at Cross Fell West and South, and 0.8 at Teeshead 1 and Cross Fell East.

The herbaceous taxa include types normally associated with woodland such as *Ilex, Hedera, Anemone, Mercurialis,* and *Stellaria holostea,* some of which are insect-pollinated and not abundant pollen producers, and also types normally associated with open non-wooded habitats such as *Lycopodium selago, Armeria,* and *Plantago maritima.* The former tend to occur in the lower and middle part of the profile, the latter in the upper parts (see, for example, the complementary occurrences of *Mercurialis* and *Lycopodium selago*). This tendency is not so clear at the other sites.

Further analyses of the data

The vegetational history shown by these diagrams differs little from that of nearby areas, but several interesting points emerge from a zone-by-zone comparison of the five sites.

Three zones are present on more than two diagrams CF4, CF5a, and CF5b. The data from each have been analysed in two ways, by Monte Carlo tests and, for CF5a and b, by principal components analysis of the mean pollen frequencies. Some interesting points also emerge from a comparison of the 5a and 5b assemblages at each site.

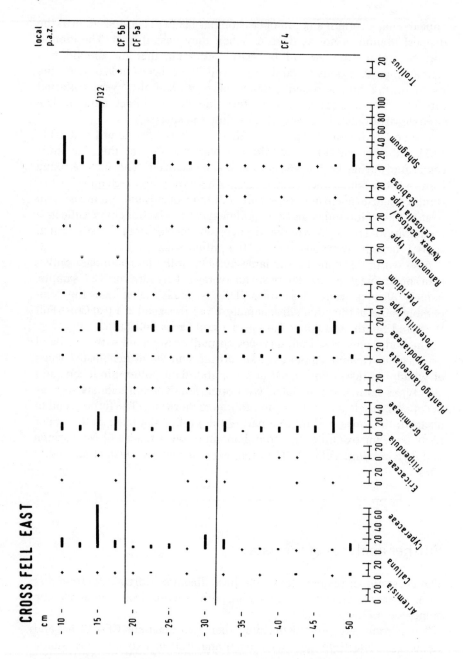

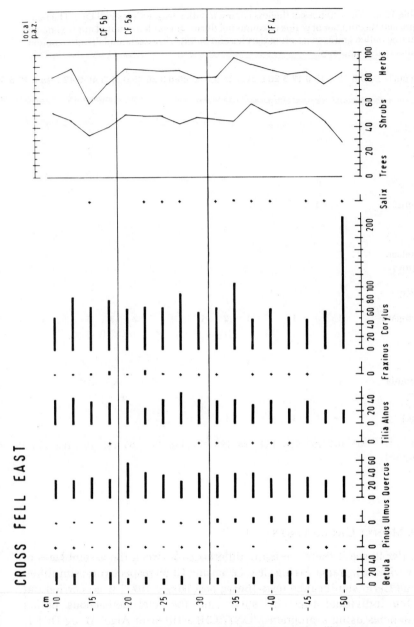

Figs. 12.10 & 12.11   Pollen diagram from Cross Fell East with pollen frequencies expressed as a percentage of the tree pollen sum (excluding *Corylus* and *Salix*).

Table 12.6.   Occurrences of the less common pollen taxa at Cross Fell East. The figures represent the number of grains encountered during the pollen count of approximately 150 tree pollen grains (excluding *Corylus* and *Salix*). The exact number of tree pollen grains is also tabulated.

Sample depth in cm	10.0	12.5	15.0	17.5	20.0	22.5	25.0	27.5	30.0	32.5	35.0	37.5	40.0	42.5	45.0	47.5	50.0
Araliaceae																	
Hedera helix								1		1							
Corylaceae																	
Carpinus																	1
Chenopodiaceae													1				
Euphorbiaceae																	
Mercurialis			1														
Gentianaceae																	
cf. Gentiana verna			1														
Gramineae																	
Cereal type		1															
Lycopodiaceae																	
Lycopodium selago																	1
Lycopodium (un.)																	2
Plantaginaceae																	
Plantago maritima					1			1									
Polypodiaceae																	
Polypodium vulgare	1		1		1	2	1	1		2	3	3	2	1	1		2
Rosaceae																	
cf. Agrimonia		1															
Rubus									1								
Salicaceae																	
cf. Populus tremula		1															
Scrophulariaceae																	
Melampyrum								4									
Umbelliferae (un.)				1													1
Total tree pollen (excluding *Corylus*)	161	161	152	156	154	184	193	157	158	157	165	167	162	162	159	157	147

THE MONTE CARLO TESTS

In order to see if there were any differences between the assemblages of the various diagrams during CF4, CF5a, and CF5b respectively, simplified Monte Carlo significance tests (Besag and Diggle 1977) were carried out on the individual tree and shrub and the total herbaceous pollen frequencies using a program MONTECHI written in Algol W by Dr J.T. Gleaves. This test is particularly useful when there is significant variation within each pollen assemblage and one wishes to know if there is also significant variation between the sites.

The program calculates a test-statistic (using the formula normally used for χ^2) to measure variation within each assemblage and then one to measure variation between assemblages. The samples are then ascribed randomly into groups each corresponding in size to those of the original assemblages and a between-group test-statistic computed. One hundred such randomized data test-statistics are calculated. If the one from the original data exceeds the highest from the randomized data (Montechi) it means that there is significant variation between the assemblages over and above that occurring within them with a probability of <0.01.

The Monte Carlo tests were first carried out on all the data from the three CF4, the six CF5a, and the five CF5b assemblages respectively. Since the results of the latter two gave significant between-site variation the data from them were further investigated by principal components analyses of the site mean frequencies. Because these principal components analyses indicated that the assemblages from certain sites were responsible for much of the variation, further Monte Carlo tests were carried out on the data omitting these sites one by one in the order of their contribution to the variation, until no significant between-site variation remained. The results of the Monte Carlo tests are shown in table 12.8.

Assemblage zone CF4. Each of the three assemblages shows significant within-assemblage variation. The test-statistic for the between-assemblage variation, 83.91 with 18 degrees of freedom, is lower than Montechi (180.10), the highest obtained in the random trials, and the probability of getting it is 0.37. There are therefore no significant differences between the three CF4 assemblages of Teeshead 2, Cross Fell South and Cross Fell East.

Assemblage zone CF5a. Each of the six sites has significant variation within its assemblage and there is also significant variation between the sites. The between-site test-statistic of 1073.26 is considerably higher than Montechi (723.84) and the probability of getting it by chance <0.01. Because the results of the PCA indicated that most of the variation of the first principal component was associated with the Summit site, the test was repeated without the Summit data. The test-statistic of 147.64 is again higher than Montechi (133.47), with $p < 0.01$. Cross Fell West has high scores on both the second and third components and so the data from this site were next omitted and although the test-statistic is lower than Montechi the probability level of 0.05 is generally regarded as significant. Teeshead 2 also has a high score on the second principal component and when data from it as well as the Summit site are omitted the test-statistic is significant at $p < 0.01$. It is only when the Cross Fell Summit, West, and either East or Teeshead 2 data are omitted that the remaining data proves homogeneous. This indicates that the assemblages from the Summit site, Cross Fell West, and both Teeshead 2 and Cross Fell East are all contributing significantly to the variation.

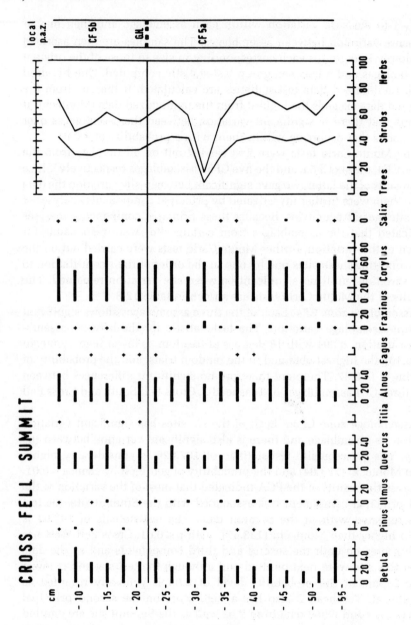

Figs. 12.12 & 12.13   Pollen diagram from Cross Fell Summit with pollen frequencies expressed as a percentage of the tree pollen sum (excluding *Corylus* and *Salix*).

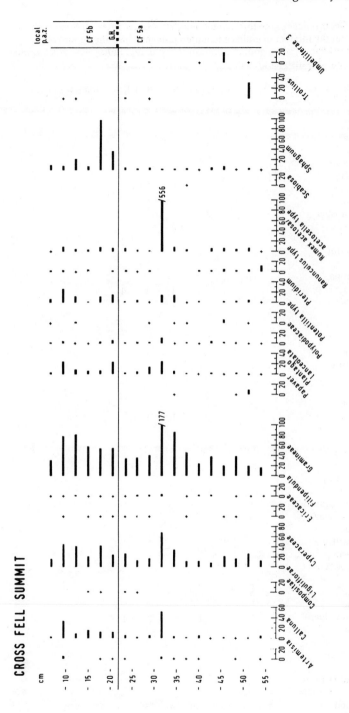

CROSS FELL SUMMIT

Table 12.7.   Occurrences of the less common pollen taxa at Cross Fell Summit. The figures represent the number of grains encountered during the pollen count of approximately 150 tree pollen grains (excluding *Corylus* and *Salix*). The exact number of tree pollen grains is also tabulated.

Sample depth in cm	7.0	9.7	12.5	15.2	18.0	20.8	23.6	26.3	29.1
Araliaceae									
Hedera helix					1		1		
Aquifoliaceae									
Ilex									
Caryophyllaceae									
Cerastium type			1						
Dianthus type									
Lychnis type									
Stellaria holostea									
Chenopodiaceae					1				
Cistaceae									
Helianthemum									1
Compositae									
Anthemis type		4	1	1					
Bidens type	1		1	1	1	1	1	1	
Cirsium type					1				
Crassulaceae									
Sedum								1	
Cruciferae									
Alliaria type						1			
Cruciferae (un.)	2	2			1				
Empetraceae									
Empetrum									
Euphorbiaceae									
Mercurialis					1	1			
Fagaceae									
Fagus			1						
Gramineae									
Cereal type		9		3	1				
Leguminosae (un.)	1								
Lycopodiaceae									
Lycopodium alpinum		1							
Lycopodium selago		3	2	1				2	
Plantaginaceae									
Plantago maritima	1		2		3	1			
Plumbaginaceae									
Armeria Type A									
Polygonaceae									
Oxyria type		3						1	1
Polypodiaceae									
Polypodium vulgare		1		1		2	1	1	
Ranunculaceae									
Anemone									
Thalictrum type									
Rhamnaceae									
Rhamnus catharticus									
Total tree pollen (excluding *Corylus*)	164	151	162	150	151	152	150	155	152

Sample depth in cm	31.9	34.6	37.4	40.2	42.9	45.7	48.5	51.2	54.0
Araliaceae									
Hedera helix									
Aquifoliaceae									
Ilex								1	
Caryophyllaceae									
Cerastium type								1	
Dianthus type							1		
Lychnis type					1				
Stellaria holostea							1		
Chenopodiaceae	1			1	1			1	
Cistaceae									
Helianthemum									
Compositae									
Anthemis type									
Bidens type			1					4	
Cirsium type									
Crassulaceae									
Sedum									
Cruciferae									
Alliaria type								1	
Cruciferae (un.)					1		2		
Empetraceae									
Empetrum								1	
Euphorbiaceae									
Mercurialis	1			1			1		
Fagaceae									
Fagus									
Gramineae									
Cereal type									
Leguminosae (un.)					1			1	
Lycopodiaceae									
Lycopodium alpinum									
Lycopodium selago			1	1					
Plantaginaceae									
Plantago maritima				1			1		
Plumbaginaceae									
Armeria Type A		1							
Polygonaceae									
Oxyria type									
Polypodiaceae									
Polypodium vulgare		1	1	2			1	2	
Ranunculaceae									
Anemone								2	
Thalictrum type							1		
Rhamnaceae									
Rhamnus catharticus			1						
Total tree pollen (excluding *Corylus*)	48	159	150	156	150	150	156	155	154

Table 12.7. (*contd*)

Sample depth in cm	7.0	9.7	12.5	15.2	18.0	20.8	23.6	26.3	29.1
Rosaceae									
Crataegus type								1	
Prunus type			1						
Rosaceae (un.)									
Rubus chamaemorus									
Rubiaceae									
Galium type	4		2	1		1	4		2
Saxifragaceae									
Saxifraga stellaris type			4						
Scrophulariaceae									
Melampyrum									1
Odontites type									1
Rhinanthus type		1	1	1		3			
Typhaceae									
Typha angustifolia type			2		1		1		
Typha latifolia type					1				
Umbelliferae									
Umbelliferae 1		1	1	1					
Umbelliferae 2	1			1				1	
Urticaceae									
Urtica type		3					1		
Total tree pollen (excluding *Corylus*)	164	151	162	150	151	152	150	155	152

Since the three sites with homogeneous assemblages during CF4 also have homogeneous assemblages during CF5a (p = 0.11), there may also have been variation on other parts of the fell during CF4.

Assemblage zone CF5b. Only the Cross Fell East site has no within-assemblage variation. The test statistic for between-site variation of 459.29 is higher than Montechi with p < 0.01. Cross Fell West is associated with the first principal component of the variation and Cross Fell Summit with the second. The data contain significant between-site variation without either of these sites, but once the data from both are omitted the remaining three sites show no significant between-site variation, as was the case during both CF4 and CF5a.

PRINCIPAL COMPONENTS ANALYSES

Principal components analyses (PCA) were carried out on the zone mean frequencies of each site in order to investigate further the nature of the

Sample depth in cm	31.9	34.6	37.4	40.2	42.9	45.7	48.5	51.2	54.0
Rosaceae									
Crataegus type									
Prunus type									
Rosaceae (un.)				1					
Rubus chamaemorus	1								
Rubiaceae									
Galium type									
Saxifragaceae									
Saxifraga stellaris type									
Scrophulariaceae									
Melampyrum				1					
Odontites type									
Rhinanthus type									
Typhaceae									
Typha angustifolia type									
Typha latifolia type									
Umbelliferae									
Umbelliferae 1				3			1		
Umbelliferae 2								1	
Urticaceae									
Urtica type									
Total tree pollen (excluding *Corylus*)	48	159	150	156	150	150	156	155	154

significant between-site variation that occurred during CF5a and CF5b. The results are given in tables 12.9 and 12.10.

Results for zone CF5a. The first principal component accounting for 59.5% of the variation is associated with high values for herb and *Betula* pollen at the Summit site and high values for *Quercus*, *Alnus*, and *Corylus* at Cross Fell South and East. The amount of herbaceous pollen at Cross Fell Summit, 40.05%, is particularly striking being nearly 20% higher than at any other site.

The second most important principal component in the data (24.1%) is associated primarily with Cross Fell West and Teeshead 2. The former has high loadings and frequencies for *Salix*, *Ulmus*, and *Pinus* and a low frequency for *Tilia*. By contrast, Teeshead 2 has a high frequency for *Tilia* and low values for the other taxa.

The third principal component accounts for only 15.0% of the variation and is associated mainly with the high *Fraxinus* frequency at Cross Fell East and the low value at Cross Fell West.

Table 12.8.   Results of Monte Carlo tests.

Zone	Sites	Montechi	test-statistic	p
CF4	Teeshead 2 Cross Fell South Cross Fell East	180.10	83.91	0.37
CF5a	Teeshead 1 Teeshead 2 Cross Fell South Cross Fell Summit Cross Fell West Cross Fell East	723.84	1073.26	<0.01
	Teeshead 1 Teeshead 2 Cross Fell South Cross Fell West Cross Foll East	133.47	147.64	<0.01
	Teeshead 1 Teeshead 2 Cross Fell South Cross Fell East	112.58	94.49	0.05
	Teeshead 1 Cross Fell South Cross Fell West Cross Fell East	103.93	129.87	<0.01
	Teeshead 1 Cross Fell South Cross Fell West	74.12	90.65	<0.01
	Teeshead 1 Teeshead 2 Cross Fell South	80.27	53.01	0.17
	Teeshead 2 Cross Fell South Cross Fell East	89.35	60.78	0.11
CF5b	Teeshead 2 Cross Fell South Cross Fell Summit Cross Fell West Cross Fell East	374.05	459.29	<0.01
	Teeshead 2 Cross Fell South Cross Fell Summit Cross Fell East	283.31	267.19	0.02
	Teeshead 2 Cross Fell South Cross Fell West Cross Fell East	350.70	330.82	0.02
	Teeshead 2 Cross Fell South Cross Fell East	272.66	108.18	0.16

Table 12.9.   Results of principal components analysis: CF5a.

Mean frequencies

	Betula	*Pinus*	*Ulmus*	*Quercus*	*Tilia*	*Alnus*	*Fraxinus*	*Corylus*	*Salix*	*Herbs*
Teeshead 1	6.55	0.51	1.70	15.67	0.11	17.91	1.10	33.96	0.28	22.22
Teeshead 2	8.37	0.27	1.59	15.14	0.16	17.00	0.60	35.70	0.22	20.94
Cross Fell South	6.19	0.65	2.09	17.34	0.14	17.55	0.58	39.57	0.29	15.61
Cross Fell Summit	10.40	0.19	1.15	10.89	0.00	11.76	1.43	23.78	0.36	40.05
Cross Fell West	9.05	0.85	2.97	14.51	0.07	16.49	0.51	36.36	0.44	18.74
Cross Fell East	5.90	0.80	2.18	18.66	0.06	19.29	1.55	35.43	0.40	15.74

	1st principal component	2nd principal component	3rd principal component
% variation	59.5	24.1	15.0
eigenvalue	5.95	2.41	1.50
Component loadings			
Betula	−0.813	0.105	−0.536
Pinus	0.751	0.650	−0.059
Ulmus	0.673	0.588	−0.444
Quercus	0.916	0.013	0.379
Tilia	0.666	−0.718	−0.197
Alnus	0.938	−0.035	0.278
Fraxinus	−0.423	0.333	0.838
Corylus	0.969	−0.130	−0.197
Salix	−0.052	0.994	−0.086
Herbs	−0.996	−0.019	0.026
Component scores			
Teeshead 1	0.139	−0.482	0.639
Teeshead 2	0.040	−1.434	−0.471
Cross Fell South	0.802	−0.503	−0.218
Cross Fell Summit	−1.952	0.247	0.167
Cross Fell West	0.319	1.220	−1.523
Cross Fell East	0.652	0.953	1.406

Quercus, Alnus, and *Corylus* pollen appear to be distributed similarly since their loadings on both the first and second principal components hardly differ.

Results for zone CF5b. The most interesting feature about these results is that they differ slightly from those for the earlier sub-zone. The first principal component of the variation (59.3%) is associated with the particularly high frequencies for *Betula* and *Salix* pollen at Cross Fell West which has correspondingly low frequencies for *Pinus* and *Corylus*. The second principal component of the variation (34.9%) is associated

Table 12.10. Results of principal components analysis: CF5b.

Mean frequencies

	Betula	*Pinus*	*Ulmus*	*Quercus*	*Tilia*	*Alnus*	*Fraxinus*	*Corylus*	*Salix*	*Herbs*
Teeshead 2	8.30	0.57	0.85	11.95	0.14	14.09	1.00	23.53	0.33	39.23
Cross Fell South	11.04	0.40	1.01	13.86	0.07	12.38	2.49	26.04	0.34	32.37
Cross Fell Summit	11.45	0.54	0.43	9.87	0.14	9.94	0.79	21.28	0.47	45.08
Cross Fell West	20.78	0.18	1.39	8.49	0.00	11.54	2.77	14.59	1.39	38.87
Cross Fell East	8.34	0.54	1.08	13.83	0.07	17.15	1.69	31.39	0.07	25.83

	1st principal component	2nd principal component	3rd principal component
% variation	59.3	34.9	4.6
eigenvalue	5.93	3.49	0.46
Component loadings			
Betula	−0.994	−0.002	0.020
Pinus	0.952	−0.278	0.126
Ulmus	−0.571	0.779	0.199
Quercus	0.756	0.569	−0.291
Tilia	0.746	−0.647	0.008
Alnus	0.535	0.718	0.445
Fraxinus	−0.672	0.674	−0.301
Corylus	0.872	0.460	−0.095
Salix	−0.984	−0.116	0.129
Herbs	−0.337	−0.934	0.008
Component scores			
Teeshead 2	0.585	−0.454	0.815
Cross Fell South	0.096	0.584	−1.609
Cross Fell Summit	0.203	−1.474	−0.324
Cross Fell West	−1.710	0.238	0.466
Cross Fell East	0.827	1.107	0.652

with the high frequencies for herbaceous pollen at Cross Fell Summit and the low values at Cross Fell East. *Fraxinus* does not have particularly high loadings on any of the principal components although its frequencies are much higher at Cross Fell South and West than at the other sites.

CHANGES DURING THE COURSE OF CF5

The changes that took place in the pollen assemblages during the course of CF5 are summarized in table 12.11 which gives the (CF5b mean frequency – CF5a mean frequency) for each taxon.

Table 12.11.   Changes in mean pollen frequencies from CF5a to CF5b.

	Betula	*Pinus*	*Ulmus*	*Quercus*	*Tilia*	*Alnus*	*Fraxinus*	*Corylus*	*Salix*	*Herbs*
Teeshead 2	−0.07	0.30	−1.34	−3.19	−0.02	−2.91	0.40	−12.17	0.11	18.29
Cross Fell South	4.85	−0.25	−1.08	−3.48	−0.07	−5.17	1.91	−12.53	0.05	16.76
Cross Fell Summit	0.66	0.35	−0.72	−1.02	0.14	−1.82	−0.64	− 2.50	0.11	5.03
Cross Fell West	11.73	−0.67	−0.58	−6.02	−0.07	−5.15	2.26	−21.77	0.95	20.13
Cross Fell East	2.44	−0.26	−1.10	−5.28	0.01	−2.14	0.14	− 4.04	−0.33	10.09

The herb pollen, which is so high in 5a at the Summit site, rises at all other sites by very large amounts. Similarly the *Betula* pollen frequency which is high in 5a at both the Summit and West sites rises at all sites on Cross Fell itself, but not at Teeshead. It attains particularly high values at Cross Fell West.

The *Quercus*, *Alnus*, and *Corylus* pollen frequencies all behave similarly, decreasing, although they do so least at the Summit site. The *Corylus* pollen frequency decreases by varying amounts at each site, the largest decrease being at Cross Fell West.

The *Fraxinus* pollen frequency, which is relatively high at Cross Fell East and the Summit site during 5a, rises at all except the Summit site, but markedly so at Cross Fell West and South.

Discussion

Although the pollen assemblages of CF5a and CF5b are similar at the five sites it has been possible to demonstrate that there are differences between them, particularly with regard to the *Betula* and herbaceous pollen frequencies. Similarly there are differences in the way the assemblages changed from 5a to 5b at each site, again the most striking being in the *Betula* and herbaceous pollen frequencies. It seems therefore that there was both temporal and spatial variation, some assemblages indicating a more open herb-rich woodland with a higher proportion of birch. The Summit diagram shows this type of woodland in 5a and by 5b all sites except Teeshead had a higher proportion of it, especially Cross Fell West.

Such variation is most easily explained by assuming that during the post-glacial climatic optimum the fell tops were carrying a low-growing open woodland which began breaking down during CF5 as the climate deteriorated, starting at the most exposed site. The woodland on the summit would have been the most vulnerable, hence the higher birch and herb frequencies during 5a at that site. Cross Fell West would have been the most exposed of the remaining sites, hence its particularly high birch and willow frequencies during 5b. The fact that the birch frequency rises at all except Teeshead between 5a and 5b makes sense in terms of Teeshead being the most sheltered and lowest altitude site.

Had Cross Fell and neighbouring summits been treeless and receiving their tree pollen from the regional pollen rain it is unlikely that this particular pattern of variation would have occurred. The results obtained are more easily explained in terms of the lightly wooded summits becoming less hospitable to trees as the climate deteriorated throughout CF5 than by long-distance transport of tree pollen from lower altitudes, a process unlikely to have produced such fine-scale local variation.

When visiting Cross Fell during the prevailing cloudy windy weather it is hard to imagine that trees ever grew there. It was much easier to do so during the exceptionally hot summer of 1976, when there was less wind and when the mean monthly temperatures were actually higher than the long-term averages by the much-quoted 1–2°C of the climatic optimum. The climatic and edaphic factors limiting tree growth on the fells today have been studied on the Moor House National Nature Reserve and it has been found that planted trees are under severe stress at 560m (Rawes 1979) although willows and rowan will grow higher, the latter to over 700m. The death of pine trees is thought to have been closely related to severe summer drought, as in 1975 and 1976, and also to severe winters such as 1979 when needles are desiccated in late frosts and heavy snow storms cause physical damage. The length of the growing season is critical for birch, mean daily maximum air temperature limiting growth in spring and the shorter day length in autumn (Millar 1965). A potassium deficiency in the soil severely limits tree growth which can be improved by adding potassium at planting (Brown, Carlisle, and White 1964, Carlisle and Brown 1969).

The climatic factors limiting growth at the tree-line have also been studied in detail in other parts of the world and results have been reviewed by Tranquillini (1979), who has pointed out that on a global scale the timberline corresponds reasonably well with the 10°C July isotherm and even better with a mean daily maximum of 11.1°C during the growing season. At Moor House the mean July temperature (1953–72) was 11.0°C and the mean daily maximum for the five months, May to September, 13.3°C, both well above the quoted temperatures. Also just above is the corresponding mean daily maximum for Great Dun Fell (834m) of 11.4°C (1938–40).

Five to seven thousand years ago the climate could only have been better. This is now reasonably well established by oxygen–isotope studies of deep-sea cores as well as by the more indirect evidence of former distribution patterns of plants and animals. Lamb (1977) in reconstructing atmospheric circulation patterns has argued that disturbances emerging from North America would have passed either side of Greenland and far enough away from north-west Europe to have given dry anticyclonic conditions with warmer summers in Britain during the mid-Flandrian. He estimates that the July and August mean temperatures would have been 2°C higher around 6500 years ago and it is unlikely that average wind speeds would have been as high in such stable anticyclonic systems. The soils on the fell tops would also have been more suitable for tree growth, not having undergone the podzolization that has occurred during the last three thousand years.

There seems therefore no reason why trees should not have become established on the fell tops during the climatic optimum and grown on the relatively deep, although stony soils which are now humus–iron podzols. This would explain why no evidence for a tree-line has been found in the FI and FII pollen data of the region. From 5000 BP or thereabouts, just before the beginning of local pollen zone CF5, as the prevailing atmospheric circulation patterns changed to give a more or less continuous flow of depressions over Britain, the forest would have gradually broken down, as described, and more peat begun to form on the fell tops, starting in small depressions and near seepage lines, probably in some of the very places where it has not yet eroded. Early during CF6, the trees probably gave way to peat-forming communities over the entire summit of Cross Fell and only subsequently did most of this peat erode exposing the stone pavement and giving rise to the present-day Festucetum and patches of *Juncus squarrosus* which overlie the now podzolized original forest soil.

Acknowledgments

I would like to thank Miss Joyce Hodgson for assistance with the field work and Dr G.A.L. Johnson for his continued encouragement and helpful comments on the geology of the region.

References

Bellamy, D.J., Bradshaw, M.E., Millington, G.R. and Simmons, I.G., 1966. Two Quaternary deposits in the Lower Tees Basin. *New Phytologist* 65: 429–42.

Besag, J. and Diggle, P.J., 1977. Simple Monte Carlo tests for spatial pattern. *Appl. Statist.* 26: 327–33.

Birks, H.J.B., 1979. Numerical methods for the zonation and correlation of biostratigraphical data. In Palaeohydrological Changes in the Temperate Zone in the last 15,000 Years, ed. B.E. Berglund, 99–123. (I.G.C.P. Project 158, University of Lund, Department of Quaternary Geology).

Brown, A.H.F., Carlisle, A. and White, E.J., 1964. Nutrient deficiencies of Scots Pine (*Pinus sylvestris* L.) on peat at 1,800 feet in the Northern Pennines. *Emp. For. Rev. 43:* 292–303.

Carlisle, A. and Brown, A.H.F., 1969. Nature Conservancy research on the nutrition of pines on high elevation peat. In *Proceedings of the Symposium on Peatland Forestry*.

Chambers, C., 1978. A radiocarbon-dated pollen diagram from Valley Bog, on the Moor House National Nature Reserve. *New Phytol. 80:* 273–80.

Eddy, A., Welch, D. and Rawes, M., 1968. The vegetation on the Moor House National Nature Reserve in the northern Pennines, England. *Vegetatio 16:* 237–84.

Faegri, K. and Iversen, J., 1964. Textbook of pollen analysis.

Godwin, H., 1975. The History of the British Flora.

Godwin, H. and Clapham, A.R., 1951. Peat deposits on Cross Fell Cumberland. *New Phytol. 50:* 167–71.

Hornung, M., 1968. The morphology, mineralogy and genesis of some of the soils on the Moor House National Nature Reserve (Ph.D. thesis, University of Durham).

Johnson, G.A.L. and Dunham, K.C., 1963. *The Geology of Moor House*.

Lamb, H.H., 1977. *Climate: Present, Past and Future*.

Manley, G., 1936. The climate of the northern Pennines: the coldest part of England. *Q. J. R. met. Soc. 62:* 103–15.

Manley, G., 1942. Meteorological observations on Dun Fell, a mountain station in northern England. *Q. J. R. met. Soc. 68:* 151–65.

Manley, G., 1943. Further climatological averages for the northern Pennines, with a note on topographical effects. *Q. J. R. met. Soc. 69:* 251–61.

Millar, A., 1965. The effect of temperature and day length on the height growth of birch (*Betula pubescens*) at 1900 feet in the northern Pennines. *J. Appl. Ecol. 2:* 17–29.

Pigott, C.D., 1956. The vegetation of Upper Teesdale in the northern Pennines. *J. Ecol. 44:* 545–86.

Precht, J., 1953. On the occurrence of the 'Upper Forest Layer' around Cold Fell, N. Pennines. *Trans. nth. Nat. Un. 2:* 44–8.

Raistrick, A. and Blackburn, K.B., 1932. The late-glacial and post-glacial periods in the northern Pennines. Part III. The post-glacial peats. *Trans. nth. Nat. Un. 1:* 79–103.

Rawes, M., 1979. Moor House. 20th Annual Report (The Nature Conservancy Council).

Squires, R. H., 1971. A contribution to the vegetational history of Upper Teesdale (Ph.D. thesis, University of Durham).

Tranquillini, W., 1979. *Physiological Ecology of the Alpine Timberline: Tree Existence at High Altitudes with Special Reference to the European Alps.* (Berlin).

Turner, J., Hewetson, V.P., Hibbert, F.A., Lowry, K.H. and Chambers, C., 1973. The history of the vegetation and flora of Widdybank Fell and the Cow

Green reservoir basin, Upper Teesdale. *Phil. Trans. R. Soc. B. 265*: 327–408.

Turner, J. and Hodgson, J., 1979. Studies in the vegetational history of the Northern Pennines I. Variations in the composition of the early Flandrian forests. *J. Ecol. 67*: 629–46.

Turner, J. and Hodgson, J., 1981. Studies in the vegetational history of the Northern Pennines II. An atypical pollen diagram from Pow Hill, Co. Durham. *J. Ecol. 69*: 171–88.

Valentine, D.H., 1978. Summarizing review and forward look. In *Upper Teesdale*, ed. A.R. Clapham, 196–209.

West, R.G., 1970. Pollen zones in the Pleistocene of Great Britain and their correlation. *New Phytol. 69*: 1179–83.

13 The Holocene vegetation of the Burren, western Ireland

W. A. Watts

Introduction

THE BURREN (fig. 13.1) is a karstic limestone region with a rich and remarkable flora and vegetation (Webb and Scannell 1983). Its flora, its dramatic scenery, and its great wealth of prehistoric remains have made it justly famous. The predominance of bare rock provokes speculation as to whether the landscape was ever forested and whether an original soil cover has been lost by deforestation and erosion. This paper attempts to survey the most important aspects of the Burren landscape and flora as a background to a presentation of new studies of its Holocene (post-glacial) vegetation history from lake sediments.

THE LANDSCAPE

Much of the Central Plain of Ireland is underlain by Carboniferous limestone. In the east it is covered by till, but in western Ireland the till cover is thinner and extensive areas of bare limestone occur. The Burren region (fig. 13.1) is the most striking example of bare karstic topography. Here, limestone hills form a plateau at about 300m, with deep north-trending fertile valleys. Strictly speaking, Burren is the name of a barony, a long disused administrative category, which coincides roughly with the hilly limestone county of north County Clare, but the name has come to be used less exactly by botanists to describe the limestone area as a whole. This includes the plateau (fig. 13.2), the 'High' Burren, and the still essentially bare 'Low' Burren which extends for a considerable distance eastward from the escarpment that bounds the plateau (map in Webb 1962).

Substantial patches of bare limestone with a diminished 'Burren' flora exist as far south as Askeaton in County Limerick and north to the isthmus between Lough Corrib and Lough Mask in County Mayo. The Aran Islands also consist of very bare karst.

The Burren was glaciated during the Last Glaciation when ice overrode the plateau (Farrington 1965). The resulting till was deposited very

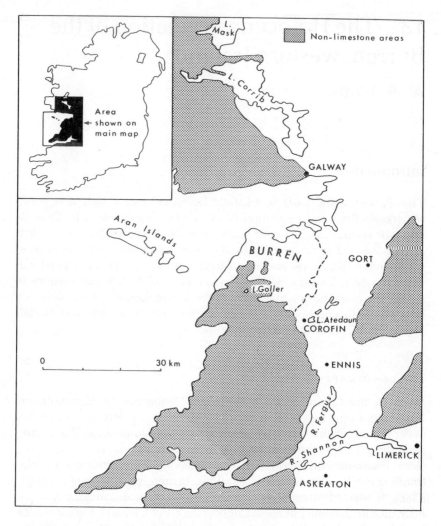

Fig. 13.1   Map of Western Ireland showing the location of the Burren and adjoining limestone regions.

irregularly. Some areas have thick deposits, in others no till was ever laid down. To the east and south the Burren is delimited by extensive drumlin fields which make a continuous cover over the limestone. Single isolated drumlins widely separated from one another by bare limestone occur in the Low Burren. A large drumlin lies west of Lough Gallaun and a drumlin-shaped till mass is plastered on the side of Mullaghmore (figs. 13.2 and 13.3).

The bare limestone is mainly flat-bedded 'pavement' in which flat rock

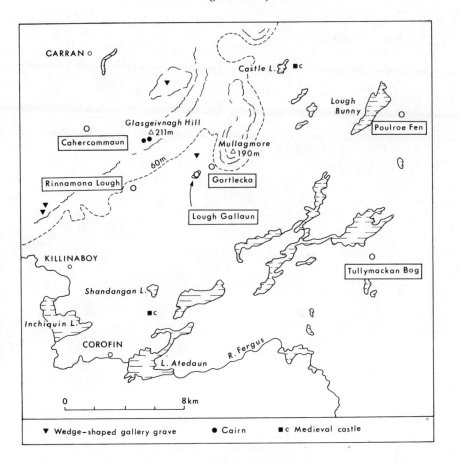

Fig. 13.2 Location map for sites referred to in the text.

surfaces alternate with deep fissures in a pattern determined by the system of joints and bedding planes. The fissures provide good micro-habitat for woodland herbs, ferns, and mosses (Dickinson *et al.* 1964). No local terminology has evolved to describe the morphology of the limestone and the 'grike' and 'clint' terminology applied to the same formation in northern England is not in use in Ireland. Throughout the region drainage is largely underground and very large cave systems are present. In the Burren plateau there are very few lakes and they show very strong fluctuations in water-level or contain water only intermittently. The Carran depression (fig. 13.2), for example, usually contains a stream, but may become a lake after heavy rainfall, especially in winter. In the Low Burren there are numerous lakes and ponds of several types. The most remarkable are 'turloughs' (Praeger 1932). These are lakes or ponds

Fig. 13.3   Mullaghmore with Lough Gallaun to its right. The photograph illustrates
the eastern margin of the Burren plateau and adjoining 'low' Burren. The landscape
is pure limestone pavement with, in foreground, ash-hazel woodland on thin till.
Woodland can be seen on cliffs at Mullaghmore. (Reproduced with permission of
the Director in Aerial Photography, University of Cambridge. Copyright reserved.)

with a strongly fluctuating water-level. They are commonly filled with
water in winter or after periods of heavy rainfall in summer, but may be
quite dry or reduced to a mud-hole after a period of low rainfall. Some are
completely vegetated with grass and herbs, in spite of long periods of
submergence. Others have a mud base, often of white carbonate mud with
a characteristic crust of blue–green algae. Castle Lough (fig. 13.2) is a
typical large turlough which may dry completely. Turloughs have a
distinctive flora and provide habitat for rare species such as *Viola
stagnina*. The most characteristic species is the moss *Cinclidotus fontina-
loides* which forms a dark-coloured zone that defines the upper limit of
normal flooding. Turloughs grade into more normal lakes. Lough Gallaun
(fig. 13.2) is a permanent lake which accumulates white carbonate mud in
its deeper water. It has a large shallow arm which behaves like a turlough.

This lake has a feature known from several other Burren lakes, a broad central shaft with vertical rock walls, in this case about 15m deep. Gortlecka and Rinn na Mona Loughs are each surrounded by swamps with abundant *Cladium mariscus* and *Schoenus nigricans*. Their water-level is relatively stable and they sediment brown algal mud.

The variety of lakes emphasize the difficulty of obtaining satisfactory cores for palaeoecological study, especially for the later Holocene. The Burren plateau has not yielded any satisfactory sites as yet. Turloughs, even where they contain sediments, may be subject to hiatuses and re-deposition. Carbonate-mud lakes preserve pollen well but are unsuitable for radiocarbon dating because of the problem of ancient carbon derived from the limestone. Peat from *Cladium* swamps is a very poor preserver of pollen, which tends to be badly oxidized. In the circumstances progress was only made when cores were obtained from open water at Gortlecka and Rinn na Mona Loughs. The brown algal muds preserve pollen well and are more satisfactory for radiocarbon dating than carbonate muds, though they cannot completely escape the possibility of error due to the presence of ancient carbon.

FLORA AND VEGETATION

The Burren flora contains both an Arctic–Alpine element and a Mediterranean element, but the species which can be referred unhesitatingly to either group are few in number. *Dryas octopetala* is very abundant locally and *Gentiana verna* frequent throughout. Other species which might be regarded as essentially arctic or high montane are *Ajuga pyramidalis*, *Saxifraga rosacea*, and *Potentilla fruticosa*. *Helianthemum canum* and *Epipactis atrorubens* belong to a group of species which are especially associated with outcrops of Carboniferous limestone in Britain as well as in Ireland. *Neotinea maculata*, an inconspicuous orchid, is frequent in the Burren in an isolated population. Its main range is Mediterranean or in south-west Europe, but it is also recorded from near Cork and has recently been found in the Isle of Man (Webb and Scannell 1983). *Adiantum capillus-veneris* reaches its northern limit in western Ireland, as does *Rubia peregrina*, another essentially southern and Mediterranean species. Many species occur in substantial populations which are rare or absent elsewhere in Ireland or, like *Teucrium scordium* and *Viola stagnina*, very rare elsewhere in both Britain and Ireland. However, as Webb (1962) observes, the interest of the Burren flora does not lie in rarity, but in species which are unusually abundant or surprisingly absent, thus bringing up interesting questions of ecological behaviour. For example, *Alnus* (alder) and *Quercus* (oak) are completely absent from the limestone region.

The upland vegetation of the Burren contains some extensive areas of scrub in which *Corylus avellana* (hazel) is dominant (Ivimey-Cook and Proctor 1966, Kelly and Kirby 1982). The scrub is often dense and luxuriant, growing to form a canopy 3–4m in height. It provides habitat for woodland herbs and ferns. Rich scrub usually grows on thin till or colonizes abandoned fields on till. Where there is little or no till open scrub can still occur, but the specimens are much smaller and more widely spaced. There are occasional small woods on till, or on thin mineral residues in valley bottoms or at the foot of cliffs. These normally include *Fraxinus excelsior* (ash) and, rarely, *Ulmus glabra* (wych elm), now rare in nature in Ireland. Both species are present in a small wood at Gortlecka, as are *Betula pubescens* (birch), *Ilex aquifolium* (holly) and *Sorbus aucuparia* (Mountain ash). Of small trees at Gortlecka, there are a few specimens each of *Malus pumila* (crab apple), *Salix cinerea* (common willow), *Salix caprea* (goat willow), *Populus tremula* (aspen) and *Sorbus hibernica* (whitebeam). Among the shrubs *Viburnum* opulus, *Euonymus europaeus*, *Cornus sanguinea* and *Rhamnus cathartica* also occur. The grazed margins of woods contain *Crataegus monogyna* (hawthorn) and *Prunus spinosa* (blackthorn) which are resistant to goat-grazing. Intensive goat-grazing by feral goats at higher elevations on Mullaghmore (fig. 13.2) destroys hazel scrub in favour of hawthorn/blackthorn scrub and ultimately removes nearly all shrubs. *Taxus baccata* (yew), the most grazing-sensitive of the Burren trees, is very heavily and preferentially grazed where accessible by goats. It survives primarily on cliff-faces. All the above species, and rare *Juniperus communis*, are present at Gortlecka. For Ireland, this is a very diverse woody flora, making the absence of *Alnus* and *Quercus* all the more conspicuous. Fig. 13.3 shows the stepped hill of Mullaghmore with woodland at the cliff-bases, on cliff-faces, and in a dissected valley.

Where woodland is open or patchy *Calluna vulgaris* and *Pteridium aquilinum* are usually present, sometimes accompanied by *Dryas octopetala*. Light grazing and trampling by cattle keeps the vegetation open and favours diversity of herbs. Where the vegetation cover is incomplete with much bare limestone, patches of herbs such as *Minuartia verna*, *Linum catharticum*, *Lotus corniculatatos*, *Sedum acre*, *Asperula cynanchica*, *Euphrasia* spp., *Thymus drucei*, *Plantago maritima*, *Geranium sanguineum*, and diverse orchids occur. *Helianthemum canum* is local in the Burren, but abundant on top of Mullaghmore. It does not occur elsewhere in Ireland.

Some of the Burren lakes have a zone of small shrubs near their winter water-level. In this position *Potentilla fruticosa*, *Salix repens* and a prostrate ecotype of *Frangula alnus* (Webb and Scannell 1983) are found.

PREHISTORY

The Burren is rich in field monuments. There is a concentration of wedge-shaped graves (Herity and Eogan 1977) at the eastern edge of the escarpment south-west of Mullaghmore (fig. 13.2). Wedge-shaped graves, of which the numerous Burren tombs are a simple variant, contain mainly Beaker pottery, but also pottery of Late Neolithic and early Bronze Age types. The archaeological material suggests that they should date to a few centuries on either side of 2000 BC. The Beaker people were predominantly pastoralists but also practised tillage. They were the first copper-using culture in Ireland and represent the transitional phase from Neolithic to Bronze Age (Herity and Eogan 1977).

Cashels are the next major class of field monuments. These consist of enclosures containing house foundations, surrounded by a high drystone wall forming the 'citadel', often with several external concentric walls. Cahercommaun (fig. 13.4 and Hencken 1938) is a fine example of this type. It was dated by artefacts to about 600 AD (Hencken 1938). It contained numerous cattle bones as well as charcoal exclusively representative of trees and shrubs common in the area today. However, cashels may have their origin in the Iron Age (Herity and Eogan 1977) and have

Fig. 13.4   Cahercommaun, a cashel in the east Burren with adjoining limestone pavement and hazel woodland. (Reproduced with permission of the Director in Aerial Photography, University of Cambridge. Copyright reserved.)

continued in use from the pre-Christian Iron Age into the Early Christian period. Apart from the cashels, and their earth equivalents (ring-forts or raths) which also occur, there are also Early Christian high crosses and churches which should date to the end of the first millennium AD. Finally, there are numerous medieval castles and churches, as well as a considerable abbey at Corcomroe in the northern part of the Burren. It seems likely that man has occupied the Burren intensively for at least the last 2500 years, and was significantly present 4000 years ago. It is likely that man was also present even before this time and in the Bronze Age in between, but evidence for that is less conclusive.

The cashel builders and perhaps also the Beaker people seem to have been pastoralists mainly interested in cattle. In spite of its stark appearance the Burren, especially its sheltered valleys, provides very good grazing for cattle ranging freely and the fact that grass tends to grow throughout the winter makes it possible to keep cattle alive in a way that would not be possible on the acid and infertile Clare shale county to the south where winter grass is inadequate. Burren 'winterages' are well known locally for this reason.

THE SITES

The difficulties of finding any site for palaeoecological studies in the Burren have already been stressed. In this study it was felt especially important to obtain a pollen record of man's impact on the landscape in the later Holocene. However, the topographic diversity of the Burren must also be taken into account. In searching for sites, it was particularly desired to obtain sites within as homogeneous an area of bare limestone as possible, away from any major till patches. This condition is not quite satisfied at Gortlecka and Rinn na Mona, but better sites probably do not exist and the two lakes meet the requirement of an open-water site with well-preserved late Holocene sediments surrounded by a wide zone of bare limestone as well as can reasonably be expected. Gortlecka is actually impounded by a small till ridge, but with that exception the entire landscape is bare limestone. One drumlin occurs about 1km to the west. Rinn na Mona is in a trough in limestone with large areas of surrounding bare limestone. Residual till patches also occur near this site and bear hazel scrub.

This study has excluded the late-glacial period because it has recently been reviewed (Watts 1977) with a pollen diagram from Poulroe (fig. 13.2). The intention of the Holocene study was to establish as far as possible what had happened specifically on the limestone. For this reason extensive investigation was made of macrofossils in the Gortlecka study and at other sites where work is still in progress.

Holocene vegetational history

GORTLECKA (fig. 13.5)

At the transition from late-glacial to Holocene a pioneer vegetation of shrubs and herbs was present, including abundant juniper, willow, grass, *Dryopteris*, *Filipendula*, and *Empetrum* with traces of aspen. Vegetation of this kind occurred universally in Ireland at this time, except for a few west coast localities on acid, infertile soils where *Empetrum* heath was present (Watts 1977). The pioneer phase ended when *Betula pubescens*, determined from macrofossils, became very abundant. Pollen of *Viburnum opulus* was already frequent at this stage. There must have been large stands of birch, but the percentage of herbs still present suggests that the landscape was open in some places at least. In the Burren the abundant large boulders and cliff-faces means that there was always habitat for light-demanding herbs, no matter how dense the forest was on flat ground. After the birch period, *Pinus sylvestris* (pine) and *Corylus avellana* invaded simultaneously, followed, after a distinct interval, by *Ulmus* and *Quercus*. The order of arrival of these four genera varies from site to site in Ireland. This may be the earliest recorded arrival of *Pinus*. At Killarney (Vokes 1966) pine was preceded by hazel by a clear margin, while in Donegal (Telford 1977) pine arrived at about 7900 BP, preceded by hazel by at least 1000 years. In Waterford, on the south coast of Ireland (Craig 1978), pine did not expand until after oak and hazel. The same is true at Scragh Bog in County Westmeath (O'Connell 1980) in the fertile till plain of the midlands. In the north-east also, pine was not frequent until after hazel, oak and elm had arrived (Singh and Smith 1973). There seems to be a case that pine was an early invader in south-western Ireland, and that in the Burren, confirmed by macrofossils, it formed one of its earliest large populations. Pine macrofossils, largely needle fragments and bud scales, show conclusively that it was growing on the limestone pavement.

The high percentages of *Quercus* throughout the early Holocene (about 10%) are surprising in view of its absence from the Burren today, while the percentages of *Ulmus* are much lower than on fertile tills of the Central Plain (O'Connell 1980). Low percentages of grass, *Calluna* and *Pteridium* suggest open ground. Probably there was woodland composed of pine, elm, hazel and birch. Elsewhere in Ireland at this time herbaceous pollen is very infrequent and there appears to have been closed deciduous forest of oak and elm on better soils.

Alnus arrived at 6680 ± 160 BP. This event seems to have taken place rather irregularly in Ireland with a wide range of dates on record. 6680 BP is acceptably within the variation already known. Like oak, alder is not at present in the Burren, and no macrofossils have been found to confirm

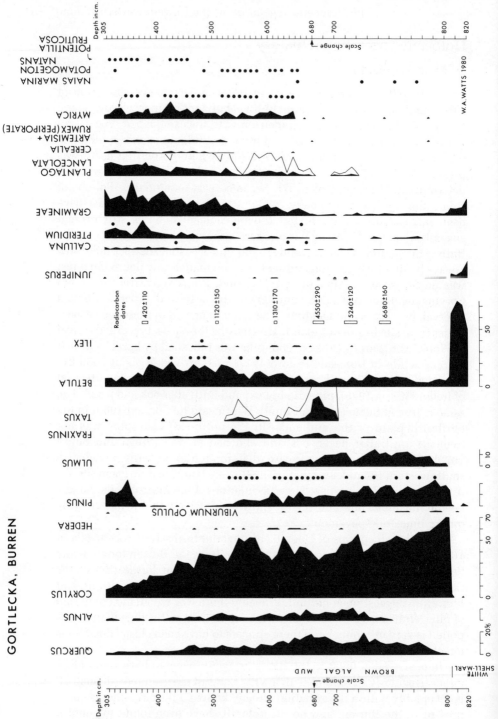

Fig. 13.5   Pollen diagram from Gortlecka. (Pollen diagrams have pollen sum of approximately 300 at all levels.)

presence in the past. The nearest non-limestone rock with alder is some 7km distant. This is regarded as 'background' or 'regional' pollen (Janssen 1966).

Elsewhere in Ireland, the elm decline at about 5100 BP is a dramatic event which must have brought about great changes in the forest. Decreases from 25% to 2% are known from Scragh Bog (O'Connell 1980). Watts (1961) argued that the scale of events was too great to be man-caused, as has been argued traditionally (Smith and Willis 1961), and that the most likely cause of the elm decline was disease. The decline is weakly expressed at Gortlecka. It is placed at 715cm where elm declines to 4%, having been as high as 15%, and the first *Plantago lanceolata* pollen is recorded. The date of 5240 ± 120 BP is consistent with a commonly accepted date of c. 5100 BP for this event. As elsewhere in Ireland elm recovers once more, rising to 12% before entering into a steady decline through the rest of the Holocene. *Fraxinus* was present before 5100 BP but, consistent with its behaviour elsewhere in Ireland, first became frequent as elm recovered. The striking evidence from other Irish sites that elm and ash behave in concert in the later Holocene, expanding or contracting their populations simultaneously (Mitchell 1956), is not replicated in the Burren.

After elm had recovered, *Taxus* appeared for the first time and briefly supplied over 20% of the pollen that was being deposited. Similar percentages are recorded from yew-woods on limestone at Killarney (Vokes 1966) where yew is the predominant canopy-former. At Killarney, yew was already present early in the Holocene. The yew peak at Gortlecka dates to 4550 BP. Subsequently there was a serious decline. The ecological role of *Taxus* today and in the past is not well understood in Ireland. In earlier studies its pollen was not identified, but it is in fact present in quantity at many Quaternary sites and favours, but is not exclusive to, limestone. Yew is slow-growing after vigorous growth as a young tree, shade- and wind-tolerant but unable to regenerate in its own shade. Observation at Killarney where natural yew-woods exist (Kelly 1981) shows that yew is more vulnerable than any other tree to browsing of its seedlings and bark by deer, goats and sheep. The role of cattle in browsing is not clear. There is plenty of evidence that yew foliage is, in some circumstances, poisonous to cattle. It seems possible that yew is well equipped to exploit a major perturbation in the forest, such as that caused by a widespread death of elms from disease, and that, once established, the main hazard to its continuation would be grazing. In the Burren it may have built up its population during and after the elm decline while the forest was still unstable. It is noticeable that its decline coincides with the time of the first major human occupancy of the eastern Burren: farmers of the Beaker Culture were present by 4000 BP and many of their wedge-shaped graves are found in the study area (Herity and Eogan 1977). *Fraxinus*, which peaks together with *Taxus* in the pollen diagram, is present throughout the Killarney yew-woods in low density.

The yew rise is associated with a decline in *Pinus* pollen, which continued until the extinction of pine at about 1500 BP. After the catastrophic decline of yew, an increase in grass, heather, bracken, and plantain took place, indicating that the woodland cover had been reduced permanently. The increase in birch, and its representation by macrofossils as well as the greater frequency of *Ilex*, suggests that a diverse open woodland and grassland was replacing more forest-like vegetation. At 550cm, probably about 2500 BP, a modest recovery of yew and ash, coupled with temporary stability in the declining pine and elm populations, suggests a slackening of human pressure on the landscape at the end of Bronze Age times, as is well established elsewhere (Mitchell 1965).

Myrica gale is present in the *Cladium/Schoenus* fen surrounding Gortlecka. It is continuously present from about 4000 BP to modern times, represented at all levels by pollen and macrofossils (fruits and leaf fragments). It is included in the pollen sum, which excludes obligate aquatics only, but it is strictly a wetland species in the Burren. Its arrival marks the moment at which the Gortlecka lake was overgrown from the margins, leaving a small central pond, separated from the upland by a broad mat of fen vegetation.

The extinction of pine is of great interest in Ireland because the date varies greatly between regions, the cause being unknown as yet. The date at Gortlecka, 1120 ± 150 BP, appears too young. Dates from two pine-stump layers at widely separate localities within the Killarney Valley show that extinction took place there at 1790 ± 95 BP and 1810 ± 95 BP. These are consistent with a date from Tullymackan Bog in the Burren (fig. 13.2) where pine was still present as a continuous curve until after 2030 ± 130 BP. Extinction took place shortly afterwards. A pine stump from a layer at Clonsast Bog in eastern Ireland dates to 1620 ± 130 BP (McAulay and Watts 1961). A further constraint in accepting a young date at Gortlecka is the occurrence of *Artemisia* in a continuous curve from the moment of extinction. After the late-glacial, *Artemisia* is absent, or present only sporadically, in Irish pollen diagrams until the late Holocene when it appears regularly in small quantity from about 1400 BP onward (Watts 1961). This seems to reflect an increased woodland disturbance for tillage in Early Christian times. All of these considerations suggest that the pine extinction at Gortlecka is likely to be somewhat older, say about 1500 BP, than the radiocarbon date suggests. The possibility must also be retained that the date is correct, in which case Gortlecka had the last known native pines.

Elsewhere in Ireland, pine extinction took place much earlier. In north-west Ireland at Glenveagh in Donegal, pine became extinct about 3220 BP (Telford 1977), while in County Mayo on the west coast, blanket bog, dated at the base to 3300 BP, contains no trace of pine, although older stump layers are common. In this region fossil fields and Neolithic tombs are buried by blanket peat (Herity M., personal communication). It is at least possible that pine forest was cleared to make way for cultivation. Why a very widespread species living in many kinds of habitat from Algeria to north of the Arctic Circle should have failed to survive in Ireland is mysterious, especially as pine survives in western Scotland in habitats similar to some of those from which it has disappeared in Ireland. It is also difficult to imagine any form of burning or exploitation which would cause complete extinction. Further, planted pines thrive today in Ireland and invade natural habitats readily. Probably the question should be seen in the context of a general loss of range in Britain also, of which the Scottish populations are only a small remnant. It will be necessary to obtain a larger number of dated pine stumps in well-recorded stratigraphic positions before much more progress can be made in Ireland, but the problem is a challenging and interesting one deserving further study.

After the pine extinction, a general decline in woodland cover took place. Herbs, especially grass, bracken, plantain and cereals, became abundant as all trees declined, especially hazel, elm, yew, and ash. This reflects, as elsewhere in Ireland, the very great extension in agriculture and in land under cultivation that seems to have characterized Early Christian times (Mitchell 1956). The reappearance of pine (without macrofossils) in the upper part of the diagram records the planting of pine from the eighteenth century onward. Extensive planting of pine in the Killarney area did not begin until the time of the Napoleonic Wars (Radcliff 1814). Before that, estate planting was largely confined to oak or other native trees.

The macrofossils at Gortlecka are not very diverse, but contain some unusual species or forms of preservation such as leaves of *Myrica* and *Calluna* and fragments of bracken fronds. Unfortunately no *Taxus* was found. Two species found as macrofossils specially deserve mention. *Naias marina*, represented by four specimens, extends from the beginning of the Holocene to after the yew decline, perhaps about 3800 BP. The species is now extinct in Ireland and has only one locality in Britain. A few other former Irish localities are known. Its preference seems to have been for carbonate-sedimenting or brackish lakes. This species merits more detailed study. If, as some authors suppose, it was warmth-demanding and became extinct because of cooling after a 'climatic optimum', then it is important to know its climatic requirements and its local extinction date. The possibility must also be entertained that it became extinct because of progressive unfavourable limnological developments in lakes.

The second species, *Potentilla fruticosa*, is represented richly by leaf fragments in late Holocene sediments, but not at any older period. One would expect this 'Arctic-Alpine' to be present, like *Dryas*, in the late-glacial sediments. The explanation may lie in recent immigration from an adjoining site. Possibly larger samples would prove it to be present in the older periods also. Evidence for the continuous presence of 'late-glacial' plants in the profile is provided by rare pollen grains of *Helianthemum* and of *Plantago maritima* that can be found by scanning slides of the middle and early Holocene sediments.

RINN NA MONA (fig. 13.6)

This lake, about 2km from Gortlecka in similar country, might be expected to yield a very similar pollen diagram (fig. 13.6), as indeed it does in many respects. There are some significant differences. The first is the apparent absence of an elm decline. The appearance of *Plantago lanceolata* pollen at 1085cm probably identifies the stratigraphic horizon which can be correlated with the elm decline elsewhere, but elm percentages are not significantly affected. It is possible that closer interval sampling would define the phenomenon more satisfactorily. However, as has already been observed, large and precipitous decreases in the elm curve are characteristic of the fertile till plains of central and eastern Ireland. The expression is weaker in the west, where elm percentages were never so high, even on limestone soils.

In common with Gortlecka, a steep peak of yew pollen follows the appearance of plantain and reduces pine pollen percentages. The pine fall is subsequently quite sudden, so that no pine pollen at all was recorded at 960cm. No macrofossil records are available as yet to define the moment of extinction, but some pine should have been locally present even with quite low percentages in the pollen diagram if Gortlecka and Rinn na Mona are strictly comparable. Later local survival at Gortlecka and earlier disappearance at Rinn na Mona are a possible explanation. The second yew peak, together with an expansion of ash, should correlate with the low second peak at Gortlecka. In this case yew was clearly more successful at Rinn na Mona. The failure to find a recent pine peak from plantations at Rinn na Mona is probably an artefact of too wide a sampling interval as yet. Further studies, especially of the macrofossil content of these cores, are still required at Rinn na Mona.

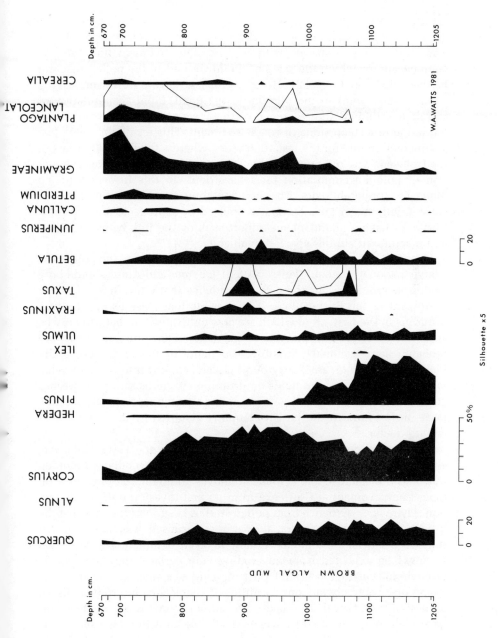

Fig. 13.6   Pollen diagram from Rinn na Mona.

Discussion

The Gortlecka lake began with a very slow sedimentation rate, to compensate for which the scale of the lower part of the pollen diagram has been doubled. In the later Holocene the lake largely grew over, leaving a small open pond in the middle. In the pond, rapid accumulation of marginal sediment due to the growth of water lilies has resulted in formation of a steep slope of coarse sediment. This results in a very fast sedimentation rate for the last 1000 years, whereas the first 5000 years of the Holocene saw only one metre of sediment form. This makes calculation of pollen-accumulation rates hazardous. In this respect Rinn na Mona is a preferable site. The sedimentation rate is less variable in a substantially larger lake, being less affected by shore factors. In contrast Gortlecka has an abundant macroflora making the link between pollen and macrofossil studies specially valuable.

At both lakes, neither of which has an inflowing stream, the pollen has a large 'regional' element (Janssen 1966). Quercus and Alnus should both belong in this class, for there is no evidence that either has ever been significant in the Burren flora. It is possible that Quercus occurred on drumlins in the past, all of which are now cultivated but that can only be speculative. Pinus (in the recent past), Cerealia, and Artemisia are probably also 'regional'. Possibly also many of the herbs, especially Plantago lanceolata, are 'background pollen', coming from the cultivable region to the east or from fields on drumlins. It can be said that at times 20% to 30% of the pollen was regional and that the representation of the local Burren flora in the pollen counts is distorted by this large 'background' factor.

The Burren appears to have been forested in the early Holocene, although there always was some open ground. The woodland included pine and the macroscopic evidence removes all doubt that pine grew on limestone pavement at Gortlecka from its first invasion until its extinction. There is no evidence from pollen studies to support the idea that the 'pavement' is a recent artefact, at least in part, caused by erosion of a thin soil cover. The lake sediments contain no layers of mineral residues such as might be expected from an eroding land surface, they are organic throughout. Farrington (1965) presents evidence that erosion of till is taking place actively at some localities. Perhaps erosion products finish up in fissures rather than in lakes. The author's view is that evidence for large-scale removal of soil cover is negligible, and that lake sediments offer no positive evidence. However, more detailed geomorphological and hydrological studies of sediment transport to the cave systems might force this view to be modified.

The expansion of Taxus at about 4500 BP is of considerable interest.

There may temporarily have been yew-woods similar to those at Killarney (Kelly 1981) and clearly they returned for a second time, at least at Rinn na Mona. A relaxation of grazing pressure today and, especially, the removal of feral goats would allow yew to play a large role in the modern vegetation. It is quite conceivable that yew woodland is the potential natural vegetation of much of the Burren.

References

Craig, A.J., 1978. Pollen percentages and influx analyses in south-east Ireland: a contribution to the ecological history of the Late-glacial period. *J. Ecol. 66:* 297–324.

Dickinson, C.H., Pearson, M.C. and Webb, D.A., 1964. Some microhabitats of the Burren, their microenvironments and vegetation. *Proc. R. Ir. Acad. 63B:* 291–302.

Farrington, A., 1965. The last glaciation in the Burren, Co. Clare. *Proc. R. Ir. Acad. 64B:* 33–9.

Hencken, H. O'N., 1938. Cahercommaun: a stone fort in County Clare. *R. Soc. Ant. Ireland, Special Vol.*

Herity, M. and Eogan, G., 1977. *Ireland in Prehistory.*

Ivimey-Cook, R.B. and Proctor, M.C.F., 1966. The plant communities of the Burren, Co. Clare. *Proc. R. Ir. Acad. 64B:* 211–301.

Janssen, C.R., 1966. Recent pollen spectra from the deciduous and coniferous-deciduous forests of north-eastern Minnesota: a study in pollen dispersal. *Ecology 47:* 804–25.

Kelly, D.L., 1981. The native forest vegetation of Killarney, south-west Ireland: an ecological account. *J. Ecol. 69:* 437–72.

Kelly, D.L. and Kirby, E.N., 1982. Irish native woodlands over limestone. *J. Life Sci. R. Dubl. Soc. 3:* 181–98.

McAulay, I.R. and Watts, W.A., 1961. Dublin radiocarbon dates I. *Radiocarbon 3:* 26–38.

Mitchell, G.F., 1956. Post-boreal pollen diagrams from Irish raised bogs. *Proc. R. Ir. Acad. 57B:* 185–251.

Mitchell, G.F., 1965. Littleton Bog, Tipperary: an Irish vegetational record. *Geol. Soc. Am., Special Paper 84:* 1–16.

O'Connell, M., 1980. The developmental history of Scragh Bog, Co. Westmeath, and the vegetational history of its hinterland. *New Phytol. 85:* 301–19.

Praeger, R.L., 1932. The flora of the turloughs: a preliminary note. *Proc. R. Ir. Acad. 41B:* 37–45.

Radcliff, T., 1814. *Report of the Agriculture and Livestock of the County of Kerry* (Dublin).

Singh, G. and Smith, A.G., 1973. Postglacial vegetational history and relative land and sea-level changes in Lecale, Co. Down. *Proc. R. Ir. Acad. 73B:* 1–51.

Smith, A.G., and Willis, E.H., 1961. Radiocarbon dating of the Fallahogy Landnam phase. *Ulster J. Archaeol. 24–5:* 16–24.

Telford, M.B., 1977. Glenveagh National Park: the past and present vegetation (Ph.D. thesis, University of Dublin (Trinity College)).

Vokes, E., 1966. The late and post-glacial vegetational history of Killarney, Co. Kerry in south-west Ireland (M.Sc. thesis, University of Dublin (Trinity College)).

Watts, W.A., 1961. Post-atlantic forests in Ireland. *Proc. Linn. Soc. Lond. 172*: 33–8.

Watts, W.A., 1977. The Late Devensian vegetation of Ireland. *Phil. Trans. R. Soc. B280*: 273–93.

Webb, D.A., 1962. Noteworthy plants of the Burren: a catalogue raisonné. *Proc. R. Ir. Acad. 62B*: 117–34.

Webb, D.A. and Scannell, M.J.P., 1983. *Flora of Connemara and the Burren* (Royal Dublin Society and Cambridge University Press).

14 Late-Quaternary pollen and plant macrofossil stratigraphy at Lochan an Druim, north-west Scotland

Hilary H. Birks

Introduction

PROFESSOR WINIFRED TUTIN made the first extensive investigation of the vegetational history of north-west Scotland (Pennington *et al.* 1972, Pennington 1975, 1977a, 1977b). She has published pollen diagrams from late- and post-glacial sediments from Loch Sionascaig, Cam Loch, Loch Borralan, and Loch Craggie in the Inverpolly area (see fig. 14.1) in conjunction with studies on sediment chemistry and diatom stratigraphy. Her work extends south to Loch Clair and Loch a'Chroisg in the Torridon area, where in addition, she made my work on Loch Maree possible (Birks H.H. 1972). To the south and west, the late- and post-glacial vegetational history of the Isle of Skye and the adjacent mainland has been reconstructed by H.J.B. Birks (1973a) and Williams (1976) (see Birks and Williams 1983). To the north of Inverpolly, Pennington (1977a) has presented a late- and post-glacial pollen diagram from Lochan an Smuraich. There are also post-glacial pollen diagrams from near Loch Assynt, Lochan an Druim (Birks H.H. unpublished data), and Duartbeg (Moar 1969a). The pioneering late-glacial pollen diagram from Loch Droma (Kirk and Godwin 1963) is associated with some macrofossil and moss records.

The present study was designed to investigate the late-glacial vegetational history of the far north-west of Scotland, thus extending Tutin's work northwards. To provide a contrast with the large lakes studied by Tutin, a small basin was chosen. It was hoped that this would provide a detailed picture of the local and nearby upland vegetation around the site. To further this aim, plant macrofossils were studied stratigraphically in conjunction with the pollen. The site chosen, Lochan an Druim beside Loch Eriboll, is of considerable botanical interest, as it is situated on Durness limestone, nearby outcrops of which today support a rich calcicolous flora, including an abundance of *Dryas octopetala* growing near sea-level (Ratcliffe 1977).

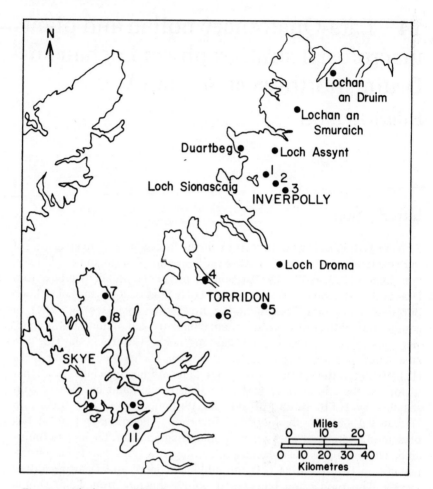

Fig. 14.1   The location of Lochan an Druim, Eriboll, and other sites with late-glacial pollen stratigraphy in north-west Scotland. (1) Cam Loch, (2) Loch Borralan, (3) Loch Craggie, (4) Loch Maree, (5) Loch a'Chroisg, (6) Loch Clair, (7) Loch Mealt, (8) Loch Fada, (9) Loch Cill Chriosd, (10) Lochan Coir'a' Ghobhainn, (11) Loch Meodal. For references to sites, see text.

Late-glacial plant macrofossils have been studied stratigraphically in conjunction with pollen at two sites on the Isle of Skye (Vasari and Vasari 1968) and at several sites in the Eastern Highlands (Birks and Mathewes 1978, Vasari and Vasari 1968, B. Huntley unpublished data). Besides providing a vegetational contrast with eastern Scotland, a site in north-west Scotland should reveal any widespread applicability for the generalizations proposed about the usefulness of plant macrofossils at Abernethy Forest (Birks and Mathewes 1978, Birks H.H. 1980).

Plant nomenclature follows Clapham, Tutin, and Warburg (1962) for vascular plants, and Corley and Hill (1981) for bryophytes.

Site description

An Druim is a small ridge running northwards from Eriboll between Loch Eriboll and the A838 road. In the shallow valley between An Druim and the road there is a small loch, to which the site name Lochan an Druim has been given (latitude 58°28′, longitude 4°42′, National Grid Reference 29/436568, altitude 25m OD). A colour photograph of the site appears in McLean (1976: 102). The bedrock is Durness limestone, and where it outcrops on the steeper slopes to the east of the road, it supports a rich *Dryas octopetala-Carex flacca* heath, containing *Campanula rotundifolia, Carex capillaris, C. pulicaris, C. rupestris, Koeleria cristata, Linum catharticum, Plantago lanceolata, P. maritima, Polygonum viviparum, Saxifraga aizoides, Selaginella selaginoides, Thymus drucei, Ctenidium molluscum, Ditrichum flexicaule,* and *Scapania aspera.* Deeper soils between outcrops support species-rich *Agrostis-Festuca* grassland, some-times with large amounts of *Pteridium aquilinum.* Around the lochan, the grasslands have been improved for grazing. These grasslands grade into *Juncus effusus*-dominated communities on damp soils near the loch, containing much *Carex nigra.* On wetter soils, *Carex nigra* predominates, associated with herbs of damp ground such as *Filipendula ulmaria, Galium palustre, Lychnis flos-cuculi, Mentha aquatica,* and *Myosotis scirpoides.* Nearer the loch, *Phragmites communis* is dominant in a dense, 2m to 3m-high, species-rich reed swamp in shallow water. As the water becomes deeper, the number of species decreases. The lochan is about 100m across, and has two small inflows and one small outflow. The open water is colonized by aquatic plants, including *Glyceria fluitans, Hippuris vulgaris,* and *Potamogeton natans.* Apart from plantations, the landscape around An Druim is virtually treeless, but in a nearby sheltered gully there are a few naturally occurring trees and bushes of *Betula pubescens, Salix aurita, S. cinerea,* and *Sorbus aucuparia,* with a herb-rich understorey.

Methods

CORING METHODS

A core was taken in August 1973 in the dense *Phragmites* fen on the west side of the lochan, using a 5cm diameter square-rod Livingstone piston corer (Wright 1980). The core segments were extruded in the field and wrapped in plastic film and aluminium foil.

SEDIMENT DESCRIPTION

The sediment lithology is described below using the notation of Troels-Smith (1955).

770–800cm   Silty fine-detritus mud, very dark, darkens on exposure to air. Nig3, strf+, sicc2, elas2, firm and rubbery; calc 0, humo 1; Ld^{1}3, Ag1, Dh +, Dg +. Upper contact very gradual.

800–812cm   Silty fine-detritus mud. Nig3, strf 0, sicc2, elas2, firm; calc 0, humo 1; Ld^{1}2, Ag1, As1, Ga +, Dh +. Upper contact not seen.

812–838cm   Clay and silty mud. Nig2, strf 0, sicc3, elas 1, very firm; calc 0; Ag2, As2, Ld^1 + , Ga +. Upper contact gradual.

838–843cm   Clay and silty fine-detritus mud. Nig2, strf 0, sicc3, elas 1, very firm; calc 0; Ld^{1}1, Ag1, As2, Ga +. Upper contact gradual over 3mm.

843–866cm   Brown silty fine-detritus mud, granular in texture. Gets greyer and paler below about 855cm. Nig2–3, strf 0, sicc3, elas 1, firm; calc 0, humo 1; Ld^{1}2, Ag2, Dh +, Dg +, Ga +. Upper contact gradual over 5mm.

866–876cm   Pale brown silty mud. Nig2, strf 0, sicc3, elas1, very firm; calc 0; Ld^{1}1, Ag3, Ga +. Large stone at 866cm. Upper contact gradual over 7mm.

876–891cm   Pale grey silty sand, speckled with black (?iron-rich) bands. Nig1, strf 1, sicc3, elas1, very firm; calc 0; Ag2, Ga2, Gs +. Upper contact gradual over 3mm. Lower contact not seen.

Samples were dried at 105°C and percentage loss on ignition was determined by combustion at 550°C for six hours, to provide a rough estimate of organic content (Dean 1974) (fig. 14.2).

POLLEN ANALYSIS

Sediment samples of 0.5cm^3 were taken with a calibrated brass sampler (Birks 1976). To each a known volume of a suspension of *Eucalyptus globulus* pollen of known concentration was added, to enable the calculation of fossil pollen concentration (Benninghoff 1962, Matthews 1969). The samples were prepared for pollen counting using standard procedures (Faegri and Iversen 1975). The residues were stained with safranin and mounted in 2000cs silicone oil.

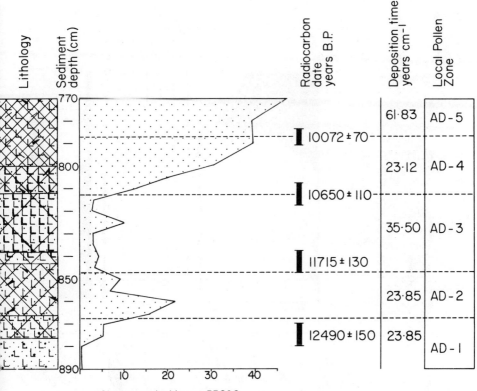

Fig. 14.2   Lochan an Druim: percentage loss-on-ignition at 550°C in relation to sediment lithology, radiocarbon dates, sediment deposition time, and local pollen zones. Symbols used in the lithology column follow Troels-Smith 1955.

The pollen was counted using a Leitz Dialux microscope with ×10 Periplan oculars and a ×40 fluorite objective. A ×95 fluorite oil-immersion objective was used for critical determinations. Regularly-spaced traverses across whole slides were made, in order to avoid errors associated with non-random distribution of pollen on the slides.

The pollen and spores were determined to the lowest possible taxon within the British flora. Naming of the morphological categories follows H.J.B. Birks (1973a) and the degree of reliability of the identifications is indicated by a standard set of conventions (Birks H.J.B. 1973a).

The percentage pollen diagram is presented in fig. 14.3. The pollen sum includes all identifiable terrestrial pollen. Percentages of pteridophyte spores, pollen of aquatic plants, indeterminable pollen, and algae have been calculated using the sums shown on the diagram. Pollen-influx curves are shown on fig. 14.4, where the influx of several taxa are

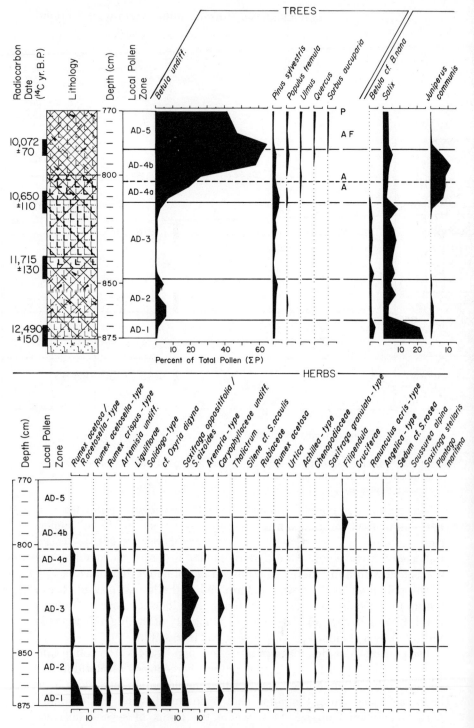

Fig. 14.3   Percentage pollen diagram from Lochan an Druim, Eriboll. Scale at base of diagram is percentages of total pollen (ΣP) for black silhouettes. Symbols used in the lithology column follow Troels-Smith 1955. Undiff. = undifferentiated; BP = before present; Indet. = indeterminable. Key to minor pollen types is in table 14.1.

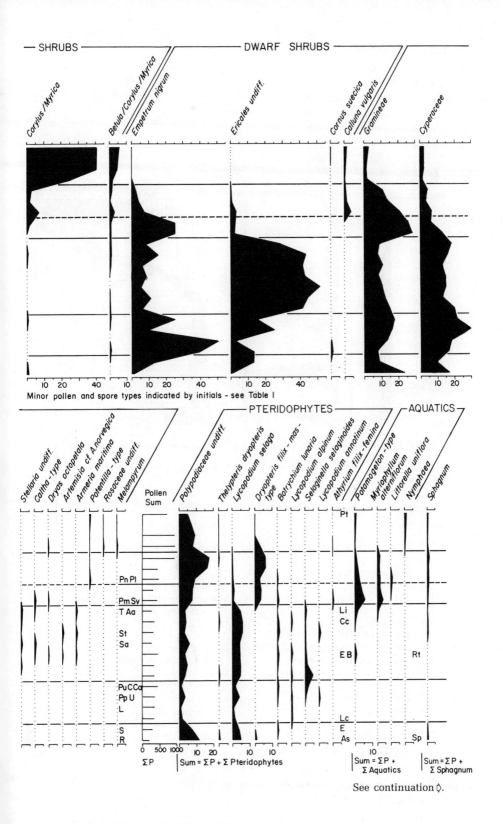

Minor pollen and spore types indicated by initials - see Table I

See continuation ◊.

Fig. 14.3 (contd)

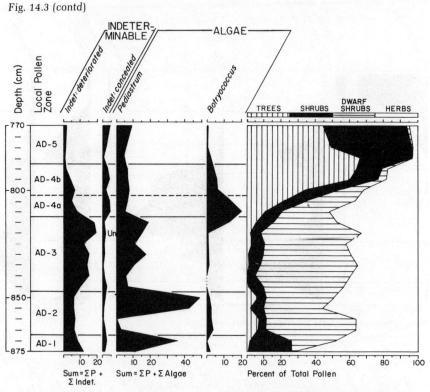

Analysed by H.H.Birks, 1974-75.

Table 14.1.   Key to minor pollen types on percentage pollen diagram, fig.14.3.

A	*Alnus glutinosa*	Sa	*Sagina*
F	*Fraxinus excelsior*	St	*Stellaria holostea*
P	*Prunus* cf. *P. padus*	Sv	*Sedum* cf. *S. villosum/S. anglicum*
V	*Vaccinium myrtillus*-type	T	*Trollius europaeus*
Aa	*Astragalus alpinus*	U	Umbelliferae undiff.
C	*Cirsium/Carduus*	As	*Asplenium*-type
Ca	*Cerastium alpinum*-type	B	*Blechnum spicant*
L	*Lotus pedunculatus*-type	Cc	*Cryptogramma crispa*
Pl	*Plantago lanceolata*	E	*Equisetum*
Pm	*Plantago media/P. major*	Lc	*Lycopodium clavatum*
Pn	*Plantago* undiff.	Li	*Lycopodium innundatum*
Pp	*Parnassia palustris*	Pt	*Pteridium aquilinum*
Pu	Papilionaceae undiff.	Rt	*Ranunculus trichophyllus*-type
R	*Rhinanthus*-type	Sp	*Sparganium*-type
S	*Silene maritima*-type	Un	Unknown

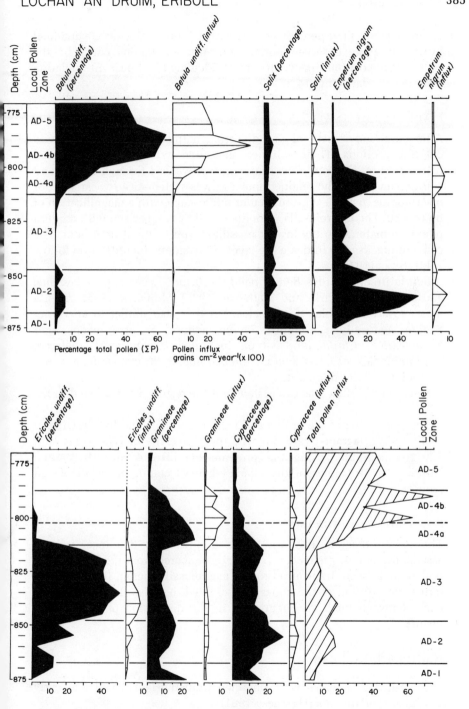

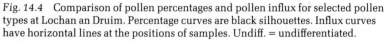

Fig. 14.4   Comparison of pollen percentages and pollen influx for selected pollen
types at Lochan an Druim. Percentage curves are black silhouettes. Influx curves
have horizontal lines at the positions of samples. Undiff. = undifferentiated.

compared with their percentage curves. The influx estimates are obtained by dividing the observed pollen concentration (grains cm^{-3}) by the sediment deposition time (years cm^{-1}) calculated by linear interpolation from the ^{14}C dates (fig. 14.2).

MACROFOSSIL ANALYSIS

After extraction of material for radiocarbon dating, the rest of the core was cut into sections. Their volumes, measured by displacement, varied between 45 and 120cm^3. The sediment was washed through 500μm and 150μm sieves and the retained material was systematically searched for macrofossils, using a Wild binocular microscope with a magnification of up to ×50. The macrofossils were identified by comparison with modern reference material to the lowest possible taxon in the British flora. The conventions used to indicate the level of taxonomic identification follow H.J.B. Birks (1973a).

Betula (tree) includes *Betula pendula, B. pubescens* spp. *pubescens*, and ssp. *odorata* (Birks and Mathewes 1978). *Minuartia* cf. *M. rubella* most closely resembles this species, but it is also similar to *M. verna*. *Sagina nodosa*-type includes *Sagina nodosa, S. procumbens*, and *S. maritima*. *Cardaminopsis*-type includes *Cardaminopsis petraea, Cardamine flexuosa, C. hirsuta, Arabis alpina, A. stricta, Draba norvegica*, and *Teesdalia nudicaulis*.

Macrofossil influx was calculated, and the results are presented in fig. 14.5.

The processing of the pollen and macrofossil data and the drawing of preliminary diagrams were done by the University of Cambridge IBM 370/165 computer using the program POLLDATA.MK5 (Birks and Huntley 1978). Copies of the tabulated fossil counts are available on request.

ZONATION OF DIAGRAMS

The pollen diagram (fig. 14.3) was divided into five local pollen-assemblage zones, prefixed by AD- and numbered from the base. These pollen zones were transferred to the macrofossil diagram (fig. 14.5). No formal description of the zones is presented, because the major differences between them are mentioned in the discussion of vegetational history.

Radiocarbon dates

Four radiocarbon dates were obtained from the late-glacial and early post-glacial sediments (Harkness 1981).

SRR-782   784–791cm   10,072 ± 70 BP

SRR-783   807.5–817.5cm   10,650 ± 110 BP

SRR-784   837.5–847.5cm   11,715 ± 130 BP

SRR-785   870–880cm   12,490 ± 150 BP

The relationship of the dates to the pollen stratigraphy is shown in figs. 14.2–14.5. The sediment-deposition time is shown on fig. 14.2.

These dates are about 600–800 years older than dates for comparable pollen-stratigraphical horizons at Abernethy Forest (Birks and Mathewes 1978) and they are about 350–500 years older than dates at Loch of Winless, Caithness (Peglar 1979). The date of 8910 ± 130 BP (I-4812) (Pennington *et al.* 1972) for the rise of *Corylus* pollen at Loch Clair is more than 1000 years younger than the date for the *Corylus* rise at Lochan an Druim. The dates for the *Corylus* rise on the Isle of Skye (Birks H.J.B. 1973a) are also consistently younger, although they are variable, and Birks discusses the difficulties of interpretation and the problems associated with the dating of minerogenic sediments which are low in organic carbon.

Lochan an Druim lies on Durness limestone. Carbonate dissolved or suspended from this may be responsible for the relatively old dates, especially at a time when soils were unstable and immature, and there was little vegetation in or around the lochan to prevent carbonate-rich silt reaching the sediments. Dates from late-glacial sediments from more sites in northern Scotland are required before a reliable chronostratigraphy can be proposed (see also Pennington 1975).

Vegetational history

ZONE AD-1 (867.5–875cm, figs. 14.3 and 14.4; 867.5–891cm, fig. 14.5)

The low pollen influx (500–700 grains $cm^{-2}yr^{-1}$) is comparable to modern values from exposed coastal tundra in Greenland (Fredskild 1973). There is also a very low influx of macrofossils, and a low organic content (loss on ignition of 0.5% in the basal grey silty mud and 5% in the overlying pale brown silty mud).

Most of the terrestrial plants represented by fossils grow today in open arctic and alpine habitats. *Salix*, Gramineae, and Cyperaceae show the highest pollen percentages, and there are relatively high percentages of *Rumex* spp. and *Oxyria digyna*. *Empetrum nigrum* and Ericales undiff. pollen increase through the zone. There is a wide variety of pollen of herb taxa, which is closely matched by the macrofossil assemblage. The pollen of *Saxifraga oppositifolia/S. aizoides* is related to seeds of *Saxifraga*

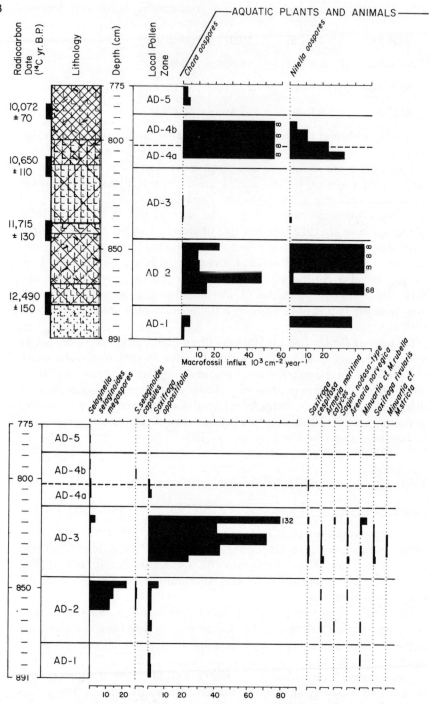

Fig. 14.5 Macrofossil influx diagram from Lochan an Druim, Eriboll. The depth of each bar indicates the stratigraphic interval of that sample. The lithological symbols follow Troels-Smith 1955. Unless otherwise indicated, the macrofossil remains are

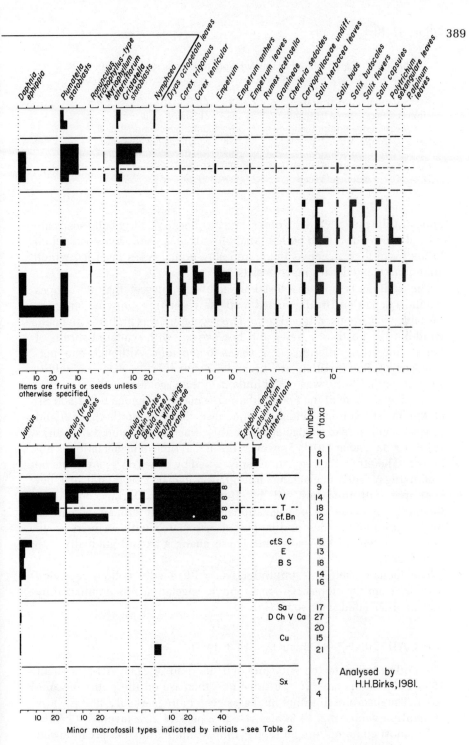

Items are fruits or seeds unless otherwise specified

Minor macrofossil types indicated by initials - see Table 2

Analysed by H.H.Birks,1981.

seeds or fruits. Undiff. = undifferentiated; BP = before present; ∞ = extremely large number; anagall. = *anagallidifolium*. Key to minor macrofossil types is given in Table 14.2.

Table 14.2.   Key to minor macrofossil types on the macrofossil influx
diagram, fig.14.5.

B	*Betula* undiff. fruit body	E	*Eleocharis* seed
cfbn	cf. *Betula nana* fruit body	cfS	cf *Sagina* seed
C	Cruciferae undiff. seed	S	*Solidago virgaurea* seed
Ca	*Cardaminopsis*-type seed	Sa	*Saxifraga aizoides* seed
Ch	*Cerastium holosteoides* seed	Sx	*Saxifraga* cf *S. aizoides* seed
Cu	*Cerastium* undiff. seed	T	*Taraxacum* seed
D	*Dryas octopetala* seed	V	*Viola* seed

oppositifolia and *S.* cf. *S. aizoides*, and *Minuartia* cf. *M. rubella* seeds are
related to *Arenaria*-type pollen. Macrofossils alone add *Dryas octopetala*
to the fossil flora, whereas there are many pollen types recorded which
are not associated with macrofossils.

The sediments and the fossil assemblages suggest that the slopes
around Lochan an Druim carried a sparse vegetation, growing on open,
unstable, calcareous, mineral soils. The vegetation was probably an open,
wind-blasted heath of grasses and low sedges, with scattered shrubs of
Salix, *Empetrum nigrum*, and *Dryas octopetala*. Although the percentages of *Salix* pollen are high, its influx (fig. 14.4) is very low,
suggesting that *Salix* was not abundant in the vegetation.

The deglaciation of the Loch Eriboll area may have ended by 12,500 BP
at the latest. Similar evidence from elsewhere in northern Scotland
suggests widespread deglaciation by this time (Pennington 1975, Birks
H.J.B. 1973a, Peglar 1979). However, the actual time of deglaciation is not
known. Therefore, the vegetation represented by Zone AD-1 may either be
colonizing recently deglaciated ground, or it may be long established, but
kept open with immature soils by climatic severity. During Zone AD-1,
there is some evidence for vegetational development with an increase in
Empetrum and Ericales undiff. pollen, associated with an increase in
organic content of the sediment. These changes extend gradually into
Zone AD-2.

The lochan itself was unproductive, with a silty bottom receiving
material from the slopes. The macrophyte vegetation was dominated by
Nitella, associated with some *Chara*.

ZONE AD-2 (847.5–867.5cm, figs. 14.3–14.5)

The pollen influx to the silty fine-detritus mud of Zone AD-2 is about
1500–2000 grains $cm^{-2}yr^{-1}$, a three-fold increase on the influx of Zone
AD-1. The macrofossil influx increases even more markedly, and the loss
on ignition values (fig. 14.2) also rise to a peak of 22% (average 13%).

The most abundant pollen type is *Empetrum nigrum*, closely followed by Cyperaceae and then Gramineae. There is a sharp decrease in *Salix* pollen percentages, that is not registered in its pollen influx (fig. 14.4), which remains as low as before. If the macrofossils are considered (fig. 14.5), *Salix herbacea* was probably abundant during Zone AD-2, where leaves, seed capsules, and other vegetative parts are frequent. A similar instance of low pollen values but frequent macrofossils of *Salix herbacea* was noted at Abernethy Forest during Zones AFP-3 and AFM-3 (Birks and Mathewes 1978). The leaves, capsules, etc. of *Salix herbacea* are shed in winter, and are blown until trapped on snow beds (Warren Wilson 1958). If these beds are adjacent to a lake, the leaves will be deposited in the water when the snow melts. According to the model of the formation of macrofossil assemblages worked out by Glaser (1981) in Alaska, most of the leaves will float to the windward shore, but others may sink and become incorporated in the sediments. In contrast, the pollen, although wind-dispersed, is produced in small amounts close to the ground, and is not dispersed any great distance. *Salix herbacea* may thus not be well represented by its pollen in lake sediments. No fossil *Salix* seeds were found, and it is possible that they are not well preserved, as they would be expected to behave in a similar way to the vegetative fragments.

The percentages of *Betula* pollen increase slightly in Zone AD-2, but the increase in influx is very small. No *Betula* macrofossils were recovered from this zone, indicating that *Betula* was absent from near the site at this time. The slight increase in pollen values probably results from long-distance pollen transport from areas at least as far south as the Isle of Skye (Birks H.J.B. 1973a) where *Betula* was locally present during Allerød times. Pollen-influx values of *Betula* determined by Pennington (1975, 1977a) in the intervening sites are all so low as to suggest that *Betula* was not locally present. The influx of herb pollen types is also low, but a wide variety of taxa is represented. Most of the taxa present in Zone AD-1 persist, often with somewhat reduced percentages but with unchanged influx, e.g. *Salix*. Several taxa appear, including *Rumex acetosa*, *Urtica*, *Achillea*-type, *Lotus pedunculatus*-type, *Parnassia palustris*, *Filipendula*, and *Lycopodium alpinum*.

Macrofossils again correspond quite well to the pollen types. *Empetrum* seeds, leaves, and anthers are virtually confined to this zone, as are *Carex* and Gramineae seeds. *Selaginella selaginoides* microspores increase in the upper half of Zone AD-2. In the same sediment, large numbers of *Selaginella* megaspores, together with portions of their capsules were recovered. The low amounts of *Saxifraga oppositifolia/S. aizoides* pollen correspond to small numbers of *Saxifraga oppositifolia* seeds and one seed of *S. aizoides*. Caryophyllaceae are not common as macrofossils, the commonest being *Cherleria sedoides*, represented by *Arenaria*-type pollen.

In contrast to this general correspondence, leaves of *Dryas octopetala* were consistently found, together with one seed, in Zone AD-2, but *Dryas* pollen was not recorded. *Dryas* is insect pollinated, whereas its leaves and seeds would be distributed by wind like *Salix herbacea*. The isolated grains of *Dryas* recorded in the subsequent zones may be washed in from soil of Zone AD-2 age, or they may have originated from *Dryas* plants growing too far away from the lochan to be represented by leaves. Alternatively, the leaves may have been prevented from reaching the sediment by the development of marginal swamp vegetation or dense aquatic vegetation.

The assemblage of terrestrial fossils in Zone AD-2 suggests that the vegetation around Lochan an Druim was dominated by dwarf shrubs, particularly *Empetrum nigrum*, but also *Dryas octopetala* and *Salix herbacea*, and perhaps a little *Juniperus* and *Cornus suecica*. These were interspersed with sedges and grasses and low-growing herbs. Small amounts of tall-herb vegetation may have occurred near the lake, with *Rumex acetosa*, *Urtica*, *Saussurea alpina*, and *Filipendula*. Sedges may have been common round the lake edge, as indicated by the frequency of both lenticular and trigonous *Carex* seeds.

Small numbers of *Polytrichum sexangulare* leaves occur throughout the zone. They indicate that areas of late snow-lie persisted near the lochan. *Salix herbacea* may also have been a component of this snow-bed vegetation. Damp runnels from the snow beds may have supported *Saxifraga stellaris*, *S. aizoides*, *Parnassia palustris*, and *Selaginella selaginoides*, although the latter occurs widely today in many damp base-rich grasslands and flush communities. *Polytrichum alpinum* leaves were found near the top of the zone. Today this moss characteristically grows in acid montane grassland, often with some snow-lie.

Fossils of plants of open calcareous grasslands and gravelly areas include *Cherleria sedoides*, *Saxifraga oppositifolia*, *Arenaria norvegica*, *Sagina nodosa*-type, *Minuartia* cf. *M. rubella*, *Saxifraga cespitosa*, and *Cardaminopsis*-type.

During Zone AD-2 the lochan was more productive, and huge numbers of *Nitella* oospores were recorded, together with large numbers of *Chara* oospores. Towards the top of the zone, *Ranunculus trichophyllus*-type seeds are found, but aquatic Ranunculi would seem to have been the only aquatic phanerogams present. *Pediastrum* colonies reach their highest values in this zone, and together with the frequency of *Boryococcus*, suggest that freshwater algae were thriving. Animals were also common, represented by high values of *Daphnia* ephippia and moderate frequencies of *Plumatella* statoblasts. The influx of silt to the lake decreased, thus allowing aquatic biota to increase in the clear water.

There is no dramatic change in the fossil flora between Zone AD-1 and

Zone AD-2. Although several taxa appear for the first time, the overall aspect of the fossil flora remains similar. However, the increased abundance of fossils together with the increased organic content of the sediment indicate that environmental conditions ameliorated sufficiently for dwarf-shrub-dominated vegetation to develop. The pollen-influx values are within the range of modern influx from sheltered dwarf-shrub heaths at the head of fjords in Greenland (Fredskild 1973), and from forest-tundra in Canada (Ritchie and Lichti-Federovich 1967). Birch trees did not colonize the area, and the evidence for late-lying snow so near sea-level suggests that the climate was too cold for them and that north-west Scotland lay beyond the northern tree-limit of this time.

ZONE AD-3 (812.5–847.5cm, figs. 14.3–14.5)

The clay- and silt-rich sediments of Zone AD-3 have an average loss on ignition of about 4% (fig. 14.2), and the sedimentation rate is lower. However, at the beginning of the zone the pollen influx rises to about 1800 grains $cm^{-2}yr^{-1}$. In contrast, the influx of macrofossils is the highest in the profile, especially in the upper part of the zone. This is primarily due to the large numbers of *Saxifraga oppositifolia* seeds (fig. 14.5).

The pollen percentages are dominated by Ericales undiff. pollen (fig. 14.3). These tetrads could not be specifically identified because of their poor state of preservation, but the majority probably belong to *Empetrum*. The values of *Empetrum nigrum* pollen fall, but those of Ericales undiff. rise considerably, and maintain high percentages until near the top of the zone. However, the influx of Ericales undiff. pollen falls steadily from the initial high values. No *Empetrum* or other Ericales macrofossils were recorded.

The pollen percentages of most of the pollen types of Zone AD-2 are reduced in Zone AD-3, and many cease to be represented at all, particularly tall-herb types. However, several types show a marked increase, particularly *Saxifraga oppositifolia/S. aizoides*, but also *Artemisia* undiff., Caryophyllaceae undiff., *Solidago*-type, and *Lycopodium selago*. In addition, several pollen and spore types appear, e.g. *Artemisia* cf. *A. norvegica*, *Armeria maritima*, *Sagina*, *Dryas octopetala*, *Astragalus alpinus*, and *Cryptogramma crispa*. *Betula* pollen is virtually absent in this zone.

The macrofossils show marked changes from the previous zone, particularly in the enormous increase in numbers of *Saxifraga oppositifolia* seeds. *Empetrum*, *Selaginella*, *Dryas*, and *Carex* spp. are completely absent after a former period of abundance in Zone AD-2. Several taxa appear for the first time, notably *Juncus*, but also several Caryophyllaceae taxa and *Saxifraga rivularis*. Other taxa become more abundant than previously, such as *Saxifraga cespitosa*, *Minuartia* cf. *M. rubella*, *Arenaria norvegica*, *Armeria maritima*, and *Polytrichum sexangulare*.

The increasing proportion of mineral matter in the sediment indicates that the surrounding soils were unstable, allowing material to be washed in. The increase in pollen influx, due largely to Ericales undiff. and also to Indeterminable:deteriorated pollen suggests that pollen was being washed in with the minerogenic material from soils which had previously supported *Empetrum* heath. As the zone progressed, the amount of inwashed pollen decreased, as the supply was depleted. The absence of any Ericales macrofossils in the presence of large amounts of poorly preserved Ericales pollen is probably due to the destruction of macrofossils in the soil and during reworking and transportation.

The aspect of the vegetation in Zone AD-3 was completely different from that of Zone AD-2. The dwarf-shrub heaths of *Empetrum* and *Dryas* were destroyed, and their place was taken by open communities containing an abundance of *Saxifraga oppositifolia* as well as *S. cespitosa*, *Artemisia norvegica*, *Lycopodium selago*, *Armeria maritima*, *Cryptogramma crispa*, *Astragalus alpinus*, and members of the Caryophyllaceae which favour open habitats, such as *Arenaria norvegica*, *Minuartia* cf. *M. rubella*, *Sagina nodosa*-type, and *Cherleria sedoides*. The extent of snow beds increased during Zone AD-3, judging from the increased numbers of *Polytrichum sexangulare* leaves and the abundance of *Salix herbacea*. The associated meltwater runnels may have supported *Saxifraga rivularis*, *Minuartia* cf. *M. stricta*, *Caltha palustris*, and *Juncus*. The species of *Juncus* was indeterminable, but there are several montane and arctic species which could have been associated with this assemblage, such as *Juncus triglumis*, *J. biglumis*, *J. castaneus*, and *J. trifidus*. Their light seeds could have been readily washed in with meltwater.

The environment at Lochan an Druim must have been bleak during Zone AD-3. The vegetation resembled high Arctic or Alpine vegetation today, with scattered herbs and a few grasses and sedges forming a discontinuous cover on open mineral gravelly soils, and long-lasting snow beds in suitable hollows, associated with meltwater swamps and runnels. During the time of Zone AD-3 glaciers re-formed on the higher mountains in north-west Scotland (Sissons 1977). His reconstructions suggest a firn line at 120–170m OD, and the development of rather small glaciers in the far north-west.

To judge from the virtual absence of pollen and macrofossils of aquatic plants and animals, the lochan was barren except for a few algae. It was presumably ice-covered for much of the year, with the sediment composed of fine silt and a small amount of organic material washed in during the summer.

ZONE AD-4 (787.5–812.5cm, figs. 14.3–14.5)

Pollen-influx values rise rapidly above all levels reached in the previous zones, with an average of 4000 grains $cm^{-2} yr^{-1}$, but exceeding 7000 grains $cm^{-2} yr^{-1}$ at the top of the zone. In contrast, the macrofossil influx falls considerably to an average of about 30–40 macrofossils $10^3 cm^{-2} yr^{-1}$. The sediment of this zone is a fine-detritus mud, with an increasing organic content indicated by loss on ignition values increasing from 13 to 39% (fig. 14.2). The sedimentation rate also increases to levels similar to those of Zone AD-2.

The pollen and spore assemblage of Zone AD-4 is characterized by high values of *Juniperus communis*, Polypodiaceae undiff., and *Dryopteris filix-mas*-type, and rising values of *Betula* undiff. In subzone AD-4a (802.5–812.5cm) percentages of *Empetrum nigrum* and Gramineae are high, but in subzone AD-4b (787.5–802.5cm) their values are lower, and *Betula* undiff. is the dominant pollen type. At the beginning of the zone there are small increases in percentages of *Rumex* spp., cf. *Oxyria digyna*, *Thalictrum*, *Achillea*-type, *Filipendula*, and *Potentilla*-type. Macrofossils of tree *Betula* and fern sporangia appear in abundance, and seeds of *Juncus* maintain their previous high levels.

During Zone AD-4 there are marked decreases in percentages of Ericales undiff. pollen, and the influx continues its decline to very low values from the previous zone. Percentages of *Saxifraga oppositifolia/S. aizoides* pollen fall, accompanied by an enormous reduction in seed influx of *Saxifraga oppositifolia*. Similarly, *Salix herbacea* macrofossils cease to be represented, and *Salix* pollen values remain low, only increasing in subzone AD-4b. The values of most of the pollen and spore types characteristic of Zone AD-3 decline, such as Cyperaceae, Caryophyllaceae undiff., *Selaginella*, *Lycopodium selago*, and Indeterminable: deteriorated. The representation of macrofossils of most taxa of Zone AD-3 ceases.

The vegetation changed quickly from an open species-rich assemblage to dwarf-shrub heath dominated by *Empetrum nigrum* and *Juniperus communis*, with an abundance of grasses and ferns. *Juncus* was common, probably on damp soils near the lochan, where it may have been associated with tall herbs such as *Rumex acetosa* and *Filipendula*. On its arrival, *Betula* quickly became dominant. There is no firm evidence for the presence of *Betula nana*. Although the birch macrofossils are generally poorly preserved, all but two could be confidently assigned to tree *Betula*. Of the two, one could possibly have been *B. nana*. *Betula* cf. *B. nana* pollen was recorded in low amounts up to Zone AD-3. The lack of macrofossils suggests that this pollen may largely be tree *Betula* pollen that falls within the morphological range of *B. nana* (Birks H.J.B. 1968).

As the birch forest developed in Zone AD-4 *Juniperus* and ferns remained common as understorey plants, but *Empetrum* and grasses declined in abundance. *Populus tremula* may have been a member of the forest canopy, although no macrofossils were recorded.

There is some evidence for the persistence of the plants of open ground near An Druim, in the form of a few seeds of *Saxifraga oppositifolia*, Caryophyllaceae undiff., and *Saxifraga cespitosa*, and single pollen grains of *Dryas octopetala*. Presumably these plants became restricted to open areas of shallow soil such as on the Durness limestone outcrops. Several of them still grow near An Druim today, notably *Dryas*, and they have probably persisted there in small quantities during the forest period of the early post-glacial, and survived subsequent forest clearance and grazing. *Saxifraga oppositifolia* failed to survive at An Druim, although it grows near sea-level elsewhere in north-west Scotland.

Major changes took place in the lochan at the opening of Zone AD-4. *Chara* became extremely abundant, and *Nitella* also became common. Other submerged aquatic plants were also present in smaller quantities, as indicated by the pollen of *Myriophyllum alterniflorum*, *Potamogeton*-type, and *Littorella uniflora*, and seeds of *Myriophyllum alterniflorum*. Algae also increased in abundance, particularly *Botryococcus*, and aquatic animals also increased, represented by fossils of *Cristatella*, *Plumatella*, and *Daphnia*. The increasing productivity of the lake is reflected by the rising organic content of the sediments (fig. 14.2).

The sudden change in both terrestrial and aquatic vegetation at the onset of Zone AD-4 must have been due to a rapid and substantial amelioration in climate at the opening of the post-glacial. The fact that birch trees arrived so soon, even in northern Scotland, suggests that climatic conditions may have been different from those of Zone AD-2, when *Betula* did not reach so far north and snow beds persisted in the vicinity of An Druim.

ZONE AD-5 (770–787.5cm, figs. 14.3 and 14.4; 775–787.5cm, fig. 14.5)

During this zone, the pollen and macrofossil influx are similar to their previous levels. The sedimentation rate also remains the same, although the sediment becomes increasingly organic (fig. 14.2).

The major change in the fossil assemblage is the expansion of *Corylus/Myrica* pollen to values equalling those of *Betula* undiff. The local presence of *Corylus* was confirmed by finding *Corylus* anthers. The values of *Juniperus*, *Empetrum nigrum*, Gramineae, *Filipendula*, and *Dryopteris filix-mas*-type all decrease substantially on the arrival of *Corylus*, and many of the tall herbs such as *Rumex* spp., *Thalictrum*, Rubiaceae, and *Sedum* cf. *S. rosea* cease to be represented.

As *Corylus* joined *Betula* and *Populus* in the canopy, shading presumably became more complete, and the understorey of *Juniperus* and ferns was reduced. *Sorbus aucuparia* may also have been present. The low pollen values of *Ulmus*, *Quercus*, and *Pinus* are probably due to long-distance transport from the south. From the fossils represented, it would appear that the understorey was species-poor, containing some ferns, *Melampyrum*, *Cornus suecica*, and *Potentilla*-type. Tall herbs and *Juncus* became considerably reduced. However, this may be a false impression created by the poor pollen dispersal of herbs beneath a tree canopy, and the presence of insect-pollinated plants which produce relatively small amounts of pollen (Birks H.J.B. 1973b). There are virtually no macrofossils of herbs in this zone, and the terrestrial macrofossil flora is dominated by tree *Betula*.

The great burst of aquatic plant productivity in Zone AD-4 declines in this zone, with a large reduction in the amounts of *Chara*, and the apparent disappearance of *Nitella*, *Myriophyllum alterniflorum*, and *Littorella uniflora*. The pollen values of *Potamogeton*-type are considerably reduced, but *Nymphaea* becomes represented by moderate quantities of both pollen and seeds. Maybe the lake had become sufficiently shallow near the coring site to allow the local growth of *Nymphaea*, which shaded out the submerged aquatic plants. Bryozoa were still present in a similar amount to before, but *Daphnia* ephippia were not recorded. Either *Daphnia* died out locally, or else ephippia were not produced, indicating that winter conditions were not sufficiently severe to initiate ephippial production. A search for *Daphnia* exoskeletons in the sediments is needed to resolve these hypotheses.

The vegetational change between Zone AD-4 and AD-5 is a result of the local establishment of *Corylus*. The causes for the change in the macrophyte vegetation are not fully understood. A short maximum of high productivity is frequently recorded in lakes during the Allerød or at the beginning of the post-glacial, when the climate ameliorates from previously severe conditions. This maximum is discussed further by H.H. Birks (1980).

Comparisons and correlations

The radiocarbon dates from Lochan an Druim indicate that the sediments were deposited during the Late Devensian stage and age (*sensu* Mitchell *et al.* 1973) (Late Weichselian *sensu* Mangerud *et al.* 1974). The interpretation of the fossil and sediment stratigraphy indicates that the climate ameliorated in Zone AD-2, reverted to severe conditions in Zone AD-3, and then became much warmer in Zone AD-4. This climatic sequence is typical of the Devensian late-glacial of Britain and the Late Weichselian elsewhere in Europe. In the terminology of Mangerud *et al.* (1974) Zone AD-1 belongs to the Older Dryas, Zone AD-2 to the Allerød interstadial, Zone AD-3 to the Younger Dryas, and Zones AD-4 and AD-5 to the Flandrian. However, the radiocarbon dates for these boundaries are all about 500 years too old to match Mangerud *et al.*'s chronozones. Probable reasons for these dating anomalies have been discussed above.

Other radiocarbon-dated sites in northern Scotland have been correlated with the Late Weichselian (Cam Loch, Pennington 1975; Loch of Winless, Peglar 1979). Because of the problems of obtaining reliable radiocarbon dates from minerogenic lake sediments poor in carbon (Sutherland 1980), the possibility that a Bølling interstadial is represented in north-west Scotland (Pennington 1975) with a short cold stadial separating it from the Allerød interstadial has not been confirmed. At most sites, including Lochan an Druim, this climatic sequence cannot be deduced from the pollen stratigraphy, even though the sediments may span the time of the Bølling chronozone (13,000–12,000 BP (Mangerud *et al.* 1974).

With the difficulties of radiocarbon dating in mind, it is possible to compare the pollen sequence from Lochan an Druim with other pollen diagrams in northern Scotland (fig. 14.1) in an attempt to illustrate any regional variations. The diagrams are correlated in table 14.3, and for each zone the most abundant pollen types are indicated.

The only site at which *Betula* trees were definitely present in the Allerød interstadial is Loch Meodal in the sheltered south of the Isle of Skye. The increases in *Betula* pollen percentages or influx at the other sites are all too low to indicate the local presence of birch trees. The recording of macrofossils would be an easy way to prove the local presence of birch, and to distinguish tree-*Betula* from *B. nana*.

At all the sites, the lowest pre-interstadial zone is dominated by Gramineae and Cyperaceae pollen associated with either *Rumex* or *Lycopodium selago*. The latter is particularly prevalent on the Isle of Skye. Apart from the presence of birch trees at Loch Meodal on Skye, the Allerød vegetation was a treeless dwarf-shrub heath, frequently dominated by *Empetrum* (presumably *E. nigrum*) and sometimes by *Juniperus*. The dwarf shrubs were associated with varying amounts of grasses, sedges, and herbs. The assemblage at Lochan Coir'a'Ghobhainn on the exposed southwest coast of Skye below the Cuillin hills is similar to that at Lochan an Druim, in suggesting that snow beds may have persisted nearby. The absence of any macrofossil studies at all the sites makes the interpretation of the pollen assemblages in vegetational terms difficult, because of important pollen 'blind spots' (*sensu* Davis 1963), in this case *Salix herbacea*, low-growing herbs, and mosses.

In the Younger Dryas, dwarf shrubs were replaced by open grass and sedge heaths. In northern Skye (Loch Mealt and Loch Fada) *Betula nana* was frequent, but at Lochan Coir'a'Ghobhainn and Loch Cill Chriosd pen vegetation and extensive snow beds occurred. At Lochan Coir'a'Ghobhainn, large amounts of Ericales undiff. pollen probably had a similar origin to those at Lochan an Druim (Birks H.J.B. 1973a). Open communities rich in herbs also occurred at Loch of Winless and Yesnaby (Orkney). In the Inverpolly area some *Empetrum* persisted, and *Betula* percentages remained relatively high. Although *Artemisia* pollen increased slightly at this time in all the sites, nowhere does it reach such high values as the Inverpolly area. This suggests that the Younger Dryas vegetation was rather different in this area, maybe as a result of the hard acid rocks.

The opening of the post-glacial is generally marked by an expansion of locally present shrubs, particularly *Empetrum* and *Juniperus*, together with an increase in Gramineae and herbs such as *Filipendula*. This is followed rapidly by the expansion of tree *Betula* and the formation of open birch woods. The exception is parts of the Isle of Skye, where *Betula* and *Corylus* expand immediately following the Younger Dryas pollen assemblages. Perhaps *Betula* and *Corylus* were already present on or near the Isle of Skye, and were able to expand there before dwarf-shrub heath could develop. Elsewhere, the time taken by *Betula* to migrate, closely followed by *Corylus*, allowed *Empetrum* and *Juniperus* to expand. Detailed chronologies are needed to test this hypothesis (cf. Lowe 1981).

The main expansion of aquatic macrophytes occurred either during the Allerød interstadial, or the early post-glacial. At Loch of Winless, macrophytes remain abundant in the Younger Dryas, but at other sites with Allerød expansions, macrophyte growth declined at this time. In the large Loch Sionascaig and Cam Loch, no macrophyte development is detected, because the water is too deep for any macrophyte growth at the coring site, which is too far from the shallow water to register any local aquatic development. Characeae often seem to be important components of pioneer aquatic vegetation, but they can only be detected fossil by their oospores. Thus the true expansion of macrophytes may differ from that indicated by pollen, if Characeae were present.

Macrofossils and vegetational reconstruction

The identification of macrofossils at Lochan an Druim has greatly enhanced the vegetational interpretations possible from pollen alone. Although detailed reconstructions can be made from careful pollen

Table 14.3.　Comparison and correlation of pollen zones at sites in northern Scotland. The most abundant pollen types (as percentages) are indicated in each zone. Types in brackets are less abundant, but characteristic of the zone. AQUATICS indicates the expansion of macrophytes. Radiocarbon dates are positioned at the appropriate zone boundaries. The authors' zone names have been used. The sites are shown in fig.14.1, except for Loch of Winless, Caithness (Peglar 1979) and Yesnaby, Orkney Mainland (Moar 1969b).

	Lochan an Druim	Loch of Winless	Yesnaby	Loch Sionascaig	Cam Loch
POST-GLACIAL	AD-5 Betula Corylus 10072	LW-4 Betula Corylus 9340		NWS-2 Betula Corylus Polypodiaceae	
POST-GLACIAL	AD-4b Betula Juniperus 10650	LW-3 Betula Tall herbs 10300	F II Betula Juniperus Empetrum AQUATICS	NWS-1 Betula Juniperus Empetrum	
POST-GLACIAL	AD-4a Juniperus Empetrum Gramineae AQUATICS 11715	LW-2c Juniperus Empetrum Gramineae AQUATICS 11250	F I Juniperus Gramineae Rumex	Transition Betula Gramineae Polypodiaceae	Cg, Ch Juniperus Empetrum Polypodiaceae
YOUNGER DRYAS	AD-3 Ericales undiff. Cyperaceae Saxifraga opp./ 　S. aizoides	LW-2b Cyperaceae Selaginella (Artemisia) AQUATICS	L III Cyperaceae Gramineae (Artemisia) Compositae Selaginella	C Empetrum Betula (Artemisia) Compositae Lycopodium selago	Ce-f Empetrum Betula Artemisia Gramineae 106?
ALLERØD	AD-2 Empetrum Gramineae Cyperaceae AQUATICS	LW-2a Empetrum Gramineae Cyperaceae Betula ? nana AQUATICS	L II Empetrum Gramineae (Cyperaceae)	B Empetrum Gramineae (Cyperaceae) (Juniperus) (Betula)	Cd, Cc, Cb Empetrum Gramineae Cyperaceae (Juniperus) (Betula) 119?
OLDER DRYAS	AD-1 Gramineae Cyperaceae Salix 12490	LW-1 Gramineae Cyperaceae 12690	L I Gramineae Cyperaceae Rumex	A Gramineae Cyperaceae Rumex	Ca Gramineae Salix Rumex Polypodiaceae 129?

Loch Craggie Loch Borralan	Lochan Coir' a'Ghobhainn	Loch Cill Chriosd	Loch Mealt Loch Fada	Loch Meodal
	LCG-6 *Betula* *Corylus*	LCC-7 *Betula* *Corylus* AQUATICS	LM-7, LF-6 *Betula* *Corylus* Filipendula	LML-5 *Betula* *Corylus* AQUATICS
NS-I, NS-II *Betula* *Juniperus* *Empetrum*	LCG-5 *Betula* *Juniperus*			
Transition *Betula ? nana* Gramineae *Rumex* AQUATICS	LCG-4 *Juniperus* *Betula* Gramineae *Rumex* AQUATICS	LCC-6 *Juniperus* *Betula* Gramineae	LM-6 *Juniperus* Gramineae Polypodiaceae	
C Cyperaceae Gramineae *Artemisia*	LCG-3 Ericales undiff. Cyperaceae *Lycopodium selago* Caryophyllaceae	LCC-5 Cyperaceae *Lycopodium selago* *Empetrum* *Selaginella*	LM-5, LF-5 *Betula nana* Gramineae Cyperaceae *Salix* Polypodiaceae	LML-4 *Betula nana* Gramineae Cyperaceae Polypodiaceae
B *Empetrum* Gramineae Cyperaceae *Rumex* (*Betula*)	LCG-2 Gramineae Cyperaceae *Lycopodium selago* (*Empetrum*) (*Juniperus*)	LCC-3, LCC-4 *Juniperus* Gramineae *Rumex* (*Betula*)	LM-2, LM-3, LM-4 LF-2, LF-3, LF-4 Gramineae Cyperaceae *Selaginella* *Juniperus* (*Betula*) AQUATICS	LML-2, LML-3 *Betula* *Juniperus* Gramineae Cyperaceae Polypodiaceae
A Gramineae Cyperaceae *Rumex* Polypodiaceae	LCG-1 Cyperaceae Gramineae *Lycopodium selago* *Oxyria digyna*	LCC-1 Cyperaceae *Lycopodium selago* Ericales undiff. *Selaginella*	LM-1, LF-1 Cyperaceae Gramineae *Lycopodium selago*	LML-1 Cyperaceae Gramineae *Rumex*

analysis combined with a knowledge of the present-day ecology of the plants concerned – such as H.J.B. Birks' (1973a) study of the Isle of Skye – macrofossils add much information, both quickly and easily, which could only be gained from a pollen study by great labour. In particular, 'blind spots' in the pollen assemblage can be filled in. At Lochan an Druim, macrofossils are particularly useful in recording the vegetational history of plants which do not produce pollen but which are important components of the vegetation, namely *Juncus*, Characeae, and *Polytrichum sexangulare*, and for plants which produce very small amounts of pollen, even though they may be quite abundant, such as *Salix herbacea*, *Dryas octopetala*, and Caryophyllaceae. The identification of macrofossils to species-level assists in the interpretation of some pollen curves. For example, *Saxifraga oppositifolia/S. aizoides* pollen is shown to belong mostly to *Saxifraga oppositifolia*, with only a very small proportion originating from *S. aizoides*, and Caryophyllaceae pollen is divided into various types, whereas the macrofossils indicate which species are involved.

Macrofossils are also useful in determining the local presence of taxa which produce widely dispersed pollen, in this case *Betula* and *Corylus*. Although absence of macrofossils cannot be taken as definite evidence for the local absence of a taxon, the absence of *Betula* macrofossils in the Allerød, combined with the evidence of low pollen percentages and influx can be taken to indicate the local absence of *Betula* at this time. Similarly, it is likely that *Betula nana* was absent from Loch Eriboll, in spite of low amounts of *B.* cf. *B. nana* pollen being recorded. In the Younger Dryas, the absence of Ericales macrofossils in the presence of large amounts of Ericales pollen supports the hypothesis that the pollen was washed in and redeposited from previously developed soils.

A small loch with a limited throughflow of water is an excellent site for macrofossil analysis (Watts 1978). It is palaeoecologically valuable because it registers the vegetation in and near the site from both pollen and macrofossils. Large deep lakes contain the pollen rain of the lake's catchment, which can make detailed vegetational reconstruction difficult. Because of their smaller numbers and poor dispersal, macrofossils seldom reach deep water in the centre of a large lake (Birks H.H. 1973), and any that do so arrive by chance and cannot readily be used for detailed vegetational reconstruction. The value of small lakes in reconstructing the local vegetation and hence the vegetational differentiation within an area is well illustrated by H.J.B. Birks' (1973a) study of the Isle of Skye. Lochan an Druim demonstrates that a small lake is an effective macrofossil trap, and that a detailed vegetational reconstruction can be made by combining both pollen and macrofossils.

More stratigraphical macrofossil studies should be made in conjunction with pollen analysis in the future, thereby laying a firm foundation for vegetational and hence environmental reconstructions. Past environments in different places can then be compared with confidence, and a detailed picture assembled of conditions in the past and how they varied, thus providing evidence upon which hypotheses about the causes of change can be proposed and tested.

Acknowledgments

I am grateful to P. Adam, H.J.B. Birks, I.C. Prentice, and J.E. Young for obtaining the core from Lochan an Druim, to D.D. Harkness for the radiocarbon dates, to Fiona Wilson for painstakingly picking out the macrofossils, and to Sylvia Peglar for preparing the pollen samples and for drawing the diagrams. I am especially grateful to H.J.B. Birks for computing the results and producing computer-drawn pollen and macrofossil diagrams, for his critical reading of the manuscript, and for continual encouragement during the project.

I wish to take this opportunity to express my thanks to Professor Winifred Tutin, F.R.S., who has provided much help and support, and many friendly discussions throughout my research on the vegetational history of north-west Scotland.

References

Benninghoff, W.S., 1962. Calculation of pollen and spore density in sediment by addition of exotic pollen in known quantities. *Pollen Spores* 4: 232.

Birks, H.H., 1972. Studies in the vegetational history of Scotland III. A radiocarbon-dated pollen diagram from Loch Maree, Ross and Cromarty. *New Phytol.* 71: 731–54.

Birks, H.H., 1973. Modern macrofossil assemblages in lake sediments in Minnesota. In *Quaternary Plant Ecology*, ed. H.J.B. Birks and R.G. West.

Birks, H.H., 1980. Plant macrofossils in Quaternary lake sediments. *Ergebn. Limnol.* Heft 15.

Birks, H.H. and Mathewes, R.W., 1978. Studies in the vegetational history of Scotland V. Late Devensian and early Flandrian pollen and macrofossil stratigraphy at Abernethy Forest, Inverness-shire. *New Phytol.* 80: 455–84.

Birks, H.J.B., 1968. The identification of *Betula nana* pollen. *New Phytol.* 67: 309–14.

Birks, H.J.B., 1973a. *Past and Present Vegetation of the Isle of Skye – A Palaeoecological Study.*

Birks, H.J.B., 1973b. Modern pollen rain studies in some arctic and alpine environments. In *Quaternary Plant Ecology*, ed. H.J.B. Birks and R.G. West.

Birks, H.J.B., 1976. Late-Wisconsinan vegetational history at Wolf Creek, central Minnesota. *Ecol. Monogr. 46:* 395–428.

Birks, H.J.B. and Huntley, B., 1978. Program POLLDATA.MK5. Documentation relating to FORTRAN IV program of 26 June 1978 (Subdepartment of Quaternary Research, University of Cambridge, internal note).

Birks, H.J.B. and Williams, W., 1983. The Late-Quaternary vegetational history of the Inner Hebrides. *Proc. R. Soc. Edinb. 83B:* 269–92.

Clapham, A.R., Tutin, T.G. and Warburg, E.F., 1962. *Flora of the British Isles.*

Corley, M.F.V. and Hill, M.O., 1981. *Distribution of Bryophytes in the British Isles* (British Bryological Society).

Davis, M.B., 1963. On the theory of pollen analysis. *Am J. Sci. 261:* 897–912.

Dean, W.E., 1974. Determination of carbonate and organic matter in calcareous sediments and sedimentary rocks by loss on ignition: comparison with other methods. *J. Sedim. Petrol. 44:* 242–8.

Faegri, K. and Iversen, J., 1975. *Textbook of Pollen Analysis* (3rd edn).

Fredskild, B., 1973. Studies in the vegetational history of Greenland. *Meddr. Grønland 198:* 1–245.

Glaser, P.H., 1981. Transport and deposition of leaves and seeds on tundra: a late-glacial analog. *Arctic Alpine Res. 13:* 173–82.

Harkness, D.D., 1981. Scottish Universities Research and Reactor Centre Radiocarbon Measurements IV. *Radiocarbon 23:* 252–304.

Kirk, W. and Godwin, H., 1963. A late-glacial site at Loch Droma, Ross and Cromarty. *Trans. R. Soc. Edinb. 65:* 225–49.

Lowe, S., 1981. Radiocarbon dating and stratigraphic resolution in Welsh lateglacial chronology. *Nature 293:* 210–12.

Mangerud, J., Andersen, S.T., Berglund, B.E. and Donner, J.J., 1974. Quaternary stratigraphy of Norden, a proposal for terminology and classification. *Boreas 3:* 109–28.

Matthews, J., 1969. The assessment of a method for the determination of absolute pollen frequencies. *New Phytol. 68:* 161–6.

McLean, A.C., 1976. *The Highlands and Islands of Scotland.*

Mitchell, G.F., Penny, L.F., Shotton, F.W. and West, R.G., 1973. Correlation of Quaternary deposits in the British Isles. *Geol. Soc. Lond. Spec. Rep. 4:* 1–99.

Moar, N.T., 1969a. A radiocarbon-dated pollen diagram from north-west Scotland. *New Phytol. 68:* 209–14.

Moar, N.T., 1969b. Two pollen diagrams from the Mainland, Orkney Islands. *New Phytol. 68:* 201–8.

Peglar, S.M., 1979. A radiocarbon-dated pollen diagram from Loch of Winless, Caithness, north-east Scotland. *New Phytol. 82:* 245–63.

Pennington, W., 1975. A chronostratigraphic comparison of Late-Weichselian and Late-Devensian subdivisions, illustrated by two radiocarbon-dated profiles from western Britain. *Boreas 4:* 157–71.

Pennington, W., 1977a. Lake sediments and the lateglacial environment in northern Scotland. In *Studies in the Scottish Lateglacial Environment,* ed. J.M. Gray and J.J. Lowe, 119–41.

Pennington, W., 1977b. The Late Devensian flora and vegetation of Britain. *Phil. Trans. R. Soc. B 280:* 247–71.

Pennington, W., Haworth, E.Y., Bonny, A.P. and Lishman, J.P., 1972. Lake sediments in northern Scotland. *Phil. Trans. R. Soc. B 264*: 191–294.

Ratcliffe, D.A., 1977. *Highland Flora* (Highlands and Islands Development Board).

Ritchie, J.C. and Lichti-Federovich, S. 1967. Pollen dispersal phenomena in arctic–subarctic Canada. *Rev. Palaeobot. Palynol. 3*: 255–66.

Sissons, J.B., 1977. The Loch Lomond Readvance in the northern mainland of Scotland. In *Studies in the Scottish Lateglacial Environment*, ed. J.M. Gray and J.J. Lowe.

Sutherland, D.G., 1980. Problems of radiocarbon dating deposits from newly deglaciated terrain: examples from the Scottish lateglacial. In *Studies in the Lateglacial of North-west Europe*, ed. J.J. Lowe, J.M. Gray and J.E. Robinson.

Troels-Smith, J., 1955. Karakterisering af løse jodarter. *Danm. Geol. Unders. Ser. IV 3(10)*: 73pp.

Vasari, Y. and Vasari, A., 1968. Late- and post-glacial macrophytic vegetation in the lochs of northern Scotland. *Acta bot. Fenn. 80*: 1–120.

Warren Wilson, J., 1958. Dirt on snow patches. *J. Ecol. 46*: 191–8.

Watts, W.A., 1978. Plant macrofossils and Quaternary paleoecology. In *Biology and Quaternary Environments*, ed. D. Walker and J. Guppy (Australian Academy of Science: Canberra).

Williams, W., 1976. The Flandrian vegetational history of the Isle of Skye and the Morar Peninsula (Ph.D. thesis, University of Cambridge).

Wright, H.E., 1980. Cores of soft lake sediments. *Boreas 9*: 107–14.

Index

(page numbers in italics refer to items in figures or tables)